7-65

CW00727597

Lecture Notes in Statistics

Edited by D. Brillinger, S. Fienberg, J. Gani, J. Hartigan, J. Kiefer, and K. Krickeberg

7

Masafumi Akahira
Kei Takeuchi

Asymptotic Efficiency of Statistical Estimators: Concepts and Higher Order Asymptotic Efficiency

Springer-Verlag
New York Heidelberg Berlin

Masafumi Akahira
Department of Mathematics
University of Electro-
Communications
Chofu, Tokyo 182,
Japan

Kei Takeuchi
Faculty of Economics
University of Tokyo
Hongo, Bunkyo-ky, Tokyo 113,
Japan

AMS Subject Classification (1970): 62F10, 62F20, 62F25

Library of Congress Cataloging in Publication Data

Akahira, Masafumi, 1945–
Asymptotic efficiency of statistical estimators.

(Lecture notes in statistics; v. 7)
Bibliography: p.
Includes index.
1. Estimation theory. 2. Asymptotic efficiencies
(Statistics) I. Takeuchi, Kei, 1933–
II. Title. III. Series: Lecture notes in statis-
tics (Springer-Verlag); 67.
QA276.8.A38 519.5′44 81–1874
AACR2

Printed in the United States of America

9 8 7 6 5 4 3 2 1

ISBN 0-387-90576-6 Springer-Verlag New York Heidelberg Berlin
ISBN 3-540-90576-6 Springer-Verlag Berlin Heidelberg New York

PREFACE

This monograph is a collection of results recently obtained by the authors. Most of these have been published, while others are awaiting publication.

Our investigation has two main purposes. Firstly, we discuss higher order asymptotic efficiency of estimators in regular situations. In these situations it is known that the maximum likelihood estimator (MLE) is asymptotically efficient in some (not always specified) sense. However, there exists here a whole class of asymptotically efficient estimators which are thus asymptotically equivalent to the MLE. It is required to make finer distinctions among the estimators, by considering higher order terms in the expansions of their asymptotic distributions.

Secondly, we discuss asymptotically efficient estimators in non-regular situations. These are situations where the MLE or other estimators are not asymptotically normally distributed, or where their order of convergence (or consistency) is not $n^{1/2}$, as in the regular cases. It is necessary to redefine the concept of asymptotic efficiency, together with the concept of the maximum order of consistency. Under the new definition as asymptotically efficient estimator may not always exist.

We have not attempted to tell the whole story in a systematic way. The field of asymptotic theory in statistical estimation is relatively uncultivated. So, we have tried to focus attention on such aspects of our recent results which throw light on the area.

While the authors have been working on this problem, Professor J.Pfanzagl of University of Cologne and his group, Professor J.K.Ghosh of Indian Statistical Institute and his group, Professor B.Efron of Stanford University among others have contributed to the development of the theory of higher order asymptotic efficiency. The authors have profited by all these works and were influenced in writing this monograph. But the authors would like to keep their own style of approach.

In the course of the investigation, we had many chances to present the partial results and to invite discussions, and were stimulated to go further by the keen interest that some Japanese mathematical statisticians have shown to us. In the summer of 1979 K.Takeuchi had a chance to visit Stanford University for the summer session and to give a course on the asymptotic theory of estimation, and a greater part of it was covered by the earlier version of this monograph. We were acknowledged to Professors T.W.Anderson, C.Stein and B.Efron for inviting him and giving him the chance to discuss with them and also with Professors J.Kiefer and P.Bickel.
Thanks are due to Miss. Gabrielle Kelly, one of the graduate students who attended the course, with the help of Mr. A. Takemura also a graduate student at Stanford, reviewed and corrected our English. We also thank Dr.K.Morimune of Kyoto University for correcting some errors in Sections 5.1 and 5.4, and Mrs. C.Tatsuno for typing and retyping the whole manuscript.

During the preparation of the manuscript the authors were partly supported by the Grand-in-Aid for Scientific Research from the Ministry of Education, Science and Culture of Japan.

Table of Contents

Chapter 1

General Discussion

1.1. The concept of asymptotic efficiency

Suppose that $X_1, X_2, \ldots, X_n, \ldots$ are a sequence of random variables. Let Θ be a parameter space, which is assumed to be an open subset of Euclidean p-space R^p. An estimator $\hat{\theta}$ $(= \hat{\theta}_n(x_1, \ldots, x_n))$ of θ is called consistent if for every $\varepsilon > 0$ and every $\theta \in \Theta$

$$\lim_{n\to\infty} P_{\theta,n}\left\{ \| \hat{\theta}_n - \theta \| > \varepsilon \right\} = 0 .$$

If the convergence is uniform in every neighborhood of every $\theta \in \Theta$, then $\hat{\theta}$ is called locally uniformly consistent. As will be discussed later, note that such pathological cases as the well-known counter-example of Hodges are avoidable by the introduction of locally uniform convergence. For a sequence of positive numbers $\{c_n\}$ (c_n tending to infinity) an estimator $\hat{\theta}$ is called consistent of order c_n (or $\{c_n\}$ -consistent for short) if for every $\varepsilon > 0$ and every $\vartheta \in \Theta$ there exist a sufficiently small positive number δ and a sufficiently large number L satisfying the following :

$$\text{(1.1.1)} \qquad \overline{\lim_{n\to\infty}} \sup_{\theta:\|\theta-\vartheta\|<\delta} P_{\theta,n}\left\{ c_n \| \hat{\theta}_n - \theta \| \geqq L \right\} < \varepsilon .$$

It should also be noted that the concept of the order of convergence is always defined in "locally uniform" sense, otherwise the maximum order of convergence could not be uniquely defined.
Let c_n be the maximum order of consistency. A $\{c_n\}$ -consistent estimator $\hat{\theta}$ is defined to be asymptotically efficient if the asymptotic distribution of $\hat{\theta}$ is most concentrated in the neighborhood of θ (in some sense) among all $\{c_n\}$ -consistent estimators.

Here we have to distinguish between the case p=1 and the case $p \geqq 2$.

First we consider the case when p=1. A $\{c_n\}$-consistent estimator $\hat{\theta}$ is called <u>asymptotically median unbiased (AMU)</u> if

$$(1.1.2) \qquad \lim_{n\to\infty} P_{\theta,n}\{\hat{\theta}_n \leqq \theta\} = \lim_{n\to\infty} P_{\theta,n}\{\hat{\theta}_n \geqq \theta\} = \frac{1}{2}$$

uniformly in a neighborhood of every θ of Ⓗ. If there exists an AMU estimator maximizing

$$(1.1.3) \qquad \lim_{n\to\infty} P_{\theta,n}\{-a < c_n(\hat{\theta}_n - \theta) < b\}$$

for all positive numbers a and b and every θ in the class of all AMU estimators, then it is called asymptotically efficient. The asymptotically efficient estimator generally may not exist. However if $X_1, \ldots, X_n$ are independently and identically distributed (i.i.d.) random variables and the support of the distribution is independent of θ and the density is continuously differentiable with respect to θ and the Fisher information is finite, then it will be shown that the maximum likelihood estimator (MLE) is asymptotically efficient with order $c_n = \sqrt{n}$. In other cases when the maximum order of consistency is $c_n = \sqrt{n \log n}$ the MLE is usually asymptotically efficient. Indeed, the most powerful test for testing the hypothesis : $\theta = \theta_0 + tc_n^{-1}$ against the alternative : $\theta = \theta_0$ is given by

$$\sum_{i=1}^{n} \frac{\partial}{\partial\theta} \log f(x_i, \theta) > \lambda \text{ or } < \lambda .$$

If the Fisher information is infinite and the distribution of $(\partial/\partial\theta) \log f(X, \theta)$ belongs to the domain of attraction of a normal law, then it follows that $c_n = \sqrt{n \log n}$, and the MLE is asymptotically efficient.

It is a very strong condition to request maximizing the probability of (1.1.3) uniformly in all positive numbers a and b also in θ. If there exists an asymptotically efficient estimator $\hat{\theta}^*$ in this sense, then $\hat{\theta}^*$ minimizes

$$\lim_{n\to\infty} E_\theta [\, L(c_n(\hat{\theta}_n - \theta))\,]$$

in the class of the all AMU estimators defined by (1.1.2), where $L(u)$ is a bounded function such that $L(u) \geqq 0$, $L(0)=0$ and $L(u)$ is monotone increasing for $u > 0$ and monotone decreasing for $u < 0$. Hence it is seen that the condition of asymptotic efficiency is stronger than that of minimizing the asymptotic variance and that the definition itself of asymptotic efficiency is independent of the normality of the asymptotic distribution. Here there are two delicate points. One point is that it is generally impossible to replace the AMU condition (1.1.2) by that of asymptotic unbiasedness

(1.1.4) $$\lim_{n\to\infty} E[c_n(\hat{\theta}_n - \theta)] = 0$$

Apart from the problem of the existence of the expectation and that of the difference between the asymptotic value of the expectation and the asymptotic variance, there does not generally exist an estimator maximizing (1.1.3) uniformly, under the condition (1.1.4). This is seen in the case when $X_1, \ldots, X_n$ are independently and normally distributed random variables with unknown mean μ and known variance σ^2. Then if $a \neq b$ we can construct an estimator $\hat{\mu}$ for which

$$P_\mu\{-a < \hat{\mu} - \mu < b\} > P_\mu\{-a < \bar{x} - \mu < b\}$$

for all μ by putting

$$\hat{\mu} = \bar{x} + \frac{b-a}{2} \quad ,$$

where U is a random variable independent of $\bar{x}$ and

$$P\{U = 1\} = 1 - \frac{1}{M} \quad ;$$

$$P\{U = 0\} = \frac{1}{M} \, ,$$

where M is a sufficiently large positive constant.

The second point is that local uniformity in θ is not required in the convergence of

$$F_\theta(b)-F_\theta(-a)=\lim_{n\to\infty} P_{\theta,n}\{-a<c_n(\hat{\theta}_n-\theta)<b\}$$

in (1.1.2). If the asymptotic distribution of $\hat{\theta}$ is normal, then local uniformity holds. However it must be remarked that the local uniformity in θ is not required in the definition of the asymptotic efficiency. In the case when the asymptotically efficient estimator does not exist in general, it can happen that there is an estimator $\hat{\theta}$ for which (1.1.3) is maximized for specified θ_0 but the asymptotic distribution of $\hat{\theta}$ in the neighborhood of θ_0 does not coincide with that at θ_0.

Indeed, Let $X_1, \ldots, X_n$ be i.i.d. random variables with the uniform distribution on $[\theta - 1/2, \theta+1/2]$. Since a sufficient statistic is given by $(\min X_i, \max X_i)$, it follows that for specified θ_0 the estimator maximizing

$$\lim_{n\to\infty} P_{\theta_0,n}\{-a<n(\hat{\theta}_n-\theta)<b\}$$

for every $a>0$ and $b>0$ can be constructed for finite n in a way as is illustrated by the Figure 1.1.1.

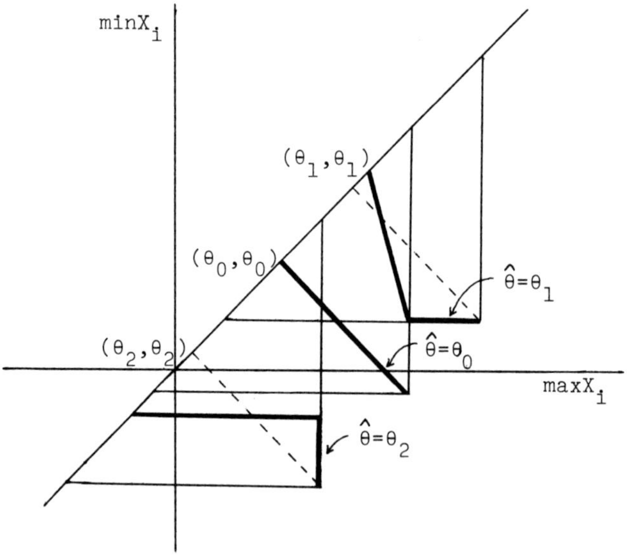

Figure 1.1.1.

Such an estimator is called <u>locally asymptotically efficient</u>.
If there exists an estimator maximizing

$$\lim_{n\to\infty} P_{\theta,n}\left\{ c_n(\hat{\theta}_n - \theta) < a \right\}$$

or

$$\lim_{n\to\infty} P_{\theta,n}\left\{ -a < c_n(\hat{\theta}_n - \theta) \right\}$$

for every $a > 0$ and every θ, then it is called <u>one-sided asymptotically efficient</u>.

For all estimators $\hat{\theta}$ for which

$$F_\theta(a) = \lim_{n\to\infty} P_{\theta,n}\left\{ c_n(\hat{\theta}_n - \theta) < a \right\}$$

exists and $F_\theta(a)$ is continuous in θ for each a, then the estimator $\hat{\theta}^*$ maximizing

$$F_\theta(a) - F_\theta(-a) = \lim_{n\to\infty} P_{\theta,n}\left\{ c_n |\hat{\theta}_n - \theta| < a \right\} \tag{1.1.5}$$

for every $a > 0$ and every θ is called <u>two-sided asymptotically efficient</u>. In the above example of uniform distribution the two-sided asymptotically efficient estimator is given by

$$\hat{\theta}^* = (\min X_i + \max X_i)/2 .$$

It should be noted that the uniformity of the convergence of (1.1.5) and the continuity of the asymptotic distribution are essential conditions.

Next, it is shown that the MLE can be asymptotically inadmissible in non-regular cases. Indeed, let $X_1, \ldots, X_n$ be i.i.d. random variables with the density

$$f(x-\theta) = \begin{cases} C \exp\left\{ -\frac{1}{2}(x-\theta)^2 \right\} & \text{if } |x-\theta| \leq 1 ; \\ 0 & \text{otherwise} , \end{cases}$$

where C is a constant.

Then the MLE $\hat{\theta}_{ML}$ is asymptotically given by

$$\hat{\theta}_{ML} = \begin{cases} \min X_i + 1 & \text{for} \quad \bar{X} > \theta \quad ; \\ \max X_i - 1 & \text{for} \quad \bar{X} < \theta \quad , \end{cases}$$

where $\bar{X} = (1/n) \sum_{i=1}^{n} X_i$.

Then the asymptotic distribution of $n(\hat{\theta}_{ML} - \theta)$ is a double exponential distribution. Put $\hat{\theta}^* = (\min X_i + \max X_i)/2$. It can be shown that the asymptotic distribution of $n(\hat{\theta}^* - \theta)$ is also a double exponential distribution but its scale parameter is one half that of the asymptotic distribution of $n(\hat{\theta}_{ML} - \theta)$. Hence the asymptotic efficiency of $\hat{\theta}_{ML}$ compared with $\hat{\theta}^*$ is exactly 50% (See section 3.3).

1.2. Multidimensional parameter case

There are two wayes to consider the case $p \geqq 2$. One way is that we concentrate our attention only a specified component ξ of θ and regard the other components as nuisance parameters and consider the estimation problem of ξ. Another way is that we consider the problem of joint estimation of all the components of θ. In the first case let $\theta = (\xi, \eta)$, where ξ is a real number. We consider the class of the all estimators $\hat{\xi}$ of ξ for which

(1.2.1) $$\lim_{n\to\infty} P_{\theta,n}\{\hat{\xi} \leqq \xi\} = \lim_{n\to\infty} P_{\theta,n}\{\hat{\xi} \geqq \xi\} = 1/2$$

locally uniformly in θ for every θ.

The estimator maximizing

$$\lim_{n\to\infty} P_{\theta,n}\{-a < c_n(\hat{\xi} - \xi) < b\}$$

for all positive numbers a and b and every θ in the above class of estimators is called the asymptotically efficient estimator of ξ.

For example, let $X_1, \ldots, X_n$ be i.i.d. random variables with the uniform distribution on $[\xi - \eta/2, \xi + \eta/2]$, where ξ and η are unknown. Then it may be shown that $\hat{\xi} = (\min X_i + \max X_i)/2$ is asymptotically efficient estimator of ξ. In the second case, if there exists an estimator maximizing

(1.2.2) $$\lim_{n\to\infty} P_{\theta,n}\left\{c_n(\hat{\theta}-\theta) \in C\right\}$$

for every convex set C containing the origin, then it seems natural to call it asymptotically efficient. However, as is stated above, it is impossible to define the estimator maximizing this probability under the condition (1.1.4) of asymptotic unbiasedness. On the other hand, it is also impossible to define the estimator maximizing (1.2.2) under the condition of coordinate-wise AMU, i.e.,

(1.2.3) $$\lim_{n\to\infty} P_{\theta,n}\left\{\hat{\theta}_\alpha \leqq \theta_\alpha\right\} = \lim_{n\to\infty} P_{\theta,n}\left\{\hat{\theta}_\alpha \geqq \theta_\alpha\right\} = \frac{1}{2}$$

for $\alpha = 1, \ldots, p$, where $\theta = (\theta_1, \ldots, \theta_p)$ and $\hat{\theta} = (\hat{\theta}_1, \ldots, \hat{\theta}_p)$. For example, let $X_1, \ldots, X_n$ be i.i.d. random vectors with a multivariate normal distribution and the components are mutually independent. Since all estimators $\hat{\underset{\sim}{\mu}}$ of mean vector $\underset{\sim}{\mu}$ of the type

$$\hat{\underset{\sim}{\mu}} = \bar{X} + \phi(A\bar{X}) \quad ,$$

where A is (p-1)xp matrix with $A\mathbf{1}=0$, satisfy (1.2.3), it is possible to have the probability of (1.2.2) for $\underset{\sim}{\mu} = \underset{\sim}{0}$ larger than that of $\hat{\underset{\sim}{\mu}} = \bar{X}$ with appropriate choice of the function ϕ.

If the stronger condition than (1.2.3) is assumed, that

(1.2.4) $$\lim_{n\to\infty} P_{\theta,n}\left\{c_n a'(\hat{\theta}-\theta) \leqq 0\right\} = \lim_{n\to\infty} P_{\theta,n}\left\{c_n a'(\hat{\theta}-\theta) \geqq 0\right\} = \frac{1}{2}$$

for all real vectors a, then there does not generally exist the estimator which satisfies (1.2.4) and maximizes (1.2.2). In order to define the estimator maximizing (1.2.2) for every convex set containing the origin we have to confine to a class of estimators with asymptotic distributions of certain types.

For example, if the asymptotic distribution of $c_n(\hat{\theta}-\theta)$ is normal with mean vector $\underset{\sim}{0}$ and variance-covariance matrix Σ, then the infimum of Σ in the matrix sense is obtained and it will be shown that the MLE attains the infimum. Thus the MLE is asymptotically efficient in the sense that it maximizes (1.2.2) in the class of the all estimators with asymptotically normal distributions. This fact coincides with the classical results of earlier papers but it is impossible to extend it to the general case.

1.3. Higher order asymptotic efficiency

The definition of higher order asymptotic efficiency is given as the direct extension of asymptotic efficiency. For the case p=1, if the estimator $\hat{\theta}$ satisfies

$$(1.3.1) \qquad \lim_{n\to\infty} d_n \left| P_{\theta,n}\{\hat{\theta}_n \leqq \theta\} - \frac{1}{2} \right| = \lim_{n\to\infty} d_n \left| P_{\theta,n}\{\hat{\theta}_n \geqq \theta\} - \frac{1}{2} \right| = 0 ,$$

for a sequence $\{d_n\}$ of positive numbers going to infinity, then it is called a <u>higher order AMU estimator</u>. If a higher order AMU estimator $\hat{\theta}^*$ satisfies

$$(1.3.2) \qquad \lim_{n\to\infty} d_n [P_{\theta,n}\{-a < c_n(\hat{\theta}_n^* - \theta) < b\} - P_{\theta,n}\{-a < c_n(\hat{\theta}_n - \theta) < b\}] \geqq 0$$

for all positive numbers a and b and every θ and for every higher order AMU estimators $\hat{\theta}_n$, then $\hat{\theta}^*$ is called <u>higher order</u> (or order d_n) <u>asymptotically efficient</u>. When $c_n=\sqrt{n}$ and $d_n=n^{(k-1)/2}$, it is called a k-th order asymptotically efficient estimator.
The following fact is useful for checking the existence of the higher order asymptotically efficient estimator. Let $X_1, \dots, X_n$ be i.i.d. random variables with a density $f(x,\theta)$. We assume that the support of $f(x,\theta)$ is independent of θ. Then it may be shown that under regularity conditions the solution $\hat{\theta}_b$ of the equation

$$(1.3.3) \qquad \sum_{i=1}^{n} f(X_i, \hat{\theta} - c_n^{-1}b) - \sum_{i=1}^{n} \log f(X_i, \hat{\theta}) = k_n(\hat{\theta})$$

satisfies (1.3.2) for $a=\infty$, where k_n is determined so that estimator $\hat{\theta}_b$ is higher order AMU. In such a way for every d_n we may obtain the estimator maximizing the asymptotic probability for specified b. If the asymptotic distribution of the estimator obtained by such a way is independent of $b(>0)$ up to order d_n^{-1}, then it may be shown that it is order d_n asymptotically efficient. If it also coincides with the asymptotic distribution of the solution $\hat{\theta}_a$ of the equation

$$(1.3.4) \qquad \sum_{i=1}^{n} \log f(X_i, \hat{\theta} + c_n^{-1}a) - \sum_{i=1}^{n} \log f(X_i, \hat{\theta}) = l_n(\hat{\theta})$$

up to order d_n^{-1}, then it is seen that the estimator is order d_n asymptotically efficient. If there is an order d_n asymptotically efficient estimator and $f(x,\theta)$ is differentiable in θ, then it asymptotically coincides with the solution of the equation

$$(1.3.5) \qquad \sum_{i=1}^{n} \frac{\partial}{\partial\theta} \log f(X_i, \hat{\theta}) = k_n^*(\hat{\theta})$$

or the estimator

$$(1.3.6) \qquad \hat{\theta}^* = \hat{\theta}_0 - k_n^*(\hat{\theta}_0) ,$$

where $\hat{\theta}_0$ is the solution of the equation

$$\sum_{i=1}^{n} \frac{\partial}{\partial\theta} \log f(X_i, \theta) = 0 .$$

Hence if there exists a higher order asymptotically efficient estimator, then it is given as the solution of the equation (1.3.5) or the modified (or adjusted) MLE (1.3.6) (See Chapter6).
It will be seen that the modified MLE is not asymptotically inadmissible in regular cases.

In the regular case it will be shown that the bound of the asymptotic distribution is given by

$$(1.3.7) \qquad P_{\theta,n}\left\{ \sqrt{n}\,(\hat{\theta}_n - \theta) \leqq t \right\}$$

$$= \Phi(\sqrt{I}\,t) + \frac{3J+2K}{6I^{3/2}\sqrt{n}} I t^2 \phi(\sqrt{I}\,t) + o\left(\frac{1}{\sqrt{n}}\right) ,$$

where $\phi(t) = \frac{1}{\sqrt{2\pi}} e^{-\frac{t^2}{2}}$, $\Phi(t) = \int_{-\infty}^{t} \phi(x)dx$,

$I = E_\theta\left[-\frac{\partial^2}{\partial\theta^2} \log f(X,\theta)\right]$, $J = E_\theta\left[\left\{\frac{\partial^2}{\partial\theta^2} \log f(X,\theta)\right\}\left\{\frac{\partial}{\partial\theta} \log f(X,\theta)\right\}\right]$,

$K = E_\theta\left[\left\{\frac{\partial}{\partial\theta} \log f(X,\theta)\right\}^3\right]$.

It follows that the estimator $\hat{\theta}^*$ satisfying the following (i), (ii) and (iii) below is second order asymptotically efficient :

(i) The asymptotic distribution of $\sqrt{n}(\hat{\theta}^* - \theta)$ is normal with mean 0 and variance matrix I^{-1} .

(ii) The distribution of $\sqrt{n}(\hat{\theta}^* - \theta)$ admits the Edgeworth expansion up to order $n^{-1/2}$;

(iii) $P_{\theta,n}\{\hat{\theta}_n^* \leq \theta\} = P_{\theta,n}\{\hat{\theta}_n^* \geq \theta\} = \frac{1}{2} + o(\frac{1}{\sqrt{n}})$.

This fact is proved since if the asymptotic distribution of $\sqrt{n}$ $(\hat{\theta}^* - \theta)$ does not satisfy the bound (1.3.7) for either $t>0$ or $t<0$, then it violates the bound for $t>0$ or $t<0$. Hence the MLE modified to be second order AMU is second order asymptotically efficient (See section 4.1). A rather straightforward result was first obtained by Takeuchi and Akahira [44] and Pfanzagl [32].

In the case when the support of $f(x,\theta)$ is independent of θ but other regularity conditions do not hold it is also possible to check the existence of the higher order asymptotically efficient estimator. For example let $X_i = (X_{1i}, X_{2i})$ $(i=1,2,\ldots)$ be random vectors with $X_{2i} = \theta X_{1i} + U_i$ $(i=1,2,\ldots)$, where U_i has the normal distribution with mean 0 and variance 1 and X_{1i} is independent of U_i and has the density $f(x)$. Since the density of $X=(X_1, X_2)$ is given by

$$\frac{1}{\sqrt{2\pi}} f(x_1) \exp\left\{ - \frac{(x_2 - \theta x_1)^2}{2} \right\}$$

it follows that the solution $\hat{\theta}_b$ of the equation (1.3.3) satisfies

$$\sum_{i=1}^{n}(X_{2i} - \hat{\theta}_b X_{1i})^2 - \sum_{i=1}^{n}(X_{2i} - (\hat{\theta}_b - c_n^{-1} b) X_{1i})^2 = k_n(\hat{\theta}_b) .$$

Then we have

$$\hat{\theta}_b - \frac{c_n k(\hat{\theta}_b)}{2b \sum_{i=1}^{n} X_{1i}^2} = \frac{\sum_{i=1}^{n} X_{1i} X_{2i}}{\sum_{i=1}^{n} X_{1i}^2} + \frac{c_n^{-1} b}{2} .$$

If $E(X_1^2) = \sigma^2 < \infty$, putting $k_n(\hat{\theta}) = n c_n^{-2} b^2 \sigma^2$ we obtain

$$\hat{\theta}_b \sim \sum_{i=1}^{n} X_{1i} X_{2i} \Big/ \sum_{i=1}^{n} X_{1i}^2$$

Hence it follows that $\hat{\theta}_b$ is asymptotically efficient estimator. In this case it is seen that $c_n = \sqrt{n}$.

Next putting $k_n(\hat{\theta}_b)=b^2\sigma^2+(k^*/n)$ we have

$$\hat{\theta}_b = \frac{\sum_{i=1}^n X_{1i}X_{2i}}{\sum_{i=1}^n X_{1i}^2} - \frac{b}{2\sqrt{n}}\left(\frac{n\sigma^2+k^*}{\sum_{i=1}^n X_{1i}^2} - 1\right).$$

Then we obtain

$$\sqrt{n}(\hat{\theta}_b-\theta) = \frac{\sqrt{n}\sum_{i=1}^n X_{1i}U_i}{\sum_{i=1}^n X_{1i}^2} - \frac{b}{2}\left(\frac{n\sigma^2+k^*}{\sum_{i=1}^n X_{1i}^2} - 1\right)$$

$$= \sqrt{\frac{n}{\sum_{i=1}^n X_{1i}^2}}\, Z - \frac{b}{2}\left(\frac{n\sigma^2+k^*}{\sum_{i=1}^n X_{1i}^2} - 1\right),$$

where Z is independent of X_{1i} and has the standard normal distribution. If $\sum_{i=1}^n (X_{1i}^2-\sigma^2)/\sqrt{n}$ has the asymptotic normal distribution, then the distribution of $\hat{\theta}_b$ is determined independently of b up to order $n^{-1/2}$. Hence $\hat{\theta}_b$ is second order asymptotically efficient. Therefore if $E(X_1^4)<\infty$ we see that the MLE $\hat{\theta}_{ML}$ of θ given by $\hat{\theta}_{ML}= \sum_{i=1}^n X_{1i}X_{2i}/\sum_{i=1}^n X_{1i}^2$ is second order asymptotically efficient. Considering the estimator $\hat{\theta}_b$ (or $\hat{\theta}_a$) we see that there does not generally exist the third order asymptotically efficient estimator even in the regular case. If we consider the estimator maximizing $\lim_{n\to\infty} P_{\theta,n}\{\sqrt{n}\,|\hat{\theta}_n-\theta|<a\}$ under the condition

$$\lim_{n\to\infty} n\left|P_{\theta,n}\{\sqrt{n}(\hat{\theta}_n-\theta)\leqq a\}-F_\theta(a)-\frac{1}{\sqrt{n}}G_\theta^1(a)-\frac{1}{n}G_\theta^2(a)\right| = 0,$$

then we see that the MLE maximizes it up to order n^{-1}, that is, the MLE is two-sided third order asymptotically efficient.

For $p\geqq 2$, we consider the case when only a specified component is estimated. Let $\theta=(\xi,\eta)$. We consider the estimator maximizing

$$P_{\theta,n}\{-a<c_n(\hat{\xi}-\xi)<b\}$$

for all positive numbers a and b and every θ, up to order d_n^{-1} in the class of all estimators $\hat{\xi}$ satisfying

$$\lim_{n\to\infty} d_n\left|P_{\theta,n}\{\hat{\xi}\leqq\xi\}-\frac{1}{2}\right| = \lim_{n\to\infty} d_n\left|P_{\theta,n}\{\hat{\xi}\geqq\xi\}-\frac{1}{2}\right| = 0.$$

Then we can discuss the higher order asymptotic efficiency in a way similar to the one-parameter case.

In the regular case it is known that $c_n = \sqrt{n}$ and the asymptotically efficient estimator whose distribution admits Edgeworth expansion up to order $n^{-1/2}$ is always second order asymptotically efficient.

1.4. The estimator in the class $\mathbb{C}$ and $\mathbb{D}$

In the regular case the distribution of the asymptotically efficient estimator admits Edgeworth expansion but it is asymptotically efficient in a wider class of estimators. However in the higher order and particularly multiparameter case the class of the estimators should be restricted. We shall discuss classes called class $\mathbb{C}$ and class $\mathbb{D}$. We consider estimators whose distributions admit Edgeworth expansion. Let $X_1, \ldots, X_n$ be i.i.d. random variables with density $f(x,\theta)$, where θ is p-dimensional real vector parameter. Put $\hat{\theta} = \hat{\theta}(X_1, \ldots, X_n)$. If the asymptotic distribution of $\sqrt{n}(\hat{\theta} - \theta)$ is normal with mean $\underset{\sim}{0}$ and variance-covariance matrix Σ, then it is defined to belong to class $\mathbb{A}$. It has been already stated that the MLE maximizes

$$\lim_{n\to\infty} P_{\theta,n}\{\sqrt{n}(\hat{\theta} - \theta) \in C\}$$

for every convex set C containing the origin in the class $\mathbb{A}$. If the asymptotic distribution of $\sqrt{n}(\hat{\theta} - \theta)$ is the multivariate normal distribution and admits Edgeworth expansion up to $n^{-1/2}$, then it is defined that $\hat{\theta}$ belongs to the class $\mathbb{B}$. Let $\mathbb{B}_0$ be the subclass of $\mathbb{B}$ whose estimators satisfy one of the following conditions : asymptotic coordinate-wise median unbiasedness of order $\sqrt{n}$, asymptotic unbiasedness and asymptotic mode unbiasedness. If the asymptotic distribution of $\sqrt{n}(\hat{\theta} - \theta)$ is multivariate normal distribution, then it will be shown that third asymptotic cumulant is given by

$$(1.4.1) \quad E[\sqrt{n}(\hat{\theta}_\alpha - \mu(\theta_\alpha))\sqrt{n}(\hat{\theta}_\beta - \mu(\theta_\beta))\sqrt{n}(\hat{\theta}_\gamma - \mu(\theta_\gamma))]$$

$$= \frac{1}{2\sqrt{n}}[(\frac{\partial}{\partial\theta_\alpha}I^{\beta\gamma} + \frac{\partial}{\partial\theta_\beta}I^{\alpha\gamma} + \frac{\partial}{\partial\theta_\gamma}I^{\alpha\beta}) + E(U_\alpha U_\beta U_\gamma)] + o(\frac{1}{\sqrt{n}}),$$

where $U_\alpha = \sum_\beta \sum_i I^{\alpha\beta} \frac{\partial}{\partial\theta} \log f(X_i, \theta)$ and $I^{\alpha\beta}$ denotes (α, β)-component of $\Sigma = I^{-1}$ (See section 5.1). Since the asymptotic expectation is determined up to order $n^{-1/2}$ by the second order asymptotic median unbiasedness condition, it follows that all asymptotically efficient estimators $\hat{\theta}^*$ belonging to the class $\mathbb{B}_0$, that is, the estimators of $\mathbb{B}_0$ with the asymptotic variance matrix I^{-1} have the same asymptotic distribution up to order $n^{-1/2}$. Hence they maximizes

$$\lim_{n\to\infty} P_{\theta,n} \left\{ \sqrt{n}(\hat{\theta} - \theta) \in C \right\}$$

for every convex set C containing the origin up to order $n^{-1/2}$ in the class $\mathbb{B}_0$.

Next we define the class $\mathbb{C}$ as those estimators $\hat{\theta}$ whose distributions admit Edgeworth expansion up to order n^{-1} and

$$(1.4.2) \qquad \sqrt{n}(\hat{\theta} - \theta) = U_\alpha + \frac{1}{\sqrt{n}} Q_\alpha + o_p\left(\frac{1}{n}\right)$$

and $E(U_\alpha Q_\beta) = o(1)$. Let $\mathbb{C}_0$ be the subclass of $\mathbb{C}$ whose elements satisfy one of the following conditions : asymptotic coordinatewise median unbiasedness of order n, asymptotic unbiasedness or asymptotic mode unbiasedness. Then it will be shown that the fourth cumulants of all asymptotically efficient estimators are equal up to order n^{-1}. Denote the asymptotic variance matrix by $I^{-1} + \frac{1}{n} Z$. It will be shown that when $\hat{\theta}$ is given by

$$(1.4.3) \qquad \hat{\theta} = \hat{\theta}_{ML} + \frac{1}{n} \varphi + o_p\left(\frac{1}{n}\right) \quad ,$$

the matrix Z is minimized, where $\hat{\theta}_{ML}$ denotes the MLE and φ is a constant depending on the parameter. If C is a symmetric convex set with respect to the origin, then it will be shown that

$$\lim_{n\to\infty} P_{\theta,n} \left\{ \sqrt{n}(\hat{\theta} - \theta) \in C \right\}$$

depends on only the asymptotic cumulants of even orders.

If the modified MLE

$$\hat{\theta}^* = \hat{\theta}_{ML} + \frac{1}{n} \varphi(\hat{\theta}_{ML})$$

belongs to the class $\mathbb{C}_0$, then it has third order (or order n) symmetric asymptotic efficiency (See sections 5.1, 5.2).

Let $\mathbb{D}$ be the subspace of $\mathbb{C}$ whose elements further satisfy

(1.4.4) $\quad E(U_\alpha Q_\beta^j)=o(1)$ $(j=1,2)$.

Let $\mathbb{D}_0$ be the subclass of $\mathbb{D}$ whos elements satisfy one of the folloing conditions : asymptotic coordinate-wise median unbiasedness of order n, the asymptotic unbiasedness and the asymptotic mode unbiasedness. The asymptotic cumulants up to fourth order of the elements of $\mathbb{D}_0$ are determined up to order n^{-1} except for the variance matrix. If the estimator of the type (1.4.2) belongs to $\mathbb{D}_0$, then it may be shown that it maximizes

$$\lim_{n\to\infty} P_{\theta,n}\left\{\sqrt{n}(\hat{\theta}-\theta) \in C\right\}$$

for every convex set C up to order n^{-1} in the class $\mathbb{D}_0$. Hence it is seen that the modified MLE is third order asymptotically efficient in the class $\mathbb{D}_0$. Further we define the class $\mathbb{E}$ by those estimators $\hat{\theta}$ whose distributions admit Edgeworth expansion up to order $n^{-3/2}$ and which satisfy

(1.4.5) $\quad \sqrt{n}(\hat{\theta}-\theta) = U_\alpha + \frac{1}{\sqrt{n}}Q_\alpha + \frac{1}{n}W_\alpha + o_p(\frac{1}{n})$.

Let $\mathbb{E}_0$ be the subclass of $\mathbb{E}$ whose elements satisfy one of the asymptotic coordinate-wise median unbiasedness of order $n^{3/2}$, the asymptotic unbiasedness and the asymptotic mode unbiasedness conditions. It will be shown that the modified MLE belonging to the class $\mathbb{E}_0$ has fourth order (or order $n^{3/2}$) symmetric asymptotic efficiency. In general we may not proceed to higher order asymptotic efficiency than fifth order. Indeed, it will be shown that there does not generally exist higher order asymptotically efficiency than fifth.

We have modified Bayes estimators with respect to the symmetric loss function as the third order asymptotically efficient estimators belonging to the class $\mathbb{D}_0$ other than the modified MLE. Let $\hat{\theta}_0$ be the Bayes estimator with respect to the symmetric loss function L(u) (i.e. L(-u)=L(u)) and the prior distribution with the smooth density $p(\theta)$. Then $\hat{\theta}_0$ belongs to $\mathbb{D}$. If the estimator $\hat{\theta}_0^*$ of the type

$$\hat{\theta}_0^* = \hat{\theta}_0 + \frac{1}{n}\Psi(\hat{\theta}_0)$$

belongs to $\mathbb{D}_0$, then $\hat{\theta}_0^*$ asymptotically coincides with the modified MLE up to order n^{-1} independently of L and p (See section 5.3). If L is asymmetric, then the Bayes estimator belongs to the class $\mathbb{C}$ but not generally to $\mathbb{D}$ (See section 5.3). Then its asymptotic distribution of order n depends on L.

From the above discussion we see that it is important to modify estimators to be AMU in the case of comparison of the asymptotic distributions of estimators up to order n^{-1}. If we compare the asymptotic values of the mean square error without such a modification, then we can not obtain any definite conclusion about the optimality of the estimators in general. From the above we see that the (modified) MLE is higher order asymptotically efficient. The asymptotic efficiency is a common property of the MLE and the Bayes estimator. This could be conjectured from the fact that the MLE is the limit of Bayes estimators in some sense (e.g. the case when $L(u)=\|u\|^{\alpha}$ and $\alpha\to\infty$ is considered). Hence the asymptotic efficiency of the Bayes estimator may be a more fundamental fact. If the likelihood function is once differentiable but not differentiable more than twice, it will be shown that the MLE has asymptotic efficiency but not higher order asymptotic efficiency in appropriate examples. In this case some kind of Bayes estimator has higher order asymptotic efficiency or it may make larger the probability of the concentration about the origin than the MLE when the higher order asymptotic distribution is considered.

1.5. More general cases

All the conclusions of the preceding sections are actually independent of the assumptions of the identity of the distributions of the observations. What is essentially required is validity of Edgeworth expansion of the asymptotic distributions of estimators, and also of the derivatives of the log-likelihood function, a fact which can be seen from the contexts of the proofs of established theorems and propositions. Therefore, in the case where the Edgeworth expansion applies, the third order asymptotic efficiency of the modified maximum likelihood estimator can be established, even when the sample variables are not identically distributed or independent. What is more difficult is to prove the validity of the formal expansion, which is derived by fitting asymptotic cumulants. It should also be noted that the above discussions exclude the case of discrete or lattice distributions, because in such cases not only the Edgeworth expansion fails to apply, but it is also impossible to have asymptotic median unbiasedness or asymptotic unbiasedness up to the order n^{-1} without resorting to randomization. Hence in discrete cases we are faced with much more complicated situation than continuous cases, and no satisfactory theory has been found to tackle them.

The case of non-identically but independently distributed samples is dealtwith.

Suppose that $X_1, X_2, \ldots, X_n, \ldots$ are independently distributed according to continuous distributions with density functions $f_i(x_i, \theta)$ $(i=1,2,\ldots)$ having the common real vector parameter θ. We assume that

$$\sum_{i=1}^{n} I_i(\theta)=O(n) \quad \text{and} \quad \sum_{i=1}^{n} \frac{I_i(\theta)}{n} \to \bar{I}(\theta) \quad (n\to\infty),$$

where

$$I_i(\theta)=E[\frac{\partial}{\partial\theta}\log f_i(x_i,\theta)\frac{\partial}{\partial\theta'}\log f_i(x_i,\theta)]$$
$$= -E[\frac{\partial^2}{\partial\theta\partial\theta'}\log f(x_i,\theta)] \quad ,$$

and that the distribution of

$$\frac{1}{\sqrt{n}}\sum_{i=1}^{n}\frac{\partial}{\partial\theta}\log f_i(x_i,\theta)$$

admits the Edgeworth expansion up to the order n^{-1} which requires some sort of the Lindeberg conditions. Then it can be proved that an estimator $\hat{\theta}$ is asymptotically efficient if $\sqrt{n}(\hat{\theta}-\theta)$ is asymptotically distributed with mean vector $\underset{\sim}{0}$ and variance matrix $\bar{I}(\theta)^{-1}$. This holds true if and only if it can be expressed as

$$\sqrt{n}(\hat{\theta}-\theta)\sim\bar{I}(\theta)^{-1}\frac{1}{\sqrt{n}}\sum_{i=1}^{n}\frac{\partial}{\partial\theta}\log f_i(x_i,\theta)$$

or

$$\sqrt{n}(\hat{\theta}_\alpha-\theta_\alpha)\sim U_\alpha \quad ,$$

where

$$U_\alpha=\sum_\beta\sum_i \bar{I}^{\alpha\beta}(\theta)\frac{\partial}{\partial\theta_\beta}\log f_i(x_i,\theta)$$

with (α,β)-element $\bar{I}^{\alpha\beta}(\theta)$ of the matrix $\bar{I}(\theta)^{-1}$.

The class $\mathbb{C}$ and the class $\mathbb{D}$ of estimators can be defined in the same way as in the i.i.d. case as the estimators of the type

$$\sqrt{n}(\hat{\theta}_\alpha-\theta_\alpha) = U_\alpha+\frac{1}{\sqrt{n}}Q_\alpha+o_p(\frac{1}{n}) \quad .$$

It is proved exactly as in the i.i.d. case that the adjusted maximum likelihood estimator maximizes

$$P_{\theta,n}\{\sqrt{n}(\hat{\theta}-\theta)\in C\}$$

up to the order n^{-1} for any symmetric convex set C among the class $\mathbb{C}$ of estimators, and for any convex set C containing the origin among the class $\mathbb{D}$ of estimators with relevant (median unbiasedness) conditions.

A more concrete, and practically important case is the multivariate normal linear regression model. $X_1, X_2, \ldots, X_n, \ldots$ are independently distributed according to the multivariate normal distribution $N(\mu_i,\Sigma)$, where the mean vectors μ_i are expressed as

$$\mu_i = \Pi z_i \quad ,$$

where Π is the matrix of regression coefficients, and z_i's are vectors of independent variables. It is further assumed that the elements of Π can be expressed as smooth functions of a real vector-valued parameter, i.e. $\Pi = \Pi(\theta)$ with the condition that $\Pi(\theta_1) \neq \Pi(\theta_2)$ if $\theta_1 \neq \theta_2$. It is assumed that

$$\frac{1}{n} \sum_{i=1}^{n} z_i z_i' = M + o(\frac{1}{n}) \quad ,$$

where M is a positive definite matrix.

In this case, the two sets of satistics $(\hat{\Pi}, \hat{\Sigma})$ given by

$$\hat{\Pi} = (\sum_{i=1}^{n} x_i z_i')(\sum_{i=1}^{n} z_i z_i')^{-1}$$

$$\hat{\Sigma} = \sum_{i=1}^{n} (x_i - \hat{\Pi} z_i)(x_i - \hat{\Pi} z_i)'/n$$

form a sufficient statistic. Hence we may restrict the class of estimators of θ to those expressible as functions of $\hat{\Pi}$ and $\hat{\Sigma}$. We call the estimator of the type

$$\hat{\theta} = g(\hat{\Pi}, \hat{\Sigma}) + \frac{1}{n} h(\hat{\Pi}, \hat{\Sigma})$$

where g and h are smooth functions independent of n, an extended regular estimator. It is noted that the class of extended regular estimators are a subset of the class $\mathbb{D}$ of estimators. This implies that among all the extended regular estimators with the condition of asymptotic (median) unbiasedness up to the order n^{-1}, an extended maximum likelihood estimator, that is, an extended regular estimator where g is equal to the maximum likelihood and h is so defined that $\hat{\theta}$ be asymptotically (median) unbiased, is third order asymptotically efficient. We have also a modified statement. Corresponding to any extended regular estimator $\hat{\theta}$ there exists an estimator of the type

$$\hat{\theta}^* = \hat{\theta}_{ML} + \frac{1}{n} h^*(\hat{\Pi}, \hat{\Sigma}) \quad ,$$

which satisfies the condition that

$$\lim_{n \to \infty} n[\, P_{\theta,n}\{\sqrt{n}(\hat{\theta}^* - \theta) \in C\} - P_{\theta,n}\{\sqrt{n}(\hat{\theta} - \theta) \in C\}\,] \geq 0$$

for any convex set C containing the origin, where $\hat{\theta}_{ML}$ denotes the MLE. Here the definition of h depends on the first estimator $\hat{\theta}$ but independent of C. Hence it follows that any extended regular estimator $\hat{\theta}$ can be uniformly improved or dominated by a modified maximum likelihood estimator of the above type.

From this fact it may be said that the class of modified MLE's is asymptotically complete up to order n^{-1} among all the extended regular estimators. As a special example of the regression model, we can discuss the linear simultaneous equation model given by

$$By_t + \Gamma Z_t = u_t \ , \ t=1, \ldots , T ,$$

where y_t and Z_t are vectors of endogeneous and exogeneous variables respectively, and u_t's are random vectors independently distributed according to the multivariate normal distribution $N(\underset{\sim}{0},\Sigma)$.
Then the above model is equivalent to

$$y_t = \Pi z_t + v_t \ ,$$

where $\Pi = B^{-1}\Gamma$. With the identifiability conditions of the parameters in B and Γ, this model is the special case of the multivariate linear regression model. It follows that the modified limited information, subsystem, full system maximum likelihood estimator in the respective information structures are third order asymptotically efficient among (the class of the) relevant extended regular estimators. They also have larger probabilities of concentration around the true value in the order n^{-1} than the two stage least squares and three stage least squares and others.

In the subsequent chapters the following are given.

In Chapter 2 the necessary conditions for the existence of consistent estimator are generally established ([1]). The bounds of the order of consistency of estimators are also obtained and the existence of estimators with the boundary order of consistency is proved ([1]).

In Chapter 3 the asymptotic efficiency of estimators is defined and discussed in the usual regular case and an autoregressive process case as a typical example of non-i.i.d. cases is discussed ([4]). Further an example of asymptotic inefficiency of the maximum probability estimator is given in a non-regular case [10].

In Chapter 4 second and higher order asymptotic efficiency of estimators is defined and discussed in one-parameter, multiparameter and autoregressive process's cases and also non-i.i.d. cases ([44], [7], [3]).

In Chapter 5 it is shown that the maximum likelihood estimator (MLE) is third order asymptotically efficient in the multiparameter exponential case ([46]), the general case ([62]) and the multivariate linear regression models. Further it is shown that the generalized Bayes estimator is second order and third order asymptotically efficient ([48]).

In Chapter 6 the discretized likelihood estimator (DLE) is defined as a solution of the discretized likelihood equation ([8]). It is shown in comparison with DLE that the MLE is second order asymptotically efficient ([8]). Further it is seen that the asymptotic efficiency (including higher order cases) may be systematically discussed by the discretized likelihood methods ([8]).

In Chapter 7 the concept of (higher order) asymptotic completeness is introduced and it is shown that the MLE is higher order asymptotically complete ([62]). Further the second order asymptotic efficiency of unbiased confidence interval is discussed ([9]).

Chapter 2

Consistency of Estimators and Order of Consistency

Summary :

Suppose that $X_1, X_2, \ldots, X_n, \ldots$ is a sequence of independent identically distributed (i.i.d.) random variables. We assume that a parameter space Θ is an open subset in a Euclidean p-space. In the textbook discussion of asymptotic theory, it is usually shown that the asymptotically best (in some sense or other) estimator $\hat{\theta}_n^*$ has an asymptotic distribution of order $\sqrt{n}$, in the sense that the distribution of $\sqrt{n}(\hat{\theta}_n^* - \theta)$ converges to some probability law (in most cases normal). There are sporadic examples where the distribution of $\sqrt{n}(\hat{\theta}_n^* - \theta)$ or $\sqrt{n \log n}(\hat{\theta}_n^* - \theta)$ converges to some law (Woodroofe [57]) when X_i's are i.i.d. random variables with an uniform distribution or a truncated distribution. The purpose of this chapter is to give a systematic treatment to the problem of whether for a given sequence $\{c_n\}$, $c_n(\hat{\theta}_n^* - \theta)$ converges to some law, and what is the possible bound for such a sequence. In the location parameter case it will be shown that such a bound can be explicitely given. The asymptotic distribution of $c_n(\hat{\theta}_n^* - \theta)$ and the bound for it in non-regular cases is discussed by Akahira [2]. Also some results in terms of the asymptotic distribution of estimators are given in Takeuchi [42]. Asymptotic sufficiency of consistent estimators is discussed by Akahira [5] in non-regular cases.

2.1. Notation and definition

Let $\mathcal{X}$ be an abstract space the element of which are denoted by x. Let $\mathcal{B}$ be a σ-field of susbet of $\mathcal{X}$.

Let Θ be a parameter space, which is assumed to be an open subset of Euclidean p-space R^p with a norm denoted by $\|\cdot\|$. We shall denote by $(\mathcal{X}^{(n)},\mathcal{B}^{(n)})$ the n-fold direct products of $(\mathcal{X},\mathcal{B})$. For each $n=1,2,\ldots$, the points of $\mathcal{X}^{(n)}$ will be denoted by $\tilde{x}_n=(x_1, \ldots, x_n)$. We consider a sequence of classes of probability measures $\{P_{\theta,n} : \theta \in \Theta\}$ $(n=1,2,\ldots)$ each defined on $(\mathcal{X}^{(n)},\mathcal{B}^{(n)})$ and where for each n and each $\theta \in \Theta$ the following holds :

$$P_{\theta,n}(B^{(n)}) = P_{\theta,n+1}(B^{(n)}\times \mathcal{X})$$

for all $B^{(n)}\in \mathcal{B}^{(n)}$.

An estimator of θ is defined to be a sequence $\{\hat{\theta}_n\}$ where $\hat{\theta}_n$ is a $\mathcal{B}^{(n)}$-measurable function from $\mathcal{X}^{(n)}$ into Θ $(n=1,2,\ldots)$.

For simplicity we denote an estimator as $\hat{\theta}_n$ instead of $\{\hat{\theta}_n\}$.

Definition 2.1.1 An estimator $\hat{\theta}_n$ is called (weakly) consistent if for every $\varepsilon > 0$ and every $\theta \in \Theta$.

$$\lim_{n\to\infty} P_{\theta,n}\{\|\hat{\theta}_n - \theta\|>\varepsilon\}= 0 \quad . \tag{2.1.1}$$

For any two points θ and θ' in Θ there exists σ-finite measure μ_n such that $P_{\theta,n}$ and $P_{\theta',n}$ are absolutely continuous with respect to μ_n. Then for any points θ and θ' in Θ we define

$$\begin{aligned} d_n(\theta,\theta') &= \int_{\mathcal{X}^{(n)}} \left| \frac{dP_{\theta,n}}{d\mu_n} - \frac{dP_{\theta',n}}{d\mu_n} \right| d\mu_n \\ &= 2 \sup_{B\in\mathcal{B}^{(n)}} \left| P_{\theta,n}(B) - P_{\theta',n}(B) \right| \end{aligned} \tag{2.1.2}$$

It is easily seen that for each n, d_n is a metric on Θ which is independent of μ_n .

2.2. Necessary conditions for consistency of estimators

In this section we shall obtain necessary conditions for the existence of consistent estimators. The following is known (e.g.[25]).

Theorem 2.2.1. If there exists a consistent estimator then for any disjoint points θ_1 and θ_2 in Θ

(2.2.1) $\lim_{n\to\infty} d_n(\theta_1, \theta_2) = 2$.

Proof. If we denote a consistent estimator by $\hat{\theta}_n$, then for every $\varepsilon > 0$

$$\lim_{n\to\infty} P_{\theta,n}\left\{ \| \hat{\theta}_n - \theta \| > \varepsilon \right\} = 0 .$$

Put $\varepsilon = \| \theta_1 - \theta_2 \| / 2$. Then it follows that for any $\delta > 0$ there exists a n_0 such that for every $n \geqq n_0$

$$P_{\theta_1,n}\left\{ \| \hat{\theta}_n - \theta_1 \| \leqq \varepsilon \right\} \geqq 1 - \frac{\delta}{4} \quad ;$$

$$P_{\theta_2,n}\left\{ \| \hat{\theta}_n - \theta_2 \| \leqq \varepsilon \right\} \geqq 1 - \frac{\delta}{4} \quad .$$

Putting $A = \left\{ \| \hat{\theta}_n - \theta_1 \| \leqq \| \hat{\theta}_n - \theta_2 \| \right\}$, we have

$$P_{\theta_1,n}(A) \geqq 1 - \frac{\delta}{4} \quad ; \quad P_{\theta_2,n}(A^c) \geqq 1 - \frac{\delta}{4} \quad , \tag{2.2.2}$$

where A^c denote the complement of the set A.

From (2.1.2) and (2.2.2) we obtain

$$d_n(\theta_1, \theta_2) = \int_{\mathcal{X}^{(n)}} \left| \frac{dP_{\theta_1,n}}{d\mu_n} - \frac{dP_{\theta_2,n}}{d\mu_n} \right| d\mu_n$$

$$\geqq \int_A \left(\frac{dP_{\theta_1,n}}{d\mu_n} - \frac{dP_{\theta_2,n}}{d\mu_n} \right) d\mu_n + \int_{A^c} \left(\frac{dP_{\theta_2,n}}{d\mu_n} - \frac{dP_{\theta_1,n}}{d\mu_n} \right) d\mu_n$$

$$= 2 - 2P_{\theta_1,n}(A^c) - 2P_{\theta_2,n}(A)$$

$$\geqq 2 - \delta \quad .$$

Letting $\delta \to 0$, we have

$$\underline{\lim}_{n\to\infty} d_n(\theta_1, \theta_2) \geqq 2 \quad .$$

On the other hand it follows from (2.1.2) that

$$\overline{\lim}_{n\to\infty} d_n(\theta_1, \theta_2) \leqq 2 \quad .$$

Hence we have

$$\lim_{n\to\infty} d_n(\theta_1, \theta_2) = 2 \quad .$$

Thus we complete the proof.

The following theorem shows that a necessary condition for the existence of a consistent estimator is that the limit of the Kullback

- Leibler information is infinite.

Theorem 2.2.2. Suppose that for each n, $\{\tilde{x}_n : dP_{\theta,n}/d\mu_n > 0\}$ does not depend on θ. If there exists a consistent estimator, then the following holds : for any two disjoint points θ_1 and θ_2

$$\lim_{n\to\infty} I_n(\theta_1, \theta_2) = \infty ,$$

where $I_n(\theta_1,\theta_2)=\int_{\mathfrak{X}^{(n)}} dP_{\theta_1,n}/d\mu_n \log (dP_{\theta_1,n}/dP_{\theta_2,n})d\mu_n$.

Proof Let $0<\delta<\frac{1}{2}$. Putting $Y_n=dP_{\theta_2,n}/dP_{\theta_1,n}$, we have from Theorem 2.2.1 for sufficiently large n,

$$\begin{aligned}(2.2.3)\qquad E_{\theta_1,n}(|Y_n-1|) &= \int_{\mathfrak{X}^{(n)}}|Y_n-1|\, dP_{\theta_1,n} \\ &= d_n(\theta_1,\theta_2) \\ &\geqq 2-2\delta .\end{aligned}$$

Putting $Y_n^+=\max\{Y_n-1, 0\}$ and $Y_n^-=\max\{1-Y_n, 0\}$, we have for each n=1,2, ... ,

$$\begin{aligned}&E_{\theta_1,n}(Y_n^+) - E_{\theta_1,n}(Y_n^-) \\ &= \int_{\mathfrak{X}^{(n)}}\left\{\frac{dP_{\theta_1,n}}{d\mu_n} - \frac{dP_{\theta_2,n}}{d\mu_n}\right\} d\mu_n . \\ &= 0 .\end{aligned}$$

and for sufficiently large n,

$$E_{\theta_1,n}(Y_n^+) + E_{\theta_1,n}(Y_n^-) = E_{\theta_1,n}(|Y_n-1|) \geqq 2-2\delta .$$

Hence we obtain for sufficiently large n,

$$(2.2.4)\qquad E_{\theta_1,n}(Y_n^+) = E_{\theta_1,n}(Y_n^-) \geqq 1-\delta .$$

Since $0\leqq Y_n^-\leqq 1$ and (2.2.4) hold, for sufficiently large n,

$$\begin{aligned}1-\delta &\leqq E_{\theta_1,n}(Y_n^-) \\ &= \int_{\{Y_n^-\geqq 1-2\delta\}} Y_n^- dP_{\theta_1,n}(\tilde{x}_n) + \int_{\{Y_n^-< 1-2\delta\}} Y_n^- dP_{\theta_1,n}(\tilde{x}_n) \\ &\leqq P_{\theta_1,n}\{Y_n^-\geqq 1-2\delta\} + (1-2\delta)P_{\theta_1,n}\{Y_n^-\leqq 1-2\delta\} \\ &= 2\delta P_{\theta_1,n}\{Y_n^-\geqq 1-2\delta\} + 1-2\delta\end{aligned}$$

Hence we have for n large,

(2.2.5) $P_{\theta_1,n}\left\{Y_n \geqq 1 - 2\delta\right\} \geqq \frac{1}{2}$.

It follows from (2.2.5) that for sufficiently large n,

$$
\begin{aligned}
I_n(\theta_1, \theta_2) &= E_{\theta_1,n}(-\log Y_n) \\
&= E_{\theta_1,n}[-\log (1+Y_n^{+})] - E_{\theta_1,n}[\log (1-Y_n^{-})] \\
&\geqq - E_{\theta_1,n}(Y_n^{+}) - \frac{1}{2} \log 2\delta \ . \\
&\geqq - 1 - \frac{1}{2}\log 2\delta \ .
\end{aligned}
$$

Letting $\delta \to 0$, we have

$$\lim_{n\to\infty} I_n(\theta_1, \theta_2) = \infty .$$

Thus we complete the proof.

2.3. Consistent estimators with order $\{c_n\}$

We shall define a consistent estimator with order $\{c_n\}$ and derive a necessary condition for its existence.

Definition 2.3.1. For an increasing sequence of positive numbers $\{c_n\}$ (c_n tending to infinity) an estimator $\hat{\theta}_n$ is called consistent with order $\{c_n\}$ (or $\{c_n\}$-consistent for short) if for every $\varepsilon > 0$ and every ϑ of Ⓗ , there exist a sufficiently small positive number δ and a sufficiently large positive number L satisfying the following :

(2.3.1) $$\overline{\lim_{n\to\infty}} \sup_{\theta:\|\theta-\vartheta\|<\delta} P_{\theta,n}\left\{ c_n \|\hat{\theta}_n - \theta\| \geqq L \right\} < \varepsilon \ .$$

It is easily seen that if $\hat{\theta}_n$ is a $\{c_n\}$-consistent estimator, then $\hat{\theta}_n$ is consistent in the sense of Definition 2.1.1.

Theorem 2.3.1. If there exists a $\{c_n\}$-consistent estimator of θ , then for each $\theta \in$ Ⓗ and every $\varepsilon > 0$ there is a positive number t such that

$$\varliminf_{n\to\infty} d_n(\theta, \theta + tc_n^{-1}\mathbf{1}) \geqq 2 - \varepsilon ,$$

where $\mathbf{1}=(1,\ldots,1)'$.

Proof. Let $\hat{\theta}_n$ be a $\{c_n\}$-consistent estimator of θ. It follows from the definition of $\{c_n\}$-consistent estimator that for every $\varepsilon>0$ and every $\theta\in\Theta$, there exist positive numbers δ and L such that

$$\varlimsup_{n\to\infty} \sup_{\vartheta:\|\vartheta-\theta\|<\delta} P_{\vartheta,n}\{c_n\|\hat{\theta}_n - \vartheta\|\geqq L\}\leqq \varepsilon/4 .$$

Let $t > 2L$ be fixed. Since there exists n_0 such that for any $n>n_0$,

$$c_n > c_{n_0} > t/\delta ,$$

and

$$\sup_{\vartheta:\|\vartheta-\theta\|<tc_{n_0}^{-1}} P_{\vartheta,n}\{c_n\|\hat{\theta}_n - \vartheta\|\geqq L\}<\varepsilon/4 ,$$

it follows that

(2.3.2) $$\varlimsup_{n\to\infty} P_{\theta+tc_n^{-1}\mathbf{1},n}\{c_n\|\hat{\theta}_n-\theta-tc_n^{-1}\mathbf{1}\|\geqq L\}<\varepsilon/4 ,$$

and

(2.3.3) $$\varlimsup_{n\to\infty} P_{\theta,n}\{c_n\|\hat{\theta}_n-\theta\|\geqq L\}<\varepsilon/4 .$$

From (2.3.2) we have

(2.3.4) $$\varlimsup_{n\to\infty} P_{\theta+tc_n^{-1}\mathbf{1},n}\{c_n\|\hat{\theta}_n-\theta-tc_n^{-1}\mathbf{1}\|\geqq t-L\}<\varepsilon/4 .$$

Since the following holds :

(2.3.5) $$\begin{aligned} &d_n(\theta, \theta+tc_n^{-1}\mathbf{1}) \\ &= 2\sup_{B\in\mathcal{B}^{(n)}} \left| P_{\theta+tc_n^{-1}\mathbf{1},n}(B) - P_{\theta,n}(B)\right| \\ &\geqq 2\left| P_{\theta+tc_n^{-1}\mathbf{1},n}\{c_n\|\hat{\theta}_n-\theta\|\geqq L\} - P_{\theta,n}\{c_n\|\hat{\theta}_n-\theta\|\geqq L\}\right| \end{aligned}$$

for all n, it is sufficient to show that the inferior limit of the last term of (2.3.5) is not smaller than $2-\varepsilon$.

Since

$$\{\tilde{x}_n : c_n\|\hat{\theta}_n(\tilde{x}_n)-\theta-tc_n^{-1}\mathbf{1}\|<t-L\}\subset\{\tilde{x}_n : c_n\|\hat{\theta}_n(\tilde{x}_n)-\theta\|\geqq L\}$$

for all n, it follows that

$$(2.3.6)\quad \varliminf_{n\to\infty} P_{\theta+tc_n^{-1}\mathbf{1},n}\left\{c_n\|\hat{\theta}_n-\theta-tc_n^{-1}\mathbf{1}\|<t-L\right\}$$

$$\leqq \varliminf_{n\to\infty} P_{\theta+tc_n^{-1}\mathbf{1},n}\left\{c_n\|\hat{\theta}_n-\theta\|\geqq L\right\}.$$

It follows from (2.3.4) and (2.3.6) that

$$(2.3.7)\quad 1-\frac{\varepsilon}{4}\leqq \varliminf_{n\to\infty} P_{\theta+tc_n^{-1}\mathbf{1},n}\left\{c_n\|\hat{\theta}_n-\theta\|\geqq L\right\}.$$

From (2.3.3) and (2.3.7) we obtain

$$2-\varepsilon\leqq \varliminf_{n\to\infty} 2\left|P_{\theta+tc_n^{-1}\mathbf{1},n}\left\{c_n\|\hat{\theta}_n-\theta\|\geqq L\right\}-P_{\theta,n}\left\{c_n\|\hat{\theta}_n-\theta\|\geqq L\right\}\right|.$$

Therefore we have

$$\varliminf_{n\to\infty} d_n(\theta,\theta+tc_n^{-1}\mathbf{1})\geqq 2-\varepsilon.$$

Similarly we also obtain

$$\varliminf_{n\to\infty} d_n(\theta,\theta-tc_n^{-1}\mathbf{1})\geqq 2-\varepsilon,$$

which completes the proof.

2.4. Order of convergence of $\{c_n\}$-consistent estimators for the location parameter case.

Before discussing order of convergence of $\{c_n\}$-consistent estimators in detail, we shall give a definition and lemmas.

Definition 2.4.1. (Generalized from Gnedenko and Kolmogorov [22])

For each $\theta\in\Theta$, the sums

$$Y_n(\theta)=X_1(\theta)+X_2(\theta)+\dots+X_n(\theta)$$

of positive independent random variables $X_1(\theta)$, $X_2(\theta)$, ... , $X_n(\theta)$, ... are said to be uniformly relatively stable for the sequence of positive constants $B_n(\theta)$ if it holds that for any $\varepsilon > 0$

$$P_{\theta,n}\left\{ \left| \frac{Y_n(\theta)}{B_n(\theta)} - 1 \right| > \varepsilon \right\} \to 0$$ as $n \to \infty$ uniformly in every compact subset of Ⓗ.

In the subsequent lemmas, we use the notation that for each k and each $\theta \in$ Ⓗ, $F_{\theta k}(x)$ is the distribution function of $X_k(\theta)$.

<u>Lemma 2.4.1.</u> (Gnedenko and Kolmogorov [22])

For each $\theta \in$ Ⓗ, let $X_1(\theta)$, $X_2(\theta)$, ... , $X_n(\theta)$, ... be a sequence of positive independent random variables. The sums

$$Y_n(\theta) = X_1(\theta) + X_2(\theta) + \dots + X_n(\theta)$$

are uniformly relatively stable for the sequence of positive constants $B_1(\theta)$, $B_2(\theta)$, ... , $B_n(\theta)$, ... , if for every $\varepsilon > 0$,

(2.4.1) $\sum_{k=1}^{n} \int_{\varepsilon B_n(\theta)}^{\infty} dF_{\theta k}(x) \to 0$ as $n \to \infty$ uniformly in every compact subset of Ⓗ, and

(2.4.2) $\frac{1}{B_n(\theta)} \sum_{k=1}^{n} \int_{0}^{\varepsilon B_n(\theta)} x\,dF_{\theta k}(x) \to 1$ as $n \to \infty$ uniformly in every compact subset of Ⓗ.

The following lemma is a generalization of Lindeberg's condition (see Gnedenko and Kolmogorov [22]) .

<u>Lemma 2.4.2.</u> For each $\theta \in$ Ⓗ. let $X_1(\theta)$, $X_2(\theta)$, ... , $X_n(\theta)$, ... be a sequence of independent random variables. The distribution laws of the sums

$$Y_n(\theta) = \frac{X_1(\theta)+X_2(\theta)+\ldots+X_n(\theta)}{B_n(\theta)}$$

converges to the normal law

$$\Phi(x) = \frac{1}{\sqrt{2\pi}} \int_{-\infty}^{x} e^{-\frac{y^2}{2}} dy$$

uniformly on every compact subset of Ⓗ, if $\lim_{n\to\infty} B_n(\theta)=\infty$ uniformly on every compact subset of Ⓗ and for every $\varepsilon > 0$,

(2.4.3) $\sum_{k=1}^{n} \int_{\{|x|>\varepsilon B_n(\theta)\}} dF_{\theta k}(x) \to 0$ as $n\to\infty$ uniformly in every compact subset of Ⓗ, and

(2.4.4) $$\frac{1}{\{B_n(\theta)\}^2} \sum_{k=1}^{n} \left\{ \int_{\{|x|<\varepsilon B_n(\theta)\}} x^2 dF_{\theta k}(x) - \left(\int_{\{|x|<\varepsilon B_n(\theta)\}} x dF_{\theta k}(x) \right)^2 \right\} \to 1 \text{ as } n\to\infty$$

uniformly in every compact subset of Ⓗ.

Now suppose that every $P_\theta(\cdot)(\theta \in Ⓗ)$ is absolutely continuous with respect to a σ-finite measure μ.
We denote the density $dP_\theta/d\mu$ by $f(\cdot:\theta)$ and by $A(\theta) \subset \mathfrak{X}$ the set of points in the space of $\mathfrak{X}$ for which $f(x:\theta) > 0$.
Then we may write $f(x:\theta) = \chi_{A(\theta)}(x) f(x:\theta)$, where $\chi_{A(\theta)}(\cdot)$ denotes the indicator of $A(\theta)$.
Let $X_1, X_2, \ldots, X_n, \ldots$ be independently and identically distributed random variables with the density $f(x:\theta)$. Then the joint density of $(X_1, X_2, \ldots, X_n)$ is given by $\prod_{i=1}^{n} f(x_i:\theta)$. We denote by $\mathop{X}_{i=1}^{n} A$ and $\mu^{(n)}$ the n-fold direct products of a set A and the measure μ, respectively.

Lemma 2.4.3.

$$\int_{\mathop{X}_{i=1}^{n}(A(\theta_1)\cap A(\theta_2))} \left\{ \prod_{i=1}^{n} \frac{f(X_i:\theta_1)}{f(X_i:\theta_2)} - 1 \right\}^2 \prod_{i=1}^{n} f(x_i:\theta_2) d\mu^{(n)} < \infty ,$$

then for any two points θ_1 and θ_2 in Ⓗ and each $n=1,2,\ldots$,

(2.4.5) $$\begin{aligned} d_n(\theta_1, \theta_2) &\leqq [1-\{P_{\theta_1}(A(\theta_2))\}^n] + [1-\{P_{\theta_2}(A(\theta_1))\}^n] \\ &+ [\{\int_{A(\theta_1)\cap A(\theta_2)} f^2(x:\theta_1)/f(x:\theta_2) d\mu\}^n - 2\{P_{\theta_1}(A(\theta_2))\}^n \\ &+ \{P_{\theta_2}(A(\theta_1))\}^n]^{1/2} . \end{aligned}$$

Proof. Since for any two points θ_1 and θ_2 in Ⓗ and each $n=1,2,\ldots$, $\frac{dP_{\theta_j}^{(n)}}{d\mu^{(n)}} = \prod_{i=1}^{n} f(x_i:\theta_j) = \prod_{i=1}^{n} \chi_{A(\theta_j)} f(x_i:\theta_j) = \chi_{\overset{n}{\underset{i=1}{X}}A(\theta_j)}(\tilde{x}_n) \prod_{i=1}^{n} f(x_i:\theta_j)$ $(j=1,2)$, from (2.1.2) we have

$$
\begin{aligned}
&d_n(\theta_1,\theta_2)\\
&=\int_{\mathfrak{X}^{(n)}} \left| \chi_{\overset{n}{\underset{i=1}{X}}A(\theta_1)}(\tilde{x}_n)\prod_{i=1}^{n} f(x_i:\theta_1) - \chi_{\overset{n}{\underset{i=1}{X}}A(\theta_2)}(\tilde{x}_n)\prod_{i=1}^{n} f(x_i:\theta_2) \right| d\mu^{(n)}\\
&=\int_{\overset{n}{\underset{i=1}{X}}A(\theta_1)-\overset{n}{\underset{i=1}{X}}A(\theta_2)} \prod_{i=1}^{n} f(x_i:\theta_1)d\mu^{(n)} + \int_{\overset{n}{\underset{i=1}{X}}A(\theta_2)-\overset{n}{\underset{i=1}{X}}A(\theta_1)} \prod_{i=1}^{n} f(X_i:\theta_2)d\mu^{(n)}\\
&\quad+\int_{\overset{n}{\underset{i=1}{X}}(A(\theta_1)\cap A(\theta_2))} \left| \prod_{i=1}^{n} f(x_i:\theta_1) - \prod_{i=1}^{n} f(x_i:\theta_2) \right| d\mu^{(n)}\\
&=1-\int_{\overset{n}{\underset{i=1}{X}}(A(\theta_1)\cap A(\theta_2))} \prod_{i=1}^{n} f(x_i:\theta_1)d\mu^{(n)} + 1 - \int_{\overset{n}{\underset{i=1}{X}}(A(\theta_1)\cap A(\theta_2))} \prod_{i=1}^{n} f(x_i:\theta_2)d\mu^{(n)}\\
&\quad+\int_{\overset{n}{\underset{i=1}{X}}(A(\theta_1)\cap A(\theta_2))} \left| \prod_{i=1}^{n} f(x_i:\theta_1) - \prod_{i=1}^{n} f(x_i:\theta_2) \right| d\mu^{(n)}\\
&=[1-\{P_{\theta_1}(A(\theta_2))\}^n] + [1-\{P_{\theta_2}(A(\theta_1))\}^n]\\
&\quad+\int_{\overset{n}{\underset{i=1}{X}}(A(\theta_1)\cap A(\theta_2))} \left| \prod_{i=1}^{n} f(x_i:\theta_1) - \prod_{i=1}^{n} f(x_i:\theta_2) \right| d\mu^{(n)} .
\end{aligned}
$$

Further it follows from the assumption and the Schwarz's inequality that

$$
\begin{aligned}
&\int_{\overset{n}{\underset{i=1}{X}}(A(\theta_1)\cap A(\theta_2))} \left| \prod_{i=1}^{n} f(x_i:\theta_1) - \prod_{i=1}^{n} f(x_i:\theta_2) \right| d\mu^{(n)}\\
&=\int_{\overset{n}{\underset{i=1}{X}}(A(\theta_1)\cap A(\theta_2))} \left| \prod_{i=1}^{n} \frac{f(x_i:\theta_1)}{f(x_i:\theta_2)} - 1 \right| \prod_{i=1}^{n} f(x_i:\theta_2)d\mu^{(n)}\\
&\leqq \left[\int_{\overset{n}{\underset{i=1}{X}}(A(\theta_1)\cap A(\theta_2))} \left\{ \prod_{i=1}^{n} \frac{f(x_i:\theta_1)}{f(x_i:\theta_2)} - 1 \right\}^2 \prod_{i=1}^{n} f(x_i:\theta_2)d\mu^{(n)}\right]^{1/2}\\
&= \left[\left\{ \int_{A(\theta_1)\cap A(\theta_2)} f^2(x:\theta_1)/f(x:\theta_2)d\mu \right\}^n - 2\{P_{\theta_1}(A(\theta_2))\}^n + \{P_{\theta_2}(A(\theta_1))\}^n \right]^{1/2} .
\end{aligned}
$$

Thus we complete the proof.

Now let $\mathcal{X} = \Theta = R^1$, and we suppose that every $P_\theta(\cdot)(\theta \in \Theta)$ is absolutely continuous with respect to a Lebesgue measure and constitutes a location parameter family. Then we denote the density dP_θ/dx by $f(\cdot : \theta)$ and $f(x:\theta)=f(x-\theta)$.

<u>For the lemmas and theorems in sections 2.4 and 2.5 we make the following assumptions.</u>

(A.2.4.1) $f(x) > 0$ for $a < x < b$,
$f(x)=0$ for $x \leqq a$, $x \geqq b$.

(A.2.4.2) $f(x)$ is twice continuously differentiable in the interval (a,b) and

$$\lim_{x \to a+0} (x-a)^{1-\alpha} f(x) = A' \quad ;$$

$$\lim_{x \to b-0} (b-x)^{1-\beta} f(x) = B' \quad ,$$

where both α and β are positive constants and A' and B' are positive finite numbers.

(A.2.4.3) $A'' = \lim_{x \to a+0} (x-a)^{2-\alpha} |f'(x)|$ and $B'' = \lim_{x \to b-0} (b-x)^{2-\beta} |f'(x)|$ are finite. For $\alpha \geqq 2$ $f''(x)$ is bounded.

For example we see that the beta distributions $Be(\alpha, \beta)$ $(0 < \alpha \leqq \beta \leqq 2$ or $3 < \alpha \leqq \beta < \infty)$ satisfy Assumptions (A.2.4.1)$\sim$(A.2.4.3).

<u>Lemma 2.4.4.</u> Suppose that a density function f satisfies Assumptions (A.2.4.1)$\sim$(A.2.4.3) with $\alpha=2$. Then the following hold : for any $\varepsilon > 0$

(2.4.6) $$n \int_{\{x : \varepsilon c_1 (n \log n) < -(\partial^2/\partial\theta^2) \log f(x-\theta)\}} f(x-\theta) dx \to 0 \text{ as } n \to \infty$$

uniformly in every compact subset of Θ and

(2.4.7) $$\frac{1}{c_1 \log n} \int_{\{x : 0 < -(\partial^2/\partial\theta^2) \log f(x-\theta) < \varepsilon c_1 n \log n\}} \{-(\partial^2/\partial\theta^2) \log f(x-\theta)\} f(x-\theta) dx \to 1 \text{ as } n \to \infty$$

uniformly in every compact subset of Θ ,

where $c_1 = \frac{1}{2}(\frac{A''^2}{A'} + \frac{B''^2}{B'})$ if $\beta = 2$, $c_1 = \frac{A''^2}{2A'}$ if $\beta > 2$.

Proof. It follows from Assumptions (A.2.4.1) $\sim$ (A.2.4.3) that there exists n_0 and η_n such that for $n \geqq n_0$ and for

(2.4.8) $\quad 0 < x-a < \eta_n$, $0 < b-x < \eta_n$, that

(2.4.9) $A' - \frac{1}{n} < (x-a)^{-1} f(x) < A' + \frac{1}{n}$, $A'' - \frac{1}{n} < |f'(x)| < A'' + \frac{1}{n}$,

$$B' - \frac{1}{n} < (b-x)^{1-\beta} f(x) < B' + \frac{1}{n},\quad B'' - \frac{1}{n} < (b-x)^{2-\beta} |f'(x)| < B'' + \frac{1}{n}.$$

Let $A_{-n} = A' - \frac{1}{n}$, $A_n = A' + \frac{1}{n}$, $B_{-n} = B' - \frac{1}{n}$, $B_n = B' + \frac{1}{n}$, $A''_{-n} = A'' - \frac{1}{n}$,

$A''_n = A'' + \frac{1}{n}$, $B''_{-n} = B'' - \frac{1}{n}$ and $B''_n = B'' + \frac{1}{n}$.

Putting

$$I_{1n} = \int_{\{x:\varepsilon c_1 n\log n < -\frac{f''(x)}{f(x)} + \{\frac{f'(x)}{f(x)}\}^2\} \cap (a, a+\eta_{n_0})} f(x)dx ,$$

$$I_{2n} = \int_{\{x:\varepsilon c_1 n\log n < -\frac{f''(x)}{f(x)} + \{\frac{f'(x)}{f(x)}\}^2\} \cap [a+\eta_{n_0}, b-\eta_{n_0}]} f(x)dx ,$$

$$I_{3n} = \int_{\{x:\varepsilon c_1 n\log n < -\frac{f''(x)}{f(x)} + \{\frac{f'(x)}{f(x)}\}^2\} \cap (b-\eta_{n_0}, b)} f(x)dx ,$$

we have

(2.4.10)
$$n\int_{\{x:\varepsilon c_1 n\log n < -\frac{\partial^2}{\partial\theta^2}\log f(x-\theta)\}} f(x-\theta)dx$$

$$= n\int_{\{x:\varepsilon c_1 n\log n < -\frac{f''(x)}{f(x)} + \{\frac{f'(x)}{f(x)}\}^2\}} f(x)dx$$

$$= n(I_{1n} + I_{2n} + I_{3n}) .$$

Since $f'(x)$ and $f''(x)$ are bounded,

(2.4.11) $\quad \lim_{n\to\infty} nI_{2n} = 0$.

Since $f''(x)$ is bounded, from (2.4.8) and (2.4.9) we have for sufficiently large n,

$$nI_{1n} = n\int_{\{x:\varepsilon c_1 n\log n < -\frac{f''(x)}{f(x)} + \{\frac{f'(x)}{f(x)}\}^2\} \cap (a, a+\eta_{n_0})} f(x)dx$$

$$\leqq n\int_{\{x:\varepsilon c_1 n\log n < \{\frac{A''_n}{A_{-n}(x-a)}\}^2\}} f(x)dx + O(n^{-1}(\log n)^{-2})$$

$$= n\int_a^{a+\frac{A''_n}{A_{-n}}(\varepsilon c_1 n \log n)^{-\frac{1}{2}}} A_n(x-a)dx + O(n^{-1}(\log n)^{-2})$$

$$= \frac{n}{2}A_n\frac{A''_n}{A_{-n}}(\varepsilon c_1 n \log n)^{-1} + O(n^{-1}(\log n)^{-2}) .$$

Hence we obtain

$$(2.4.12)\quad \overline{\lim_{n\to\infty}}\, nI_{1n}$$

$$\leqq \overline{\lim_{n\to\infty}}\, \frac{n}{2}A_n \frac{A''_n}{A_{-n}} (\varepsilon c_1 n \log n)^{-1}$$

$$= \frac{1}{2}A''/(\varepsilon c_1)^{-1} \lim_{n\to\infty}(\log n)^{-1}$$

$$= 0 .$$

Repeating a similar argument, we have

$$(2.4.13)\quad \lim_{n\to\infty} nI_{3n}=0 .$$

It follows from (2.4.10), (2.4.11) and (2.4.12) that (2.4.6) holds.

Putting

$$I'_{1n}=\int_{\left\{x:0<-\frac{f''(x)}{f(x)}+\left\{\frac{f'(x)}{f(x)}\right\}^2<\varepsilon c_1 n \log n\right\}\cap(a,a+\eta_{n_0})} [-\frac{f''(x)}{f(x)}+\left\{\frac{f'(x)}{f(x)}\right\}^2]f(x)dx ,$$

$$I'_{2n}=\int_{\left\{x:0<-\frac{f''(x)}{f(x)}+\left\{\frac{f'(x)}{f(x)}\right\}^2<\varepsilon c_1 n \log n\right\}\cap[a+\eta_{n_0},b-\eta_{n_0}]} [-\frac{f''(x)}{f(x)}+\left\{\frac{f'(x)}{f(x)}\right\}^2]f(x)dx ,$$

$$I'_{3n}=\int_{\left\{x:0<-\frac{f''(x)}{f(x)}+\left\{\frac{f'(x)}{f(x)}\right\}^2<\varepsilon c_1 n \log n\right\}\cap(b-\eta_{n_0},b)} [-\frac{f''(x)}{f(x)}+\left\{\frac{f'(x)}{f(x)}\right\}^2]f(x)dx ,$$

we have

$$(2.4.14)\quad \frac{n}{c_1 n \log n}\int_{\left\{x:0<-\frac{\partial^2}{\partial\theta^2}\log f(x-\theta)<\varepsilon c_1 n \log n\right\}} \left\{-\frac{\partial^2}{\partial\theta^2}\log f(x-\theta)\right\} f(x-\theta)dx$$

$$= \frac{1}{c_1 \log n}\int_{\left\{x:0<-\frac{f''(x)}{f(x)}+\left\{\frac{f'(x)}{f(x)}\right\}^2<\varepsilon c_1 n \log n\right\}} [-\frac{f''(x)}{f(x)}+\left\{\frac{f'(x)}{f(x)}\right\}^2]f(x)dx$$

$$= \frac{1}{c_1 \log n}(I'_{1n}+I'_{2n}+I'_{3n}) .$$

Since f'(x) and f"(x) are bounded,

(2.4.15) $$\lim_{n\to\infty} \frac{1}{c_1 \log n} I'_{2n} = 0 ,$$

Since f"(x) is bounded, from (2.4.8) and (2.4.9) we have for sufficiently large n,

$$\frac{1}{c_1 \log n} I'_{1n}$$

$$= \frac{1}{c_1 \log n} \int_{\{x:0< -\frac{f''(x)}{f(x)} + \{\frac{f'(x)}{f(x)}\}^2 < \varepsilon c_1 n \log n\} \cap (a, a+\eta_{n_0})} [-\frac{f''(x)}{f(x)} + \{\frac{f'(x)}{f(x)}\}^2] f(x) dx$$

$$\leqq \frac{1}{c_1 \log n} \int_{\{x: \frac{A''^2_{-n}}{A^2_n} \cdot \frac{1}{(x-a)^2} < \varepsilon c_1 n \log n\} \cap (a, a+\eta_{n_0})} [-\frac{f''(x)}{f(x)} + \{\frac{f'(x)}{f(x)}\}^2] f(x) dx$$

$$\leqq \frac{1}{c_1 \log n} \int_{\frac{A''_{-n}}{A_n}(\varepsilon c_1 n \log n)^{-1/2}}^{\eta_{n_0}} \frac{A''^2_n}{A_{-n}} \cdot \frac{1}{x} dx + O(n^{-1})$$

$$= \frac{1}{c_1 \log n} \left\{ \frac{A''^2_n}{A_{-n}} (\log \eta_{n_0} - \log \frac{A''_{-n}}{A_n} + \frac{1}{2} \log \varepsilon c_1 + \frac{1}{2} \log n + \frac{1}{2} \log \log n) \right\}$$

$$+ O(n^{-1}) .$$

Hence we obtain

(2.4.16) $$\overline{\lim_{n\to\infty}} \frac{1}{c_1 \log n} I'_{1n} \leqq \frac{A''^2}{2c_1 A'} .$$

Further since for sufficiently large n,

$$\frac{1}{c_1 \log n} I'_{1n}$$

$$\geqq \frac{1}{c_1 \log n} \int_{\{x: \frac{A''^2_n}{A^2_{-n}(x-a)^2} < \varepsilon c_1 n \log n\} \cap (a, a+\eta_{n_0})} [-\frac{f''(x)}{f(x)} + \{\frac{f'(x)}{f(x)}\}^2] f(x) dx$$

$$\geq \frac{1}{c_1 \log n} \int_{\frac{A''_n}{A_{-n}}(c_1 n \log n)^{1/2}}^{\eta_{n_0}} \frac{A''^2_{-n}}{A_n} \cdot \frac{1}{x} dx$$

$$= \frac{1}{c_1 \log n} \left\{ \frac{A''^2_{-n}}{A_n} \left(\log \eta_{n_0} - \log \frac{A''_n}{A_{-n}} + \frac{1}{2}\log \varepsilon c_1 + \frac{1}{2}\log n + \frac{1}{2}\log \log n\right) \right\}$$

we have

$$\underline{\lim}_{n\to\infty} \frac{1}{c_1 \log n} I'_{1n} \geq \frac{A''^2}{2c_1 A'} \quad . \tag{2.4.17}$$

It follows from (2.4.16) and (2.4.17) that

$$\lim_{n\to\infty} \frac{1}{c_1 \log n} I'_{1n} = \frac{A''^2}{2c_1 A'} \quad . \tag{2.4.18}$$

Repeating a similar argument, we have

$$\lim_{n\to\infty} \frac{1}{c_1 \log n} I'_{3n} = \begin{cases} \dfrac{B''}{2c_1 B'} & \text{for } \beta = 2 \quad , \\ 0 & \text{for } \beta > 2 \quad . \end{cases} \tag{2.4.19}$$

Hence it follows from (2.4.15), (2.4.18) that (2.4.7) holds, and $c_1 = \frac{1}{2}\left(\frac{A''^2}{A'} + \frac{B''^2}{B'} \right)$ for $\beta = 2$, $c_1 = \frac{A''^2}{2A'}$ for $\beta > 2$.

Lemma 2.4.5. Suppose a density function f satisfies Assumptions (A.2.4.1) $\sim$ (A.2.4.3) with $\alpha = 2$. Then the following hold : for any $\varepsilon > 0$

$$n \int_{\left\{x: \left|\frac{\partial}{\partial\theta}\log f(x-\theta)\right| > \varepsilon c_2 (n \log n)^{1/2}\right\}} f(x-\theta)dx \to 0 \text{ as } n \to \infty$$

uniformly in $\theta \in \text{Ⓗ}$, and

$$\frac{1}{c^2 \log n}\left[\int_{\left\{x: \left|\frac{\partial}{\partial\theta}\log f(x-\theta)\right| < \varepsilon c_2 (n \log n)^{1/2}\right\}} \left\{\frac{\partial}{\partial\theta}\log f(x-\theta)\right\}^2 f(x-\theta)dx \right.$$

$$\left. - \left(\int_{\left\{x: \left|\frac{\partial}{\partial\theta}\log f(x-\theta)\right| < \varepsilon c_2 (n \log n)^{1/2}\right\}} \left\{\frac{\partial}{\partial\theta}\log f(x-\theta)\right\} f(x-\theta)dx \right)^2 \right] \to 1$$

as $n \to \infty$ uniformly in $\theta \in (H)$, $c_2 = \left\{\frac{1}{2}\left(\frac{A''^2}{A'} + \frac{B''^2}{B'}\right)\right\}^{\frac{1}{2}}$

if $\beta = 2$, $c_2 = \frac{A''}{\sqrt{2A'}}$ if $\beta > 2$.

The proof is omitted as it is similar to that of lemma 2.4.4.

Lemma 2.4.6. Suppose that a density function f satisfies Assumptions (A.2.4.1)$\sim$(A.2.4.3) with $\alpha > 2$. Then

$$\int_a^b \frac{f'(x)^2}{f(x)} dx < \infty .$$

Proof. Since $f(a)=0$ and $\lim_{x \to a+} f(x)=0$, by the second mean value theorem, in a neighborhood of a we have

$$\frac{\{f'(x)\}^2}{f(x)} = \frac{2f'(\xi)f''(\xi)}{f'(\xi)} = 2f''(\xi)$$

where $a < \xi < x$. Since $f''(x)$ is continuous and bounded, $\{f'(x)\}^2/f(x)$ is bounded in a neighborhood of a. By a similar argument $\{f'(x)\}^2/f(x)$ is bounded in a neighborhood of b. Therefore the integral is bounded. This completes the proof.

In the following theorem we shall show that for density functions in a family satisfying Asssumptions (A.2.4.1)$\sim$(A.2.4.3), there exist consistent estimators of different orders.

Theorem 2.4.1. Let $X_1, X_2, \ldots, X_n, \ldots$ be a sequence of independent identically distributed random variables with a density function satisfying Assumptions (A.2.4.1)$\sim$(A.2.4.3). For each value of α there exists a consistent estimator with the order given in Table 2.4.1 below.

α	order c_n	$\{c_n\}$-consistent estimator
$0<\alpha<2$	$n^{1/\alpha}$	$\{\min X_i+\max X_i-(a+b)\}/2$
$\alpha=2$	$(n \log n)^{1/2}$	MLE
$\alpha>2$	$n^{1/2}$	MLE

Table 2.4.1.

where MLE denotes the maximum likelihood estimator of θ. The existence of the MLE is guaranteed since f is continuous and bounded.

Proof. 1) $0<\alpha<2$. Let $\hat{\theta}_n = \{\min X_i+\max X_i-(a+b)\}/2$. It follows from Assumptions (A.2.4.1) and (A.2.4.2) that there are positive constants C and γ such that $C \leqq (x-a)^{1-\alpha} f(x)$ for all $x \in (a, a+\gamma)$; $C \leqq (b-x)^{1-\beta} f(x)$ for all $x \in (b-\gamma, b)$.
In order to show that $\hat{\theta}_n$ is $\{n^{1/\alpha}\}$-consistent, it is sufficient to know that every $\varepsilon>0$ we can choose L satisfying $L > \max\{(1/2)((\alpha/C)\log(2/\varepsilon))^{1/\alpha}, 0\}$. Indeed, we have for each n

$$(6.4.20)\quad P_{\theta,n}\{\hat{\theta}_n-\theta > Ln^{-1/\alpha}\} = P_{\theta,n}\{\hat{\theta}_n-\theta > Ln^{-1/\alpha} \text{ and } \max_i x_i \leqq b+\theta\}$$

$$\leqq \left\{1-\int_a^{a+2Ln^{-1/\alpha}} f(x)\,dx\right\}^n .$$

Similarly we obtain for each n

$$(6.4.21)\quad P_{\theta,n}\{\hat{\theta}_n-\theta < -Ln^{-1/\alpha}\} = \left\{1-\int_{b-2Ln^{-1/\alpha}}^{b} f(x)\,dx\right\}^n .$$

It follows from (2.4.20) and (2.4.21) that for each n

$$P_{\theta,n}\{|\hat{\theta}_n-\theta| > Ln^{-1/\alpha}\}$$

$$\leqq \left\{1-\int_a^{a+2Ln^{-1/\alpha}} f(x)\,dx\right\}^n + \left\{1+\int_{b-2Ln^{-1/\alpha}}^{b} f(x)\,dx\right\}^n .$$

Hence we have uniformly in $\theta \in \Theta$,

$$\overline{\lim_{n\to\infty}} P_{\theta,n}\left\{|\hat{\theta}_n-\theta|>Ln^{-1/\alpha}\right\}$$

$$\leqq \lim_{n\to\infty}\left\{1-\int_a^{a+2Ln^{-1/\alpha}} f(x)dx\right\}^n + \lim_{n\to\infty}\left\{1-\int_{b-2Ln^{-1/\alpha}}^{b} f(x)dx\right\}^n$$

$$\leqq 2\exp\left\{-\frac{C(2L)^\alpha}{\alpha}\right\}<\varepsilon .$$

Therefore it is seen that $\hat{\theta}_n$ is $\{n^{1/\alpha}\}$-consistent.

2) $\alpha=2$. Since the MLE is a consistent estimator (Wald [52]) and it is a root of

(2.4.22) $$\sum_{i=1}^{n}\frac{\partial}{\partial\theta}\log f(x_i-\theta) = 0 ,$$

there a least exists a consistent solution of (2.4.22). We denote it by $\hat{\theta}_n^*$.

Let $A_n=(n\log n)^{1/2}$ and put $L_n(\theta, \tilde{x}_n)=\prod_{i=1}^{n} f(x_i-\theta)$ for $\theta+a<x_i<\theta+b$ $(i=1,2,\ldots,n)$. Using the mean value theorem, we have

(2.4.23) $$-\frac{1}{c^2A_n^2}\left[\frac{\partial^2}{\partial\theta^2}\log L_n\right]_{\theta=\theta_n^*} cA_n(\hat{\theta}_n^*-\theta)=\frac{1}{cA_n}\left[\frac{\partial}{\partial\theta}\log L_n\right]_{\theta=\theta} ,$$

where $|\theta-\theta_n^*|\leqq|\theta-\hat{\theta}_n^*|$ and $c=\left\{\frac{1}{2}\left(\frac{A''^2}{A'}+\frac{B''^2}{B'}\right)\right\}^{1/2}$ if $\beta=2$, $c=\frac{A''}{\sqrt{2A'}}$ if $\beta>2$. $-(\partial^2/\partial\theta^2)\log L_n$ is the sums of positive i.i.d. random variables $-(\partial^2/\partial\theta^2)\log f(X_1-\theta)$, $-(\partial^2/\partial\theta^2)\log f(X_2-\theta)$, ... , $-(\partial^2/\partial\theta^2)\log f(X_n-\theta)$. If $c^2A_n^2$ is taken as $B_n(\theta)$ in lemma 2.4.1, then it follows from lemma 2.4.4 that the conditions (2.4.1) and (2.4.2) hold. From lemma 2.4.1 we conclude that $-(\partial^2/\partial\theta^2)\log L_n$ is uniformly relatively stable for constants $c^2A_n^2$. Since $\hat{\theta}_n^*$ is uniformly consistent in any compact subset of Ⓗ (Wald [52]), θ_n^* converges in probability to θ uniformly in every compact subset of Ⓗ. Furthermore since $(\partial^2/\partial\theta^2)\log L_n(\theta, \tilde{x}_n)$ is uniformly continuous in any compact subset of Ⓗ, $(-1/c^2A_n^2)$ $[(\partial^2/\partial\theta^2)\log L_n]_{\theta=\theta_n^*}$ converges in probability to 1 uniformly in every compact subset of Ⓗ. $(\partial/\partial\theta)\log L_n$ is the sums of i.i.d.

random variables $f_\theta(X_1-\theta)/f(X_1-\theta)$, $f_\theta(X_2-\theta)/f(X_2-\theta)$, ... , $f_\theta(X_n-\theta)/f(X_n-\theta)$, where $f_\theta(X-\theta)=(\partial/\partial\theta)f(X-\theta)$. If cA_n is taken as $B_n(\theta)$ in lemma 2.4.2 then it follows from lemma 2.4.5 that conditions (2.4.3) and (2.4.4) are satisfied. From lemma 2.4.2 we see that the distribution law of $(1/cA_n)\{(\partial/\partial\theta)\log L_n\}$ converges to the normal law $\Phi(x)=(1/\sqrt{2\pi})\int_{-\infty}^{x} e^{-y^2/2}\,dy$ uniformly in every compact subset of Ⓗ.

Since from (2.4.23)

$$cA_n(\hat{\theta}_n^*-\theta) = \frac{(1/cA_n)[(\partial/\partial\theta)\log L_n]_{\theta=\theta}}{(-1/c^2A_n^2)[(\partial^2/\partial\theta^2)\log L_n]_{\theta=\theta_n^*}},$$

it follows that the distribution law of $cA_n(\hat{\theta}_n^*-\theta)$ converges to the normal law $\Phi(x)$ uniformly in every compact subset of Ⓗ.

In order to prove that $\hat{\theta}_n^*$ is an $\{A_n\}$-consistent estimator, it is sufficient to show that for any $\varepsilon>0$ we can choose an L satisfying $\int_{-cL}^{cL}(1/\sqrt{2\pi})e^{-x^2/2}\,dx>1-\varepsilon$ and that (2.3.1) holds.

Since

$$\begin{aligned} &P_{\theta,n}\{A_n|\hat{\theta}_n^*-\theta|\geqq L\}\\ &= P_{\theta,n}\{cA_n|\hat{\theta}_n^*-\theta|\geqq cL\}\\ &= 1-P_{\theta,n}\{cA_n|\hat{\theta}_n^*-\theta|<cL\}, \end{aligned}$$

it follows that for every $\vartheta\in$ Ⓗ there exists $\delta>0$ such that

$$\begin{aligned} &\overline{\lim_{n\to\infty}}\ \sup_{\theta:|\theta-\vartheta|<\delta} P_{\theta,n}\{A_n|\hat{\theta}_n^*-\theta|\geqq L\}\\ &= 1-\int_{-cL}^{cL}(1/\sqrt{2\pi})e^{-\frac{x^2}{2}}dx\\ &<\varepsilon\ . \end{aligned}$$

<u>3) $\alpha>2$.</u> It follows from Assumption (A.2.4.3) that $E_\theta(Z_\theta)=0$ and $E_\theta(Z_{\theta\theta})+E_\theta(Z_\theta^2)=0$, where $Z_\theta=(\partial/\partial\theta)\log f(X-\theta)$ and $Z_{\theta\theta}=(\partial^2/\partial\theta^2)\log f(X-\theta)$. Further it is seen from lemma 2.4.6 that $E_\theta(Z_\theta^2)<\infty$. Hence the distribution law of $\sqrt{I\,n}\,(\hat{\theta}_n^*-\theta)$ converges to the normal law $\Phi(x)$ uniformly in every compact subset of Ⓗ, where $I=E_\theta(Z_\theta^2)$

(Cramér [17]). Therefore it may be shown in the same way as the case $\alpha=2$ that $\hat{\theta}_n^*$ is a $\{n^{1/2}\}$-consistent estimator. This completes the proof.

2.5. Bounds for the order of convergence of consistent estimators

In this section we shall show that for each α, there does not exist a consistent sequence of estimators with order greater than the values as given in Table 2.4.1 of Theorem 2.4.1. That is, the order given by Table 2.4.1 is an upper bound for the order of convergence of consistent estimators. Before proceeding to the next theorem, we shall prove the following lemmas.

Lemma 2.5.1. Let f be a density function satisfying Assumption (A.2.4.1) of section 2.4. Suppose that for $0<\Delta<b-a$, there exists a measurable function $g(\cdot)$ on $\mathfrak{X}$ such that $g(x)>0$ if $a-\Delta<x<b$, $g(x)=0$ otherwise and $\int_{\mathfrak{X}} g(x)dx=1$. If $\int_{a-\Delta}^{b}\{f(x+\Delta)-g(x)\}^2/g(x)dx<\infty$, $\int_{a-\Delta}^{b}\{f(x)-g(x)\}^2/g(x)dx<\infty$, then

$$(2.5.1)\qquad d_n(\theta-\Delta,\theta)\leqq\left[\left\{\int_{a-\Delta}^{b}\frac{(f(x+\Delta)-g(x))^2}{g(x)}dx+1\right\}^n-1\right]^{1/2}$$
$$+\left[\left\{\int_{a-\Delta}^{b}\frac{(f(x)-g(x))^2}{g(x)}dx+1\right\}^n-1\right]^{1/2}.$$

Proof. First we have

$$(2.5.2)\qquad d_n(\theta-\Delta,\theta)$$
$$=\int_{\mathfrak{X}^{(n)}}\left|\prod_{i=1}^n f(x_i-\theta+\Delta)-\prod_{i=1}^n f(x_i-\theta)\right|\prod_{i=1}^n dx_i$$
$$\leqq\int_{\mathfrak{X}^{(n)}}\left|\prod_{i=1}^n f(x_i-\theta+\Delta)-\prod_{i=1}^n g(x_i-\theta)\right|\prod_{i=1}^n dx_i+\int_{\mathfrak{X}^{(n)}}\left|\prod_{i=1}^n f(x_i-\theta)-\prod_{i=1}^n g(x_i-\theta)\right|\prod_{i=1}^n dx_i$$
$$=\int_{a-\Delta}^{b}\cdots\int_{a-\Delta}^{b}\left|\prod_i f(x_i+\Delta)-\prod_i g(x_i)\right|\prod_i dx_i$$

$$+\int_{a-\Delta}^{b}\cdots\int_{a-\Delta}^{b}\left|\prod_i f(x_i)-\prod_i g(x_i)\right|\prod_i dx_i$$

$$=\int_{a-\Delta}^{b}\cdots\int_{a-\Delta}^{b}\left|\prod_i\frac{f(x_i+\Delta)}{g(x_i)}-1\right|\prod_i g(x_i)\prod_i dx_i$$

$$+\int_{a-\Delta}^{b}\cdots\int_{a-\Delta}^{b}\left|\prod_i\frac{f(x_i)}{g(x_i)}-1\right|\prod_i g(x_i)\prod_i dx_i$$

$$\leqq\left[\int_{a-\Delta}^{b}\cdots\int_{a-\Delta}^{b}\left\{\prod_i\frac{f(x_i+\Delta)}{g(x_i)}-1\right\}^2\prod_i g(x_i)\prod_i dx_i\right]^{1/2}$$

$$+\left[\int_{a-\Delta}^{b}\cdots\int_{a-\Delta}^{b}\left\{\prod_i\frac{f(x_i)}{g(x_i)}-1\right\}^2\prod_i g(x_i)\prod_i dx_i\right]^{1/2}.$$

Furthermore we have

$$(2.5.3)\quad\int_{a-\Delta}^{b}\cdots\int_{a-\Delta}^{b}\left\{\prod_i\frac{f(x_i+\Delta)}{g(x_i)}-1\right\}^2\prod_i g(x_i)\prod_i dx_i$$

$$=\int_{a-\Delta}^{b}\cdots\int_{a-\Delta}^{b}\left\{\prod_i\frac{f(x_i+\Delta)}{g(x_i)}\right\}^2\prod_i g(x_i)\prod_i dx_i$$

$$-2\int_{a-\Delta}^{b}\cdots\int_{a-\Delta}^{b}\prod_i f(x_i+\Delta)\prod_i dx_i+\int_{a-\Delta}^{b}\cdots\int_{a-\Delta}^{b}\prod_i g(x_i)\prod_i dx_i$$

$$=\left[\int_{a-\Delta}^{b}\left\{\frac{f(x+\Delta)}{g(x)}\right\}^2 g(x)dx\right]^n-1$$

$$=\left[\int_{a-\Delta}^{b}\frac{\{(f(x+\Delta)-g(x))+g(x)\}^2}{g(x)}dx\right]^n-1$$

$$=\left[\int_{a-\Delta}^{b}\frac{\{f(x+\Delta)-g(x)\}^2}{g(x)}dx+2\int_{a-\Delta}^{b}\{f(x+\Delta)-g(x)\}dx\right.$$

$$\left.+\int_{a-\Delta}^{b}g(x)dx\right]^n-1$$

$$=\left[\int_{a-\Delta}^{b}\frac{\{f(x+\Delta)-g(x)\}^2}{g(x)}dx+1\right]^n-1.$$

Similarly we have

$$(2.5.4)\quad\int_{a-\Delta}^{b}\cdots\int_{a-\Delta}^{b}\left\{\prod_i\frac{f(x_i)}{g(x_i)}-1\right\}^2\prod_i g(x_i)\prod_i dx_i$$

$$=\left[\int_{a-\Delta}^{b}\frac{\{f(x)-g(x)\}^2}{g(x)}dx+1\right]^n-1.$$

It follows from (2.5.2), (2.5.3) and (2.5.4) that (2.5.1) holds. Thus we complete the proof.

If the assumptions of lemma 2.5.1 hold, we can define an information number by

$$I = \int_{a-\Delta}^{b} \frac{\{f(x+\Delta)-g(x)\}^2}{g(x)}\,dx$$

For the remainder of section 2.5 for $0<\Delta<b-a$, we put $g(x) = \frac{1}{2}\{f(x+\Delta)+f(x)\}$. Then it is easily seen that $g(\cdot)$ satisfies the assumption of lemma 2.5.1.

Since

$$f(x+\Delta)- g(x) = \frac{1}{2}\{f(x+\Delta) - f(x)\}$$

and

$$f(x) - g(x) = \frac{1}{2}\{f(x) - f(x+\Delta)\},$$

it follows from (2.5.1) that

$$(2.5.5) \qquad d_n(\theta-\Delta,\theta) \leqq 2[\{\int_{a-\Delta}^{b} \frac{\{f(x+\Delta)-g(x)\}^2}{g(x)}\,dx+1\}^n -1]^{1/2}$$

$$= 2\{(I+1)^n - 1\}^{1/2}.$$

For the remainder of this section we suppose that $f(x)$ satisfies Assumptions (A.2.4.1) $\sim$ (A.2.4.3).

Then there exist numbers K_i, K_i' $(i=1,2,3)$ and $\varepsilon>0$ such that

$$(2.5.6) \qquad 0<K_1 \leqq (x-a)^{1-\alpha}f(x) \leqq K_2 \quad \text{for } a<x<a+\varepsilon,$$

$$(2.5.7) \qquad 0<K_1' \leqq (b-x)^{1-\beta}f(x) \leqq K_2' \quad \text{for } b-\varepsilon<x<b,$$

$$(2.5.8) \qquad (x-a)^{2-\alpha}|f'(x)| \leqq K_3 \quad \text{for } a<x<a+\varepsilon,$$

$$(2.5.9) \qquad (b-x)^{2-\beta}|f'(x)| \leqq K_3' \quad \text{for } b-\varepsilon<x<b,$$

$$0<\varepsilon<\min\left\{1, \frac{b-a}{2}\right\}.$$

Let $0<\Delta<\frac{\varepsilon}{2}$ and consider the case when Δ approaches to zero.

Now we divide I into six parts I_0, I_1, I_2, I_3, I_4 and I_5, that is,

$$I = \sum_{i=0}^{5} I_i ,$$

where $I_0 = \int_{a-\Delta}^{a} \frac{\{f(x+\Delta)-g(x)\}^2}{g(x)}\,dx$, $I_1 = \int_{a}^{a+\Delta} \frac{\{f(x+\Delta)-g(x)\}^2}{g(x)}\,dx$,

$I_2 = \int_{a+\Delta}^{a+\varepsilon} \frac{\{f(x+\Delta)-g(x)\}^2}{g(x)}\,dx$, $I_4 = \int_{a+\varepsilon}^{b-\varepsilon} \frac{\{f(x+\Delta)-g(x)\}^2}{g(x)}\,dx$,

$I_4 = \int_{b-\varepsilon}^{b-\Delta} \frac{\{f(x+\Delta)-g(x)\}^2}{g(x)}\,dx$, $I_5 = \int_{b-\Delta}^{b} \frac{\{f(x+\Delta)-g(x)\}^2}{g(x)}\,dx$,

Lemma 2.5.2 For $\alpha > 0$, the orders of I_0, I_1, I_2, I_3, I_4, I_5 and I are given by Table 2.5.1.

α	I_0	I_1	I_2	I_3	I_4	I_5	I
$0<\alpha<2$	$O(\Delta^\alpha)$	$O(\Delta^\alpha)$	$O(\Delta^\alpha)$				$O(\Delta^\alpha)$
$\alpha = 2$	$O(\Delta^2)$	$O(\Delta^2)$	$O(\Delta^2\lvert\log\Delta\rvert)$	$O(\Delta^2)$	$O(\Delta^2)$ if $\beta\neq 2$ $O(\Delta^2\lvert\log\Delta\rvert)$ if $\beta=2$	$O(\Delta^\beta)$	$O(\Delta^2\lvert\log\Delta\rvert)$
$\alpha > 2$	$O(\Delta^\alpha)$	$O(\Delta^\alpha)$	$O(\Delta^2)$				$O(\Delta^2)$

Table 2.5.1.

Proof. i) I_0 and I_1. It follows from (2.5.6) that

$$I_0 = \int_{a-\Delta}^{a} \frac{\{f(x+\Delta)-g(x)\}^2}{g(x)}\,dx = \int_{a-\Delta}^{a} \frac{f(x+\Delta)}{2}\,dx = O(\Delta^\alpha) . \tag{2.5.10}$$

Since

$$I_1 = \int_{a}^{a+\Delta} \frac{\{f(x+\Delta)-g(x)\}^2}{g(x)}\,dx$$

$$= \int_{a}^{a+\Delta} \frac{\{(f(x+\Delta)-f(x))/2\}^2}{\{(f(x+\Delta)+f(x))/2\}^2}\,(f(x+\Delta)+f(x))/2\,dx$$

$$\leqq \frac{1}{2}\int_{a}^{a+\Delta} \{f(x+\Delta)+f(x)\}\,dx ,$$

it follows from (2.5.6) that

(2.5.11) $\quad I_1 = O(\Delta^{\alpha})$.

ii) I_2. It follows by the mean value theorem that

$$(2.5.12) \quad I_2 = \int_{a+\Delta}^{a+\varepsilon} \frac{\{f(x+\Delta)-g(x)\}^2}{g(x)}\, dx$$

$$= \int_{a+\Delta}^{a+\varepsilon} \frac{\{(f(x+\Delta)-f(x))/2\}^2}{(f(x+\Delta)+f(x))/2}\, dx$$

$$\leqq \frac{1}{2} \int_{a+\Delta}^{a+\varepsilon} \frac{\{f(x+\Delta)-f(x)\}^2}{f(x)}\, dx$$

$$= \frac{1}{2} \int_{a+\Delta}^{a+\varepsilon} \Delta^2 \frac{\{f'(\xi(x,\Delta))\}^2}{f(x)}\, dx ,$$

where $a+\Delta < x < \xi(x,\Delta) < x+\Delta < a+\Delta+\varepsilon$.

If $0 < \alpha < 2$, then it follows from (2.5.6), (2.5.8) and (2.5.12) that

$$(2.5.13) \quad I_2 \leqq \int_{a+\Delta}^{a+\varepsilon} \Delta^2 c_1 \frac{(\xi-a)^{2\alpha-4}}{(x-a)^{\alpha-1}}\, dx$$

$$\leqq c_1 \Delta^2 \int_{a+\Delta}^{a+\varepsilon} \frac{(x-a)^{2\alpha-4}}{(x-a)^{\alpha-1}}\, dx$$

$$= c_1 \Delta^2 \int_{\Delta}^{\varepsilon} x^{\alpha-3}\, dx$$

$$= \frac{c_1}{\alpha-2} \varepsilon^{\alpha-2} \Delta^2 - \frac{c_1}{\alpha-2} \Delta^{\alpha},$$

where c_1 is some positive constant.

If $\alpha=2$, then it follows from (2.5.8) that $f'(x)$ is bounded on $(a, a+\varepsilon)$. From (2.5.6) and (2.5.12) we have

$$(2.5.14) \quad I_2 \leqq C_2 \Delta^2 \int_{a+\Delta}^{a+\varepsilon} \{1/(x-a)\}\, dx$$

$$= C_2 \Delta^2 (\log \varepsilon - \log \Delta) ,$$

where C_2 is some positive constant.

If $\alpha > 2$, then it follows from (2.5.6), (2.5.8) and (2.5.12) that

(2.5.15) $$I_2 \leqq C_3 \int_{a+\Delta}^{a+\varepsilon} \Delta^2 \frac{(\xi - a)^{2\alpha - 4}}{(x-a)^{\alpha - 1}} dx$$

$$\leqq C_3 \Delta^2 \int_{a+\Delta}^{a+\varepsilon} \frac{(x-a+\Delta)^{2\alpha - 4}}{(x-a)^{\alpha - 1}} dx$$

$$= C_2 \Delta^2 \int_{\Delta}^{\varepsilon} x^{\alpha - 3} (1+\frac{\Delta}{x})^{2\alpha - 4} dx$$

$$\leqq 2^{2\alpha - 4} C_3 \Delta^2 \int_{\Delta}^{\varepsilon} x^{\alpha - 3} dx$$

$$\leqq 2^{2\alpha - 4} C_3 \varepsilon^{\alpha - 2} \Delta^2 - \frac{2^{2\alpha - 4}}{\alpha - 2} C_3 \Delta^{\alpha} ,$$

where C_3 is some positive constant.

Hence it follows from (2.5.13), (2.5.14) and (2.5.15) that

(2.5.16) $$I_2 = \begin{cases} O(\Delta^{\alpha}) & \text{if} \quad 0 < \alpha < 2 , \\ O(\Delta^2 |\log \Delta|) & \text{if} \quad \alpha = 2 , \\ O(\Delta^2) & \text{if} \quad \alpha > 2 . \end{cases}$$

<u>iii) I_3.</u> Since $f(x)$ and $f'(x)$ are continuous functions on (a,b), it follows that

(2.5.17) $$I_3 = \int_{a+\varepsilon}^{b-\varepsilon} \frac{\{f(x+\Delta)-g(x)\}^2}{g(x)} dx$$

$$\leqq \frac{1}{2} \int_{a+\varepsilon}^{b-\varepsilon} \frac{\{f(x+\Delta)-f(x)\}^2}{f(x)} dx$$

$$= \frac{1}{2} \int_{a+\varepsilon}^{b-\varepsilon} \frac{\Delta^2 \{f'(\xi(x,\Delta))\}^2}{f(x)} dx$$

$$\leqq C_4' \Delta^2 \int_{a+\varepsilon}^{b-\varepsilon} \{1/f(x)\} dx$$

$$= C_4 \Delta^2 ,$$

where $a+\varepsilon < x < \xi(x,\Delta) < x+\Delta < b-(\varepsilon/2)$, and C_4' and C_4 are some positive constants.

Hence we have

(2.5.18) $\quad I_3=O(\Delta^2)$.

<u>iv) I_4.</u> It follows by the mean value theorem that

$$(2.5.19)\quad I_4=\int_{b-\varepsilon}^{b-\Delta}\frac{\{f(x+\Delta)-g(x)\}^2}{g(x)}dx$$

$$\leqq \frac{1}{2}\int_{b-\varepsilon}^{b-\Delta}\frac{\{f(x+\Delta)-f(x)\}^2}{f(x)}dx$$

$$=\frac{1}{2}\int_{b-\varepsilon}^{b-\Delta}\frac{\Delta^2\{f'(\xi(x,\Delta))\}^2}{f(x)}dx,$$

where $b-\varepsilon<x<\xi(x,\Delta)<x+\Delta<b-(\varepsilon/2)$.

If $0<\beta<2$, then it follows from (2.5.7), (2.5.9) and (2.5.19)

$$(2.5.20)\quad I_4\leqq C_5\Delta^2\int_{b-\varepsilon}^{b-\Delta}\frac{(b-\xi)^{2\beta-4}}{(b-x)^{\beta-1}}dx$$

$$\leqq C_5\Delta^2\left(\frac{\varepsilon}{2}\right)^{2\beta-4}\int_{\Delta}^{\varepsilon}x^{1-\beta}\,dx$$

$$=\frac{C_5}{2-\beta}\,\frac{\varepsilon^{\beta-2}}{2^{2\beta-4}}\,\Delta^2-\frac{C_5}{2-\beta}\left(\frac{\varepsilon}{2}\right)^{2\beta-4}\Delta^{4-\beta},$$

where C_5 is some positive constant.

If $2\leqq\beta$, then it follows from (2.5.7), (2.5.9) and (2.5.19) that

$$(2.5.21)\quad I_4\leqq\int_{b-\varepsilon}^{b-\Delta}C_6\Delta^2\frac{(b-\xi)^{2\beta-4}}{(b-x)^{\beta-1}}dx$$

$$\leqq C_6\Delta^2\int_{b-\varepsilon}^{b-\Delta}\frac{(b-x)^{2\beta-4}}{(b-x)^{\beta-1}}dx$$

$$=C_6\Delta^2\int_{\Delta}^{\varepsilon}x^{\beta-3}\,dx$$

$$=\begin{cases}C_6\Delta^2(\log\varepsilon-\log\Delta) & \text{if } \beta=2,\\ C_6\Delta^2\dfrac{1}{\beta-2}(\varepsilon^{\beta-2}-\Delta^{\beta-2}) & \text{if } \beta>2,\end{cases}$$

where C_6 is some positive constant.

Hence it follows from (2.5.20) and (2.5.21) that

$$(2.5.22) \qquad I_4 = \begin{cases} O(\Delta^2) & \text{if } \beta \neq 2 \\ O(\Delta^2 |\log\Delta|) & \text{if } \beta = 2 . \end{cases}$$

v) I_5. It follows from (2.5.8) that

$$(2.5.23) \qquad I_5 = \int_{b-\Delta}^{b} \frac{\{f(x+\Delta)-g(x)\}^2}{g(x)}\,dx$$

$$= \int_{b-\Delta}^{b} \frac{f(x)}{2}\,dx$$

$$= O(\Delta^{\beta}) .$$

since $I = \sum_{i=0}^{5} I_i$, it follows from (2.5.10), (2.5.11), (2.5.16), (2.5.18), (2.5.22) and (2.5.23) that

$$I = \begin{cases} O(\Delta^{\alpha}) & \text{if } 0<\alpha<2 , \\ O(\Delta^2|\log\Delta|) & \text{if } \alpha = 2 , \\ O(\Delta^2) & \text{if } \alpha > 2. \end{cases}$$

Thus we complete the proof.

Remark : We also define another information number I^* by

$$I^* = \int_a^b \frac{\{f(x+\Delta)-f(x)\}^2}{f(x)}\,dx$$

Since

$$\frac{\{f(x+\Delta)-g(x)\}^2}{g(x)} \leqq \frac{1}{2} \frac{\{f(x+\Delta)-f(x)\}^2}{f(x)} \quad \text{for } a<x<b ,$$

it follows that $I_i \leqq I_i^*$ $(i=1,2,3,4,5)$ where $I^* = \sum_{i=1}^{5} I_i^*$,

$$I_1^* = \int_a^{a+\Delta} \frac{\{f(x+\Delta)-f(x)\}^2}{f(x)}\,dx , \quad I_2^* = \int_{a+\Delta}^{a+\varepsilon} \frac{\{f(x+\Delta)-f(x)\}^2}{f(x)}\,dx ,$$

$$I_3^* = \int_{a+\varepsilon}^{b-\varepsilon} \frac{\{f(x+\Delta)-f(x)\}^2}{f(x)}\,dx , \quad I_4^* = \int_{b-\varepsilon}^{b-\Delta} \frac{\{f(x+\Delta)-f(x)\}^2}{f(x)}\,dx ,$$

$$I_5^* = \int_{b-\Delta}^{b} \frac{\{f(x+\Delta)-f(x)\}^2}{f(x)}\,dx .$$

It follows from the proof of lemma 2.5.2 that for each $\alpha > 0$ the orders of I_i^* (i=2,3,4,5) coincide with the orders of I_i (i=2,3,4,5) respectively, given by Table 2.5.1. Furthermore if $0<\alpha \leqq 1$, then it follows from (2.5.6) that there exists a positive constant C_7 such that

$$(2.5.24) \quad 0<\frac{f(x+\Delta)}{f(x)} \leqq \frac{K_2(x+\Delta-a)^{\alpha-1}}{K_1(x-a)^{\alpha-1}} = \frac{K_2}{K_1}(1+\frac{\Delta}{x-a})^{\alpha-1} \leqq C_7 \text{ for } a<x<a+\Delta$$

and the following hold :

$$(2.5.25) \quad \int_a^{a+\Delta} f(x)dx=O(\Delta^{\alpha}) ,$$

$$(2.5.26) \quad \int_a^{a+\Delta} f(x+\Delta)dx = O(\Delta^{\alpha}) .$$

From (2.5.24) we have

$$(2.5.27) \quad I_1^* = \int_a^{a+\Delta} \frac{\{f(x+\Delta)-f(x)\}^2}{f(x)} dx$$

$$= \int_a^{a+\Delta} \frac{\{f(x+\Delta)\}^2}{f(x)} dx - 2\int_a^{a+\Delta} f(x+\Delta)dx + \int_a^{a+\Delta} f(x)dx .$$

$$\leqq (C_7^2 + 1)\int_a^{a+\Delta} f(x)dx - 2\int_a^{a+\Delta} f(x+\Delta)dx$$

It follows from (2.5.25), (2.5.26) and (2.5.27) that $I_1^* = O(\Delta^2)$. Hence if $0<\alpha\leqq 1$, the order of I_1^* is equal to the order of I_1.

From lemmas 2.5.1 and 2.5.2 and (2.5.5) we get the following lemma.

<u>Lemma 2.5.3.</u>

$$d_n(\theta-\Delta,\theta) = \begin{cases} 2[\{1+O(\Delta^{\alpha})\}^n - 1]^{1/2} & \text{if } 0<\alpha<2 , \\ 2[\{1+O(\Delta^2 |\log\Delta|)\}^n - 1]^{1/2} & \text{if } \alpha=2 , \\ 2[\{1+O(\Delta^2)\}^n - 1]^{1/2} & \text{if } \alpha>2 . \end{cases}$$

Theorem 2.5.1. Let $X_1, X_2, \ldots, X_n, \ldots$ be a sequence of independent identically distributed random variables with a density function satisfying Assumptions (A.2.4.1) $\sim$ (A.2.4.3) of section 2.4. For each α, the order given by Table 2.4.1 of Theorem 2.4.1 is the bound of the order of convergence of consistent estimators ; that is, there does not exist a consistent estimator with order greater than values as given in Table 2.4.1.

Proof.

1) $0<\alpha<2$. From lemma 2.5.3 we obtain for sufficiently large n and every $t>0$,

$$d_n(\theta - tc_n^{-1}, \theta) \leqq 2[\{1+O((tc_n^{-1})^{\alpha})\}^n - 1]^{1/2} .$$

If order c_n is greater than order $n^{1/\alpha}$, then $\lim_{n\to\infty} d_n(\theta - tc_n^{-1}, \theta)=0$ for all $t>0$ and all $\theta \in \textcircled{H}$. Hence it follows from Theorem 2.3.1 that there does not a consistent estimator with the order greater than order $n^{1/\alpha}$.

2) $\alpha=2$. From Lemma 2.5.3 we obtain for sufficiently large n and every $t>0$,

$$d_n(\theta - tc_n^{-1}, \theta) \leqq 2[\{1+O((tc_n^{-1})^2 \,|\log tc_n^{-1}|)\}^n - 1]^{1/2} .$$

If order c_n is greater than order $(n \log n)^{1/2}$, then $\lim_{n\to\infty} d(\theta - tc_n^{-1}, \theta)=0$ for all $t>0$ and $\theta \in \textcircled{H}$. Hence it follows from Theorem 2.3.1 that there does not exist a consistent estimator with the order greater than order $(n \log n)^{1/2}$.

3) $\alpha > 2$. From Lemma 2.5.3 we have for sufficiently large n and every $t>0$,

$$d_n(\theta - tc_n^{-1}, \theta) \leqq 2[\{1+O((tc_n^{-1})^2)\}^n - 1]^{1/2} .$$

If order c_n is greater than $n^{1/2}$, then $\lim_{n\to\infty} d_n(\theta - tc_n^{-1}, \theta)=0$ for all $t>0$ and all $\theta \in \textcircled{H}$. Hence it follows from Theorem 2.3.1 that there does not exist a consistent estimator with the order greater than order $n^{1/2}$. Thus we complete the proof.

Remark: Since $A(\theta)=(a+\theta\ ,\ b+\theta\ )$, it follows from Assumptions (A.2.4.1) and (A.2.4.2) that for every $t>0$ and sufficiently large n

$$\left\{P_\theta(A(\theta - tc_n^{-1}))\right\}^n = \left\{1-\int_{b+\theta-tc_n^{-1}}^{b+\theta} f(x-\theta)dx\right\}^n$$

$$= \exp\left[\, n \log\left\{1 - \int_{b-tc_n^{-1}}^{b} f(x)dx\right\}\right]$$

$$= \exp\left[\, n\left\{-\frac{K}{\beta}t^\beta c_n^{-\beta} + O(c_n^{-2\beta})\right\}\right] \ ;$$

$$\left\{P_{\theta - tc_n^{-1}}(A(\theta))\right\}^n = \exp\left[\, n\left\{-\frac{K}{\alpha}t^\alpha c_n^{-\alpha} + O(c_n^{-2\alpha})\right\}\right] \ ,$$

where K is some positive constant.

From Lemma 2.4.3 we obtain the following results.

If $0<\alpha<2$ and $\alpha<\beta$, then every $\theta \in \text{Ⓗ}$ and every $t>0$,

$$\lim_{n\to\infty} L(\theta - tn^{-\frac{1}{\alpha}}, \theta) = 1 - e^{-\frac{K}{\alpha}t^\alpha} ,$$

$$\lim_{n\to\infty} R(\theta - tn^{-\frac{1}{\alpha}}, \theta) = 0 ,$$

$$\lim_{n\to\infty} M(\theta - tn^{-\frac{1}{\alpha}}, \theta) = \begin{cases} (e^{K't^\alpha} - 2e^{-\frac{K}{\alpha}t^\alpha} + 1)^{1/2} & \text{for } 0<\alpha\leqq 1 , \\ \infty & \text{for } 1<\alpha<2 , \end{cases}$$

$$\overline{\lim_{n\to\infty}}\, d_n(\theta - tn^{-\frac{1}{\alpha}}, \theta) \leqq 1-e^{-\frac{K}{\alpha}t^\alpha} + (e^{K't^\alpha} - 2e^{-\frac{K}{\alpha}t^\alpha} + 1)^{1/2} \quad \text{for } 0<\alpha\leqq 1,$$

$$\overline{\lim_{n\to\infty}}\, d_n(\theta - tn^{-\frac{1}{\alpha}}, \theta) \leqq 2(e^{ct^\alpha} - 1)^{1/2} \quad \text{for } 1<\alpha<2 ,$$

where c is some positive constant and K' is some constant.

Here we write

$$L(\theta_1, \theta_2) = 1 - \left\{P_{\theta_1}(A(\theta_2))\right\}^n \ ;$$

$$R(\theta_1, \theta_2) = 1 - \left\{P_{\theta_2}(A(\theta_1))\right\}^n \ ;$$

$$M(\theta_1, \theta_2) = \left[\left\{\int_{A(\theta_1)\cap A(\theta_2)} f^2(x-\theta_1)/f(x-\theta_2)d\mu\right\}^n - 2\left\{P_{\theta_1}(A(\theta_2))\right\}^n + \left\{P_{\theta_2}(A(\theta_1))\right\}^n\right]^{1/2} ,$$

and we note that

$$M(\theta_1,\theta_2) \geqq \int_{\overset{n}{\underset{i=1}{X}}(A(\theta_1)\cap A(\theta_2))} \left| \prod_{i=1}^{n} f(x_i-\theta_1) - \prod_{i=1}^{n} f(x_i-\theta_2) \right| d\mu^{(n)}.$$

If $0<\alpha<2$ and $\alpha=\beta$, then for every $\theta \in (H)$ and every $t>0$,

$$\lim_{n\to\infty} L(\theta - tn^{-\frac{1}{\alpha}}, \theta) = \lim_{n\to\infty} R(\theta - tn^{-\frac{1}{\alpha}}, \theta) = 1-e^{-\frac{K}{\alpha}t^{\alpha}},$$

$$\lim_{n\to\infty} M(\theta - tn^{-\frac{1}{\alpha}}, \theta) = \begin{cases} (e^{K't^{\alpha}} - e^{-\frac{K}{\alpha}t^{\alpha}})^{1/2} & \text{for } 0<\alpha<1, \\ 0 & \text{for } \alpha=1, \\ \infty & \text{for } 1<\alpha<2, \end{cases}$$

$$\overline{\lim_{n\to\infty}}\, d_n(\theta - tn^{-\frac{1}{\alpha}}, \theta) \leqq 2(1-e^{-\frac{K}{\alpha}t^{\alpha}}) + (e^{K't^{\alpha}} - e^{-\frac{K}{\alpha}t^{\alpha}})^{1/2} \quad \text{for } 0<\alpha<1,$$

$$\overline{\lim_{n\to\infty}}\, d_n(\theta - tn^{-1}, \theta) \leqq 2(1-e^{-Kt}) \quad \text{for } \alpha=1,$$

$$\overline{\lim_{n\to\infty}}\, d_n(\theta - tn^{-\frac{1}{\alpha}}, \theta) \leqq 2(e^{ct^{\alpha}}-1)^{1/2} \quad \text{for } 1<\alpha<2.$$

If $\alpha=2$, then for every $\theta \in (H)$ and every $t>0$,

$$\lim_{n\to\infty} L(\theta - t(n\log n)^{-\frac{1}{2}}, \theta) = \lim_{n\to\infty} R(\theta - t(n\log n)^{-\frac{1}{2}}, \theta) = 0,$$

$$\lim_{n\to\infty} M(\theta - t(n\log n)^{-\frac{1}{2}}, \theta) = \infty,$$

but

$$\overline{\lim_{n\to\infty}}\, d_n(\theta - t(n\log n)^{-\frac{1}{2}}, \theta) \leqq 2(e^{c}-1)^{\frac{1}{2}},$$

where c is some positive constant.

If $\alpha>2$, then for every $\theta \in (H)$ and every $t>0$,

$$\lim_{n\to\infty} L(\theta - tn^{-\frac{1}{2}}, \theta) = \lim_{n\to\infty} R(\theta - tn^{-\frac{1}{2}}, \theta) = 0,$$

$$\lim_{n\to\infty} M(\theta - tn^{-\frac{1}{2}}, \theta) = \infty,$$

but

$$\overline{\lim_{n\to\infty}}\, d_n(\theta - tn^{-\frac{1}{2}}, \theta) \leqq 2(e^{c'}-1)$$

where c' is some positive constant.

The above argument shows that L, R and M are not sufficient to determine the exact bound of order of convergence of $\{c_n\}$-consistent estimators.

The conclusions of this section and the previous one can be readily generalized to the case where either $a=-\infty$ or $b=\infty$, with appropriate but obvious modifications of regularity conditions. The case where $a=0, b=\infty$ and $\alpha=2$ coincides with the result obtained by Woodroofe [57].

2.6. Order of convergence of consistent estimators in an autoregressive process

We consider an autoregressive model as a typical example of non-i.i.d. cases.

Let X_t $(t=1,2,\dots)$ be defined recursively by

$$X_t = \theta X_{t-1} + U_t \quad (t=1,2,\dots) ,$$

where $X_0=0$ and $\{U_t : t=1,2,\dots\}$ is a sequence of independent identically distributed random variables having a density function f(w.r.t. a Lebesgue measure) with mean 0, variance σ^2 and finite fourth moment. Assume that $|\theta| < 1$ and $f(u) > 0$ for all u.

Since the joint density of $X_1, \dots, X_n$ is given by $\prod_{t=1}^{n} f(x_t - \theta x_{t-1})$, it follows that for every θ_1, θ_2 with $\theta_1 \neq \theta_2$

$$I_n(\theta_1,\theta_2) = \int \cdots \int \prod_{t=1}^{n} f(x_t - \theta_1 x_{t-1}) \sum_{t=1}^{n} \log \frac{f(x_t - \theta_1 x_{t-1})}{f(x_t - \theta_2 x_{t-1})} \prod_{t=1}^{n} dx_t .$$

Putting $\theta_2 = \theta_1 + \Delta$, we have

$$I_n(\theta_1,\theta_2) = \int \cdots \int \prod_{t=1}^{n} f(u_t) \sum_{t=1}^{n} \log \frac{f(u_t)}{f(u_t - \Delta x_{t-1})} \prod_{t=1}^{n} du_t$$

$$= \sum_{t=1}^{n} E\left[\int f(u_t) \log \frac{f(u_t)}{f(u_t - \Delta X_{t-1})} du_t \right] .$$

For small $|\Delta|$ we obtain

$$\int f(u) \log \frac{f(u)}{f(u-\Delta)} du = O(\Delta^2) .$$

If for large $|\Delta|$

$$\log \frac{f(u)}{f(u-\Delta)} = O(\Delta^2) ,$$

then we have

$$\sup_{0<|\Delta|<\infty} \frac{1}{\Delta^2} \int f(u) \log \frac{f(u)}{f(u-\Delta)} du < \infty .$$

Thus we have

$$I_n(\theta_1,\theta_2) = O\left(\Delta^2 \sum_{t=1}^{n} E(X_{t-1}^2)\right) . \tag{2.6.1}$$

Since for each t

$$E(X_t)=\theta E(X_{t-1})= \dots = \theta^t E(X_0)=0 ;$$

$$\begin{aligned} E(X_t^2) &= \theta^2 E(X_{t-1}^2) + E(U_t^2) \\ &= (\theta^{2(t-1)} + \dots + \theta^2 + 1)\sigma^2 \\ &= \left(\frac{1-\theta^{2t}}{1-\theta^2}\right)\sigma^2 , \end{aligned}$$

we obtain

$$(2.6.2) \qquad \sum_{t=1}^{n} E(X_{t-1}^2) = \left\{ \frac{n}{1-\theta^2} - \frac{1-\theta^{2n}}{(1-\theta^2)^2} \right\}\sigma^2 .$$

In order to show that the order of convergence of consistent estimators is $\sqrt{n}$, we need the following theorem. The notation is that of section 2.1.

Theorem 2.6.1. Suppose that for each n, $\{ \tilde{x}_n : dP_{\theta,n}/d\mu_n > 0 \}$ does not depend on θ . If there exists a $\{ c_n \}$-consistent estimators, then the following holds : for every $\theta \in$ Ⓗ and every $a \neq 0$

$$\lim_{L\to\infty} \underline{\lim}_{n\to\infty} I_n(\theta, \theta + aLc_n^{-1}) = \infty ,$$

where $I_n(\theta_1, \theta_2)$ is the Kullback-Leibler information given by Theorem 2.2.2.

The proof is omitted since it is quite similar to the proof of Theorem 2.2.2.

Letting $\Delta = Lc_n^{-1}$, in order to get

$$\lim_{L\to\infty} \underline{\lim}_{n\to\infty} I_n(\theta_1, \theta_1 + Lc_n^{-1}) = \infty ,$$

from (2.6.1), (2.6.2) and Theorem 2.6.1 we have to obtain $c_n = O(\sqrt{n})$. Hence, it is seen that the order of convergence of consistent estimators is $\sqrt{n}$.

Chapter 3

Asymptotic Efficiency

3.1. Definitions and a sufficient condition for the existence of an asymptotically efficient estimator

Summary :

In this section we define the class of asymptotically median unbiased estimators. We then define the asymptotically efficient estimator, which belongs to this class. We use an approach similar to Bahadur [12] dealing with the bound of asymptotic variances. We obtain a sufficient condition for the existence of an asymptotically efficient estimator.

Throughout this section we assume θ is a real-valued parameter and $\hat{\theta}_n$ is a $\{c_n\}$-consistent estimator.

Definition 3.1.1. $\hat{\theta}_n$ is asymptotically median unbiased (or AMU for short) if for any $\vartheta \in \Theta$ there exists a positive number δ such that

$$\limsup_{n\to\infty} \sup_{\theta: |\theta-\vartheta|<\delta} \left| P_{\theta,n}\left\{ \hat{\theta}_n \leqq \theta \right\} - \frac{1}{2} \right| = 0 \quad ;$$

$$\limsup_{n\to\infty} \sup_{\theta: |\theta-\vartheta|<\delta} \left| P_{\theta,n}\left\{ \hat{\theta}_n \geqq \theta \right\} - \frac{1}{2} \right| = 0 \quad .$$

Definition 3.1.2. For $\hat{\theta}_n$ asymptotically median unbiased a distribution function $F_\theta(t)$ is called an asymptotic distribution of $c_n(\hat{\theta}_n - \theta)$ (or $\hat{\theta}_n$ for short) if for each t, $F_\theta(t)$ is continuous in θ and for every continuity points t of $F_\theta(t)$

$$\lim_{n\to\infty} \left| P_{\theta,n}\left\{ c_n(\hat{\theta}_n - \theta) \leqq t \right\} - F_\theta(t) \right| = 0 \; .$$

Since $\hat{\theta}_n$ is a $\{c_n\}$-consistent estimator, it follows that $F_\theta(-\infty)=0$ and $F_\theta(\infty)=1$.

Let $\hat{\theta}_n$ be a AMU estimator. Then it follows that $F_{\theta-0}(0)\leqq 1/2$ and $F_{\theta+0}(0)\geqq 1/2$ for all $\theta \in \textcircled{H}$.

For $\hat{\theta}_n$ AMU, that G_θ^+ and G_θ^- are defined as follows :

(3.1.1) $\quad G_\theta^+(t)=\overline{\lim_{n\to\infty}} P_{\theta,n}\left\{c_n(\hat{\theta}_n-\theta)\leqq t\right\}$ for all $t\geqq 0$,

(3.1.2) $\quad G_\theta^-(t)=\underline{\lim_{n\to\infty}} P_{\theta,n}\left\{c_n(\hat{\theta}_n-\theta)\leqq t\right\}$ for all $t<0$.

Let $\theta_0(\in \textcircled{H})$ be arbitrary but fixed. Consider the problem of testing the hypothesis $H^+ : \theta = \theta_0+tn^{-1/2}(t>0)$ against alternative $K:\theta=\theta_0$. We define $\beta^+_{\theta_0}(t)$ as follows :

(3.1.3) $\quad \beta^+_{\theta_0}(t)=\sup_{\{\phi_n\}\in\Phi_{1/2}} \overline{\lim_{n\to\infty}} E_{\theta_0,n}(\phi_n)$,

where $\Phi_{1/2}=\left\{\{\phi_n\} : \lim_{n\to\infty} E_{\theta_0+tc_n^{-1},n}(\phi_n)=1/2,\ 0\leqq\phi_n(\tilde{x}_n)\leqq 1\right.$ for all $\tilde{x}_n$ $\left.(n=1,2,\ldots)\right\}$.

Putting $A_{\hat{\theta}_n,\theta_0}=\left\{c_n(\hat{\theta}_n-\theta_0)\leqq t\right\}$ we have for $t>0$

$P_{\theta_0+tc_n^{-1},n}(A_{\hat{\theta}_n,\theta_0})=P_{\theta_0+tc_n^{-1},n}\left\{c_n(\hat{\theta}_n-\theta_0-tc_n^{-1})\leqq 0\right\}\to 1/2\ (n\to\infty)$.

Since a sequence $\left\{\chi_{A_{\hat{\theta}_n,\theta_0}}\right\}$ of the indicators (or characteristic functions) of $A_{\hat{\theta}_n,\theta_0}(n=1,2,\ldots)$ belongs to $\Phi_{1/2}$, it follows from (3.1.1) and (3.1.3) that

(3.1.4) $\quad G^+_{\theta_0}(t)\leqq\beta^+_{\theta_0}(t)$

for all $t>0$.

Consider next the problem of testing the hypothesis $H^-:\theta=\theta_0+tc_n^{-1}$ $(t<0)$ against alternative $K:\theta=\theta_0$. Then we define $\beta^-_{\theta_0}(t)$ as follows :

(3.1.5) $\quad \beta^-_{\theta_0}(t)=\inf_{\{\phi_n\}\in\Phi_{\frac{1}{2}}} \underline{\lim_{n\to\infty}} E_{\theta_0,n}(\phi_n)$.

Note that

(3.1.6) $\quad \beta^-_{\theta_0}(t)=1-\sup_{\{\phi_n\}\in\Phi_{\frac{1}{2}}} \overline{\lim_{n\to\infty}} E_{\theta_0,n}(\phi_n)$.

In a similar way as the case $t>0$, we have from (3.1.2) and (3.1.5)

(3.1.7) $\quad G^-_{\theta_0}(t)\geqq\beta^-_{\theta_0}(t)$

for all $t<0$. Since θ_0 is arbitrary, the bounds of the asymptotic distributions of AMU estimators are given by :

$$G^+_\theta(t) \leqq \beta^+_\theta(t) \quad \text{for all} \quad t>0 \quad ;$$
$$G^-_\theta(t) \geqq \beta^-_\theta(t) \quad \text{for all} \quad t<0 \quad .$$

<u>Definition 3.1.3</u> An AMU estimator $\hat{\theta}_n$ is asymptotically efficient if for each $\theta \in \text{Ⓗ}$

$$(3.1.8) \qquad F_\theta(t) = \begin{cases} \beta^+_\theta(t) & \text{for all } t>0 \quad , \\ \beta^-_\theta(t) & \text{for all } t<0 \quad . \end{cases}$$

[Note that for t=0 we have $F_\theta(0)=\beta^+_\theta(0)=\beta^-_\theta(0)$ from the condition of asymptotically median unbiasedness.]

It can be easily seen that for any AMU estimator $\hat{\theta}_n$ and any $\theta \in \text{Ⓗ}$

$$\overline{\lim_{n\to\infty}} P_{\theta,n}\left\{c_n(\hat{\theta}_n-\theta)<t\right\} \leqq \beta^+_\theta(t) \quad \text{for all} \quad t>0 \quad ;$$
$$\underline{\lim_{n\to\infty}} P_{\theta,n}\left\{c_n(\hat{\theta}_n-\theta)<t\right\} \geqq \beta^-_\theta(t) \quad \text{for all} \quad t<0 \quad .$$

Hence if $\hat{\theta}^*_n$ satisfies the above condition of the asymptotic efficiency, we have for any AMU estimator $\hat{\theta}_n$

$$\underline{\lim_{n\to\infty}} [P_{\theta,n}\{-a<c_n(\hat{\theta}^*_n-\theta)<b\} - P_{\theta,n}\{-a<c_n(\hat{\theta}_n-\theta)<b\}] \geqq 0 \ .$$

for all positive numbers a and b all $\theta \in \text{Ⓗ}$.

<u>Remark:</u> The above definition of asymptotic efficiency is very strong in that it requires uniformity both in θ and t, which may appear to be too strict. However, it is shown by Takeuchi and Akahira [51] that the above definition of asymptotic efficiency works in most of situations.

We also note that we require the existence of the asymptotic distribution $F_\theta(t)$ of $\hat{\theta}_n$ which may be omitted in some cases, and the omission may yield different results in certain situations as is discussed in Chapter 1. Since, however, we are interested in more regular cases, we retain this assumption.

Let $X_1, X_2, \ldots, X_n \ldots$ be independently and identically distributed (i.i.d.) random variables with a density function $f(x,\theta)$ (w.r.t. a σ-finite measure μ) satisfying the following (A.3.1.1), (A.3.1.2) and (A.3.1.3).

(A.3.1.1) $\{x : f(x, \theta) > 0\}$ does not depend on θ.

(A.3.1.2) For almost all $x[\mu]$, $f(x,\theta)$ is twice continuously differentiable in θ.

(A.3.1.3) For each $\theta \in \textcircled{H}$

$$0 < I(\theta) = E_\theta[\{\frac{\partial}{\partial\theta}\log f(x,\theta)\}^2]$$
$$= -E_\theta[\frac{\partial^2}{\partial\theta^2}\log f(x,\theta)]$$
$$< \infty .$$

Let θ_0 be arbitrary but fixed in $\textcircled{H}$. Consider the problem of testing hypothesis $H^+ : \theta = \theta_0 + tn^{-1/2}$ $(t > 0)$ against alternative $K : \theta = \theta_0$. Then the rejection region of the most powerful test is given by

$$T_n = \sum_{i=1}^n Z_{ni} > c \quad ,$$

where $Z_{ni} = \log\{f(X_i, \theta_0)/f(X_i, \theta_0 + tn^{-\frac{1}{2}})\}$.

Since

$$T_n = \sum_{i=1}^n Z_{ni} = -\frac{t}{\sqrt{n}}\sum_{i=1}^n \frac{\partial}{\partial\theta}\log f(X_i,\theta)$$
$$- \frac{t^2}{2n}\sum_{i=1}^n \frac{\partial^2}{\partial\theta^2}\log f(X_i,\theta) + o_p(\frac{1}{\sqrt{n}})$$

For $\theta = \theta_0$, T_n is asymptotically normal with mean $t^2I(\theta_0)/2$ and variance $t^2I(\theta_0)$ and for $\theta = \theta_0 + (tn^{-1/2})$, T_n is asymptotically normal with mean $-t^2I(\theta_0)/2$ and variance $t^2I(\theta_0)$.

Hence it follows that

$$\beta^+_{\theta_0}(t) = \Phi(t\sqrt{I(\theta_0)}) ,$$

where $\Phi(u) = \int_{-\infty}^u \frac{1}{\sqrt{2\pi}} e^{-\frac{x^2}{2}} dx$.

From (3.1.4) we have

$$F_{\theta_0}(t) \leqq \Phi(t\sqrt{I(\theta_0)}) \text{ for all } t>0 .$$

In a similar way as the case $t>0$, we obtain from (3.1.7) for all $t<0$

$$F_{\theta_0}(t) \geqq 1 - \Phi(|t|\sqrt{I(\theta_0)}) = \Phi(t\sqrt{I(\theta_0)}) .$$

Since θ_0 is arbitrary we have now established the following well-known theorem.

Theorem 3.1.1. Under conditions (A.3.1.1), (A.3.1.2) and (A.3.1.3), if the asymptotic distribution of $\sqrt{n}(\hat{\theta}_n - \theta)$ is normal with mean 0 and variance $1/I(\theta)$, then $\hat{\theta}_n$ is asymptotically efficient.

The following theorem is well known and established ;

Theorem 3.1.2. Under Assumption (A.3.1.1), (A.3.1.2) and (A.3.1.3), if the maximum likelihood estimator is consistent, then it is asymptotically efficient.

3.2. Asymptotic efficiency in an autoregressive process

We consider asymptotic efficiency in an autoregressive process as a typical example of non-i.i.d. cases.

Let X_t $(t=1,2,...)$ be defined recursively by

$$X_t = \theta X_{t-1} + U_t \quad , \; t=1,2,... \; ,$$

where $X_0=0$ and $\{U_t : t=1,2,...\}$ is a sequence of independent identically distributed real random variables having a density function f with mean 0 and variance σ^2.
We assume that $|\theta| < 1$. In this section we shall obtain the bound of the asymptotic distributions of AMU estimators of θ using the asymptotic normality of the likelihood ratio statistic statistic and the sufficient condition that an AMU estimator be asymptotically efficient. We shall also give a necessary and

sufficient condition that the least squares estimator of θ be asymptotically efficient. In fact, Theorem 3.2.2 shows that under some regularity conditions the bound of the asymptotic distributions of AMU estimators of θ is a normal distribution with mean 0 and variance $(1-\theta^2)/\sigma^2 I$, where I is the Fisher information of f. Thus it is easily seen that an AMU estimator is asymptotically efficient if it has an asymptotic normal distribution with variance equal to the above bound. The least squares estimator $\hat{\theta}_{LS}$ of θ is given by $(\sum_{t=1}^{n} X_{t-1}X_t)/\sum_{t=1}^{n} X_{t-1}^2$. It is shown by Anderson [11] that $\sqrt{n}(\hat{\theta}_{LS}-\theta)$ has a limiting normal distribution with mean 0 and variance $1-\theta^2$. Then Theorem 3.2.4 shows that under some conditions the limiting distribution of $\sqrt{n}(\hat{\theta}_{LS}-\theta)$ attains the bound of the asymptotic distributions given by Theorem 3.2.2 if and only if f is a normal density function with mean 0 and variance σ^2. In the subsequent discussion in this section it is enough to consider only the case $c_n=\sqrt{n}$ (See section 2.6).

Throughout this section we assume that $\mathcal{X}=R^1$ and Ⓗ is an open interval (-1,1) and consider the autoregressive process $\{X_t\}$ given above.

The following lemma is given in Diananda [18].

<u>Lemma 3.2.1.</u> Let $\{Z_n : n=1,2,...\}$ be a sequence of random variables satisfying the following :

(i) $Z_n=Z_{n,N}+R_{n,N}$ $(n>N)$;

(ii) For each fixed N, the asymptotic distribution of $Z_{n,N}$ is normal with mean 0 and variance σ_N^2 ;

(iii) $\lim_{N\to\infty}\sigma_N^2=\sigma^2$;

(iv) $R_{n,N}$ converges in probability to 0 uniformly in n.

Then Z_n has a limiting normal distribution with mean 0 and variance σ^2 .

Let

(3.2.1) $Y_1, Y_2, \ldots$

be a sequence of random variables.

If for some function $g(n)$ the inequality $s-r > g(n)$ implies that the two sets

$$(Y_1, Y_2, \ldots, Y_r), (Y_s, Y_{s+1}, \ldots, Y_n)$$

are independent, then the sequence (3.2.1) is said to be $g(n)$-dependent ([24]). When $g(n)=m$, we say that (3.2.1) is m-dependent.

Let (3.2.1) be m-dependent and such that $E(Y_i)=0$, $E(Y_i^2)<\infty$ $(i=1,2,\ldots)$. Then we define

$$A_i=E(Y_{i+m}^2) + 2\sum_{j=1}^{m} E(Y_{i+m-j}Y_{i+m}) \quad (i=1,2,\ldots) .$$

The following lemma is given by Hoeffding and Robbins [24].

Lemma 3.2.2. Let $Y_1, Y_2, \ldots$ be an m-dependent sequence of random variables such that

(a) $E(Y_i)=0$, $E(|Y_i|^3)\leqq R^3<\infty$ $(i=1,2,\ldots)$,

(b) $\lim_{p\to\infty} p^{-1}\sum_{h=1}^{p} A_{i+h}=A$ exists, uniformly for all $i=0,1,\ldots$.

Then $\sum_{i=1}^{n} Y_i$ is asymptotically normal with mean 0 and variance nA.

Throughout the remainder of this section we assume the following :

(A.3.2.1). f is once differentiable and $f(u)>0$ for all u and $\lim_{u\to\pm\infty} f(u)=0$.

Then we get the following lemma.

Lemma 3.2.3. Under Assumption (A.3.2.1), if $E[|U_t|^3]<\infty$ and $E\left[\left|\frac{f'(U_t)}{f(U_t)}\right|^3\right]<\infty$, then the limiting distribution of $\frac{1}{\sqrt{n}}\sum_{t=1}^{n}\frac{f'(U_t)}{f(U_t)}X_{t-1}$ is normal with mean 0 and variance $\frac{\sigma^2 I}{1-\theta^2}$, where $I=\int\frac{\{f'(u)\}^2}{f(u)}du$.

Proof. Putting $V_t = \frac{f'(U_t)}{f(U_t)}$, $(t=1,2,\ldots)$, we have

$$\sum_{t=1}^{n} V_t X_{t-1} = \sum_{t=2}^{n} V_t \sum_{i=1}^{t-1} \theta^{t-1-i} U_i = \sum_{j=1}^{n-1} \theta^{j-1} \sum_{t=1}^{n-j} V_{t+j} U_t .$$

Put $W_{j,t} = V_{t+j} U_t$. Then for any fixed j, $\{W_{j,t} : t=1,2,\ldots\}$ is a j-dependent sequence and $E(W_{j,t})=0$ $(t=1,2,\ldots)$. Since $E(U_t^2)=\sigma^2$ and $E(V_{t+j}^2)=I(t=1,2,\ldots)$, we obtain $E(W_{j,t}^2)=\sigma^2 I(t=1,2,\ldots)$.

Put $Z_n = (1/\sqrt{n}) \sum_{j=1}^{n-1} \theta^{j-1} \sum_{t=1}^{n-j} W_{j,t}$ $(n=1,2,\ldots)$. Then for $n \geqq N+2$

$$Z_n = Z_{n,N} + R_{n,N} ,$$

where $Z_{n,N} = (1/\sqrt{n}) \sum_{j=1}^{N} \theta^{j-1} \sum_{t=1}^{n-j} W_{j,t}$ and $R_{n,N} = (1/\sqrt{n}) \sum_{j=N+1}^{n-1} \theta^{j-1} \sum_{t=1}^{n-j} W_{j,t}$.

We now show that $Z_{n,N}$ has an asymptotic normal distribution with mean 0 and variance $\left(\frac{1-\theta^{2N}}{1-\theta^2}\right)\sigma^2 I$. Since $E(W_{j,t+j-i+h} W_{j,t+j+h})=0$ $(i=1,\ldots,j\ ;\ t,h=1,2,\ldots)$ and $\frac{1}{p} \sum_{h=1}^{p} E(W_{j,t+j+h}^2)=\sigma^2 I$, it follows by Lemma 3.2.2 that $W_{j,t}$ is asymptotically normal with mean 0 and variance $(n-j)\sigma^2 I$. Hence $\theta^{j-1} \sum_{t=1}^{n-j} W_{j,t}$ is asymptotically normal with mean 0 and variance $(n-j)^{2(j-1)}\sigma^2 I$.

Put $W_j(n) = (1/\sqrt{n})\theta^{j-1} \sum_{t=1}^{n-j} W_{j,t}$ $(j=1,2,\ldots,N)$.

Then it can be shown that for each fixed N, $W_1(n), W_2(n), \ldots, W_N(n)$ are asymptotically jointly normal with mean 0 and variance $\left(\frac{1-\theta^{2N}}{1-\theta^2}\right)\sigma^2 I$. Hence for each fixed N, $Z_{n,N}$ has a limiting normal distribution with mean 0 and variance $\left(\frac{1-\theta^{2N}}{1-\theta^2}\right)\sigma^2 I$.

Next we show that $R_{n,N}$ converges in probability to 0 uniformly in n. We have

$$E_\theta(R_{n,N}^2) = \frac{1}{n} E_\theta\Big[\Big(\sum_{j=N+1}^{n-1} \theta^{j-1} \sum_{t=1}^{n-j} W_{j,t} \Big)^2 \Big]$$

$$= \frac{1}{n} E \Big[\sum_{j=N+1}^{n-1} \theta^{2(j-1)} \Big(\sum_{t=1}^{n-j} W_{j,t} \Big)^2$$

$$+ 2 \sum_{j<j'} \theta^{j+j'-2} \Big(\sum_{t=1}^{n-j} W_{j,t} \Big) \Big(\sum_{t=1}^{n-j'} W_{j',t} \Big) \Big]$$

$$= \frac{1}{n} \sum_{j=N+1}^{n-1} \theta^{2(n-j)} \sum_{t=1}^{n-j} E(W_{j,t}^2)$$

$$= \frac{1}{n} \sum_{j=N+1}^{n-1} \theta^{2(j-1)} (n-j) \sigma^2 I$$

$$\leqq \frac{1}{n} \sum_{k=1}^{n-N-1} \theta^{2(n-k-1)} k \sigma^2 I$$

$$= \frac{1}{n} \left\{ \sigma^2 I \, \theta^{2(n-2)} \sum_{k=1}^{n-N-1} k \, \theta^{-2(k-1)} \right\}$$

$$\leqq \sigma^2 I \, \theta^{2(N-2)} \sum_{k=1}^{n-N-1} k \, \theta^{-2(k-1)}$$

$$\leqq \sigma^2 I \, \theta^{2(N-2)} \sum_{k=1}^{\infty} k \, \theta^{-2(k-1)}$$

$$= \frac{\sigma^2 I \, \theta^{2(N-2)}}{(1-\theta^{-2})^2}$$

Hence we see that $\lim_{N \to \infty} E_\theta(R_{n,N}^2) = 0$ uniformly in n. Using Chebyshev's inequality we obtain $R_{n,N}$ converges in probability to 0 uniformly in n.

Since $\lim_{N \to \infty} \frac{1-\theta^{2N}}{1-\theta^2} \sigma^2 I = \frac{\sigma^2 I}{1-\theta^2}$ it follows by Lemma 3.2.1 that Z_n has a limiting normal distribution with mean 0 and variance $\frac{\sigma^2 I}{1-\theta^2}$.

Thus we complete the proof.

Next it will be shown that the bound of the asymptotic distributions of AMU estimators of θ is obtained using the best test statistics and that the least squares estimator of θ is asymptotically efficient if and only if f a normal density with mean 0 and variance σ^2.

Let θ_0 be arbitrary but fixed in Ⓗ. Putting $\theta_1 = \theta_0 + \lambda n^{-\frac{1}{2}} (\lambda > 0)$ we define Z_{nt} as follows :

$$Z_{nt} = \log \frac{f(X_t - \theta_0 X_{t-1})}{f(X_t - \theta_1 X_{t-1})} .$$

We assume that f is twice continuously differentiable. If $\theta = \theta_0$, then we have

$$(3.2.2) \quad \sum_{t=1}^{n} Z_{nt} = \sum_{t=1}^{n} \log \frac{f(U_t)}{f(U_t - \lambda n^{-1/2} X_{t-1})}$$

$$= \sum_{t=1}^{n} \left\{ \log f(U_t) - \log f(U_t - \lambda n^{-1/2} X_{t-1}) \right\}$$

$$= \lambda n^{-1/2} \sum_{t=1}^{n} \frac{f'(U_t)}{f(U_t)} X_{t-1} - \frac{\lambda^2}{2} n^{-1} \sum_{t=1}^{n} \frac{d^2 \log f(U_t^*)}{dU_t^2} X_{t-1} ,$$

where for each t, U_t^* lies between U_t and $U_t - \lambda n^{-1/2} X_{t-1}$.

If $\theta = \theta_1$, then we have

$$(3.2.3) \quad \sum_{t=1}^{n} Z_{nt} = \sum_{t=1}^{n} \log \frac{f(U_t + \lambda n^{-1/2} X_{t-1})}{f(U_t)}$$

$$= \sum_{t=1}^{n} \left\{ \log f(U_t + \lambda n^{-1/2} X_{t-1}) - \log f(U_t) \right\}$$

$$= \lambda n^{-1/2} \sum_{t=1}^{n} \frac{f'(U_t)}{f(U_t)} X_{t-1} + \frac{\lambda^2}{2} n^{-1} \sum_{t=1}^{n} \frac{d^2 \log f(U_t^{**})}{dU_t^2} X_{t-1} ,$$

where for each t, U_t^{**} lies between U_t and $U_t + \lambda n^{-1/2} X_{t-1}$.

Throughout the subsequent discussion we make the following aassumptions :

(A.3.2.2) f is three times differentiable on the real line and $\lim_{u \to \pm\infty} f'(u) = 0$.

(A.3.2.3) $d^2 \log f(u)/du^2$ is a bounded function and $E(|U_t|^4) < \infty$.

(A.3.2.4) For each $\theta_0 \in$ Ⓗ the following hold :

(a) $\lim_{n\to\infty} n^{-3/2} \sum_{t=1}^{n} E_{\theta_j}[\, |X_{t-1}|^3 \sup_{0<|\eta|<\lambda n^{-1/2}|X_{t-1}|} |g'(U_t+\eta)|\,] = 0 \quad (j=0,1)$;

(b) $\lim_{n\to\infty} n^{-3} E_{\theta_j}[\{ \sum_{t=1}^{n} |X_{t-1}|^3 \sup_{0<|\eta|<\lambda n^{-1/2}|X_{t-1}|} |g'(U_t+\eta)| \}^2] = 0 \quad (j=0,1)$,

where $g(u) = -\dfrac{d^2 \log f(u)}{du^2}$.

Putting $T_n = \sum_{t=1}^{n} X_{t-1}^2 \, g(U_t)$, $T_n^* = \sum_{t=1}^{n} X_{t-1}^2 g(U_t^*)$ and $T_n^{**} = \sum_{t=1}^{n} X_{t-1}^2 g(U_t^{**})$,

we get the following lemmas and theorem.

Lemma 3.2.4. Under Assumptions (A.3.2.1)$\sim$(A.3.2.4) the following hold :

$$(3.2.4)\quad E_{\theta_j}(T_n)=\sum_{t=1}^{n} E_{\theta_j}[X_{t-1}^2 \left\{ \frac{f'(U_t)}{f(U_t)} \right\}^2] = \sigma^2 I\left\{\frac{n-1}{1-\theta_j^2} - \frac{\theta_j^4(1-\theta_j^{2(n-1)})}{(1-\theta_j^2)^2}\right\}$$

$(j=0,1)$;

$$(3.2.5)\quad \lim_{n\to\infty} \left| E_{\theta_0}(T_n^*/n) - E_{\theta_0}(T_n/n) \right| = 0 \quad ;$$

$$(3.2.6)\quad \lim_{n\to\infty} \left| E_{\theta_1}(T_n^{**}/n) - E_{\theta_1}(T_n/n) \right| = 0 \quad ;$$

$$(3.2.7)\quad \lim_{n\to\infty} \left| E_{\theta_0}(T_n^{*2}/n^2) - E_{\theta_0}(T_n^2/n^2) \right| = 0 \ ;$$

$$(3.2.8)\quad \lim_{n\to\infty} \left| E_{\theta_1}(T_n^{**2}/n^2) - E_{\theta_1}(T_n^2/n^2) \right| = 0 \ .$$

Proof. Since

$$X_{t-1} = \sum_{i=1}^{t-1} \theta_j^{t-1-i} U_i \quad (j=0,1) ,$$

we have

$$\begin{aligned}
E_{\theta_j}(T_n) &= E_{\theta_j}[\sum_{t=1}^{n} X_{t-1}^2 g(U_t)] \\
&= \sum_{t=1}^{n} E_{\theta_j}[X_{t-1}^2 \left\{- \frac{d^2\log f(U_t)}{dU_t^2} \right\}] \\
&= -\sum_{t=1}^{n} E_{\theta_j}[X_{t-1}^2 \frac{f''(U_t)}{f(U_t)} - X_{t-1}^2 \left\{ \frac{f'(U_t)}{f(U_t)} \right\}^2] \\
&= \sum_{t=1}^{n} E_{\theta_j}[X_{t-1}^2 \left\{ \frac{f'(U_t)}{f(U_t)} \right\}^2] \\
&= \sum_{t=1}^{n} E_{\theta_j}(X_{t-1}^2) E[\left\{ \frac{f'(U_t)}{f(U_t)} \right\}^2] \\
&= \sum_{t=1}^{n} E_{\theta_j}(X_{t-1}^2) I \\
&= I \sum_{t=1}^{n} E_{\theta_j}[\left\{ \sum_{i=1}^{t-1} \theta_j^{t-1-i} U_i \right\}^2] \\
&= I \sum_{t=2}^{n} \sum_{i=1}^{t-1} \theta_j^{2(t-1-i)} E(U_i^2) \\
&= I \sum_{t=2}^{n} \sigma^2 \frac{1-\theta_j^{2t}}{1-\theta_j^2}
\end{aligned}$$

$$= \sigma^2 I \left\{ \frac{n-1}{1-\theta_j^2} - \frac{\theta_j^4(1-\theta_j^{2(n-1)})}{(1-\theta_j^2)^2} \right\} \qquad (j=0,1) .$$

Hence (3.2.4) holds.

Using the mean value theorem we have

$$\left| E_{\theta_0}(T_n^*/n) - E_{\theta_0}(T_n/n) \right|$$
$$\leqq (1/n) \sum_{t=1}^{n} E_{\theta_0}[X_{t-1}^2 \; | g(U_t^*) - g(U_t)| \;]$$
$$\leqq |\lambda| \, n^{-3/2} \sum_{t=1}^{n} E_{\theta_0}[\; |X_{t-1}|^3 \sup_{0<|\eta|<\lambda n^{-1/2}|X_{t-1}|} |g'(U_t+\eta)| \;] .$$

It follows from (a) of Assumption (A.3.2.4) that (3.2.5) holds. Similarly (3.2.6) holds. By Schwarz's inequality we have

$$\text{(3.2.9)} \quad \left| E_{\theta_0}(T_n^{*2}/n^2) - E_{\theta_0}(T_n^2/n^2) \right|$$
$$\leqq \left\{ E_{\theta_0}(T_n^*/n - T_n/n)^2 \right\}^{\frac{1}{2}} \left\{ E_{\theta_0}(T_n^*/n + T_n/n)^2 \right\}^{\frac{1}{2}} .$$

It follows from Assumption (A.3.2.3) and that

$$\overline{\lim_{n\to\infty}} E_{\theta_0}(T_n^{*2}/n^2) = \overline{\lim_{n\to\infty}} n^{-2} E_{\theta_0}[\left\{ \sum_{t=1}^{n} X_{t-1}^2 g(U_t^*) \right\}^2]$$
$$\leqq \overline{\lim_{n\to\infty}} n^{-2} E_{\theta_0}[\left\{ \sum_{t=1}^{n} X_{t-1}^2 \sup_{|U_t - U_t^*|<|\lambda| n^{-\frac{1}{2}}|X_{t-1}|} g(U_t^*) \right\}^2] .$$

Similarly it follows that

$$\overline{\lim_{n\to\infty}} E_{\theta_0}(T_n^2/n^2) < \infty .$$

Hence we have

$$\text{(3.2.10)} \qquad \overline{\lim_{n\to\infty}} \left\{ E_{\theta_0}(T_n^*/n + T_n/n)^2 \right\}^{\frac{1}{2}} < \infty .$$

On the other hand it follows from (b) of Assumption (A.3.2.4) that

$$\text{(3.2.11)} \quad \overline{\lim_{n\to\infty}} E_{\theta_0}(T_n^*/n - T_n/n)^2$$
$$= \overline{\lim_{n\to\infty}} n^{-2} E_{\theta_0}[\left\{ \sum_{t=1}^{n} X_{t-1}^2 (g(U_t^*) - g(U_t)) \right\}^2]$$
$$\leqq \overline{\lim_{n\to\infty}} n^{-2} E_{\theta_0}[\lambda^2 n^{-1} \left\{ \sum_{t=1}^{n} |X_{t-1}|^3 \sup_{0<|\eta|<|\lambda| n^{-\frac{1}{2}}|X_{t-1}|} |g'(U_t+\eta)| \right\}^2]$$

$$= \overline{\lim_{n\to\infty}} \lambda^2 n^{-3} E_{\theta_0}\left[\left\{\sum_{t=1}^{n} |X_{t-1}|^3 \sup_{0<|\eta|<\lambda n^{-\frac{1}{2}}|X_{t-1}|} |g'(U_t+\eta)|\right\}^2\right]=0 .$$

From (3.2.9), (3.2.10) and (3.2.11) we have

$$\lim_{n\to\infty} \left| E_{\theta_0}(T_n^{*2}/n^2) - E_{\theta_0}(T_n^2/n^2) \right| = 0 .$$

In a similar way it follows from Assumption (A.3.2.3) and (b) of Assumption (A.3.2.4) and (3.2.4) that

$$\lim_{n\to\infty} \left| E_{\theta_1}(T_n^{**2}/n^2) - E_{\theta_1}(T_n^2/n^2) \right| = 0 .$$

Thus we complete the proof.

For Lemma 3.2.5 and Theorems 3.2.1, 3.2.2 and 3.2.3 we assume the following :

(A.3.2.5) $$E\left[\left| \frac{f'(U_t)}{f(U_t)} \right|^4 \right] < \infty .$$

Lemma 3.2.5. Under Assumptions (A.3.2.1) $\sim$ (A.3.2.5), both of the sequences $T_n^*/E_{\theta_0}(T_n^*)$ and $T_n^{**}/E_{\theta_1}(T_n^{**})$ converge in probability to 1.

Proof. From (3.2.5) $\sim$ (3.2.8) of Lemma 3.2.4 we have

$$\lim_{n\to\infty} \left| \mathrm{Var}_{\theta_0}(T_n^*/n) - \mathrm{Var}_{\theta_0}(T_n/n) \right| = 0 \quad ;$$

$$\lim_{n\to\infty} \left| \mathrm{Var}_{\theta_1}(T_n^{**}/n) - \mathrm{Var}_{\theta_1}(T_n/n) \right| = 0 \quad ;$$

$$\lim_{n\to\infty} \left| \left\{ E_{\theta_0}(T_n^*/n) \right\}^2 - \left\{ E_{\theta_0}(T_n/n) \right\}^2 \right| = 0 \quad ;$$

$$\lim_{n\to\infty} \left| \left\{ E_{\theta_1}(T_n^{**}/n) \right\}^2 - \left\{ E_{\theta_1}(T_n/n) \right\}^2 \right| = 0 \quad ,$$

where Var designates variance. Hence in order to prove that both of the sequence $T_n^*/E_{\theta_0}(T_n^{**})$ and $T_n^{**}/E_{\theta_1}(T_n^{**})$ converge in probability to 1, it is enough to show that

(3.2.12) $$\lim_{n\to\infty} \frac{\mathrm{Var}_{\theta_j}(T_n)}{\left\{ E_{\theta_j}(T_n) \right\}^2} = 0 \ , \ (j=0,1) \quad .$$

Indeed, it follows from (3.2.4) of Lemma 3.2.4 that

$$\left\{ E_{\theta_j}(T_n) \right\}^2 = O(n^2) \qquad (j=0,1) \quad .$$

Also it follows from Assumption (A.3.2.3) that

$$\mathrm{Var}_{\theta_j}(T_n) = \sum_{t=1}^{n} \mathrm{Var}_{\theta_j}(X_{t-1}^2 g(U_t))$$

$$+ \sum_{t \neq t'}\sum \mathrm{Cov}(X_{t-1}^2 g(U_t),\ X_{t-1}^2 g(U_{t'})) = O(n) \qquad (j=0,1) \ ,$$

where Cov designates covariance. Hence (3.2.12) holds. Thus we complete the proof.

In the following theorem we shall show that the best test statistics have limiting normal distributions.

Theorem 3.2.1. Suppose that Assumptions (3.2.1) $\sim$ (A.3.2.5) hold. If $\theta = \theta_0$, then $\sum_{t=1}^{n} Z_{nt}$ has a limiting normal distribution with mean $\dfrac{\lambda^2\sigma^2 I}{2(1-\theta_0^2)}$ and variance $\dfrac{\lambda^2\sigma^2 I}{1-\theta_0^2}$. If $\theta = \theta_1$, then $\sum_{t=1}^{n} Z_{nt}$ has a limiting normal distribution with mean $-\dfrac{\lambda^2\sigma^2 I}{2(1-\theta_0^2)}$ and variance $\dfrac{\lambda^2\sigma^2 I}{1-\theta_0^2}$.

Proof. If $\theta = \theta_0$, then if follows from (3.2.2) that

$$\text{(3.2.13)} \qquad \sum_{t=1}^{n} Z_{nt} = \lambda n^{-1/2} \sum_{t=1}^{n} \frac{f'(U_t)}{f(U_t)} X_{t-1} + \frac{\lambda^2}{2} n^{-1} T_n^*$$

$$= n^{-1} E_{\theta_0}(T_n^*) \left\{ \lambda \frac{n^{-1/2} \sum_{t=1}^{n} (f'(U_t)/f(U_t)) X_{t-1}}{n^{-1} E_{\theta_0}(T_n^*)} + \frac{\lambda^2}{2} \frac{T_n^*}{E_{\theta_0}(T_n^*)} \right\}.$$

If $\theta = \theta_1$, then it follows from (3.2.3) that

$$\text{(3.2.14)} \qquad \sum_{t=1}^{n} Z_{nt} = \lambda n^{-1/2} \sum_{t=1}^{n} \frac{f'(U_t)}{f(U_t)} X_{t-1} - \frac{\lambda^2}{2} n^{-1} T_n^{**}$$

$$= n^{-1} E_{\theta_1}(T_n^{**}) \left\{ \lambda \frac{n^{-1/2} \sum_{t=1}^{n} (f'(U_t)/f(U_t)) X_{t-1}}{n^{-1} E_{\theta_1}(T_n^{**})} - \frac{\lambda^2}{2} \frac{T_n^{**}}{E_{\theta_1}(T_n^{**})} \right\}$$

It follows from (3.2.4) and (3.2.5) of Lemma 3.2.4 that

$$\lim_{n\to\infty} n^{-1} E_{\theta_j}(T_n^*) = \lim_{n\to\infty} n^{-1} E_{\theta_j}(T_n^{**}) = \frac{\sigma^2 I}{1-\theta_0^2} \qquad (j=0,1) \ .$$

Hence it is seen from Lemma 3.2.3 that both of the sequences of

$$\frac{n^{-1/2}\sum_{t=1}^{n}(f'(U_t)/f(U_t))X_{t-1}}{n^{-1}\,E_{\theta_0}(T_n^{*})} \quad \text{and} \quad \frac{n^{-1/2}\sum_{t=1}^{n}(f'(U_t)/f(U_t))X_{t-1}}{n^{-1}\,E_{\theta_1}(T_n^{**})}$$

have a limiting normal distribution with mean 0 and variance $\dfrac{1-\theta_0^2}{\sigma^2 I}$.

Therefore it follows from (3.2.13), (3.2.14) and Lemma 3.2.5 that Z_{nt} has limiting normal distributions with means $\dfrac{\lambda^2\sigma^2 I}{2(1-\theta_0^2)}$ and $-\dfrac{\lambda^2\sigma^2 I}{2(1-\theta_0^2)}$ and common variances $\dfrac{\lambda^2\sigma^2 I}{1-\theta_0}$ for $\theta=\theta_0$ and for $\theta=\theta_1$, respectively. This completes the proof.

Theorem 3.2.2. Under Assumptions (A.3.2.1) ~ (A.3.2.5), the bound of the asymptotic distributions of AMU estimators $\hat{\theta}_n$ is given as follows : for each $\theta \in \text{Ⓗ}$

(3.2.15) $$\overline{\lim_{n\to\infty}}\, P_{\theta,n}(\{\sqrt{n}(\hat{\theta}_n-\theta)\leqq\lambda\})\leqq\Phi\left(\frac{\lambda\sigma\sqrt{I}}{\sqrt{1-\theta^2}}\right) \text{ for all } \lambda\geqq 0\ ;$$

(3.2.16) $$\underline{\lim_{n\to\infty}}\, P_{\theta,n}(\{\sqrt{n}(\hat{\theta}_n-\theta)\geqq\lambda\})\geqq\Phi\left(\frac{\lambda\sigma\sqrt{I}}{\sqrt{1-\theta^2}}\right) \text{ for all } \lambda< 0\ ,$$

where Φ is a normal distribution with mean 0 and variance 1.

Proof. Let θ_0 be arbitrary but fixed in Ⓗ. Let λ be arbitrary positive number. Then we consider the problem of testing hypothesis $H^+ : \theta=\theta_0+\lambda n^{-1/2}$ against alternative $K: \theta=\theta_0$. If we choose a sequence $\{k_n\}$ such that $\lim_{n\to\infty} P_{\theta_0+\lambda n^{-1/2},n}(\{\sum_{t=1}^{n} Z_{nt}>k_n\})=1/2$, then it follows by Theorem 3.2.1 that $\lim_{n\to\infty} k_n=-\dfrac{\lambda^2\sigma^2 I}{2(1-\theta_0^2)}$

Furthermore we have from Theorem 3.2.1

$$\lim_{n\to\infty} P_{\theta_0,n}\left(\left\{\sum_{t=1}^{n} Z_{nt}>k_n\right\}\right)=\lim_{n\to\infty} P_{\theta_0,n}\left(\left\{\frac{\sum_{t=1}^{n} Z_{nt}-J/2}{\sqrt{I^*}}>\frac{k_n-J/2}{\sqrt{I^*}}\right\}\right)$$

$$=1-\Phi(-\sqrt{I^*})=\Phi(\sqrt{I^*})\ ,$$

where $I^*=\dfrac{\lambda^2\sigma^2 I}{1-\theta_0^2}$. Hence it follows by (3.1.3) and the fundamental lemma of Neyman and Pearson that for each $\lambda>0$

$$\beta^{+}_{\theta_0}(\lambda) = \Phi\left(\frac{\lambda\sigma\sqrt{I}}{\sqrt{1-\theta_0^2}}\right) .$$

From (3.1.1) and (3.1.4) we obtain for every $\lambda > 0$

$$\overline{\lim_{n\to\infty}} P_{\theta_0,n}\left\{\sqrt{n}\,(\hat{\theta}_n - \theta_0) \leqq \lambda\right\} \leqq \Phi\left(\frac{\lambda\sigma\sqrt{I}}{\sqrt{1-\theta_0^2}}\right) .$$

Since $\hat{\theta}_n$ is AMU, $\beta^{+}_{\theta_0}(0) = \Phi(0) = 1/2$. Hence since θ_0 is arbitrary, it follows that (3.2.15) holds.

Now let λ be an arbitrary negative number. Then we consider the problem of testing hypothesis $H^- : \theta = \theta_0 + \lambda n^{-1/2}$ against alternative $K : \theta = \theta_0$. By a similar way as the case $\lambda > 0$, we have from (3.1.6)

$$\beta^{-}_{\theta_0}(\lambda) = 1 - \Phi\left(-\frac{\lambda\sigma\sqrt{I}}{\sqrt{1-\theta_0^2}}\right) = \Phi\left(\frac{\lambda\sigma\sqrt{I}}{\sqrt{1-\theta_0^2}}\right)$$

for all $\lambda < 0$. Hence it follows from (3.1.2) and (3.1.7) that for each $\lambda < 0$

$$\underline{\lim_{n\to\infty}} P_{\theta_0,n}\left\{\sqrt{n}(\hat{\theta}_n - \theta_0) \leqq \lambda\right\} \geqq \Phi\left(\frac{\lambda\sigma\sqrt{I}}{\sqrt{1-\theta_0^2}}\right) .$$

Since θ_0 is arbitrary, (3.2.16) holds. Thus we complete the proof.

From Theorem 3.2.2 and Definitions 3.1.2 and 3.1.3 we get the following theorem.

Theorem 3.2.3. Under Assumptions (A.3.2.1) $\sim$ (A.3.2.5), an AMU estimator $\hat{\theta}_n$ is asymptotically efficient if and only if the limiting distribution of $\sqrt{n}(\hat{\theta}_n - \theta)$ is normal with mean 0 and variance $(1-\theta^2)/\sigma^2 I$.

The least squares estimator $\hat{\theta}_{LS}$ of θ is given by $(\sum_{t=1}^{n} X_{t-1}X_t)/\sum_{t=1}^{n} X_{t-1}^2$. It is shown by Anderson [11] that if $E(U_t^2) < \infty$ then for $|\theta| < 1$, $\sqrt{n}(\hat{\theta}_n - \theta)$ has a limiting normal distribution with mean 0 and variance $1-\theta^2$. It is seen that under Assumptions (A.3.2.1) $\sim$ (A.3.2.3) $\hat{\theta}_{LS}$ is asymptotically median unbiased.

Throughout the remainder of this section we assume the following :

(A.3.2.6) $\lim_{u\to\pm\infty} uf(u)=0$.

Then it will be proved that the least squares estimator of θ is asymptotically efficient if and only if $f'(u)/f(u)=cu$, where c is some constant. Indeed, since

$$\sigma^2 I=\left\{\int u^2 f(u)du\right\}\left\{\int \left(\frac{f'(u)}{f(u)}\right)^2 f(u)du\right\} \geqq \left\{\int uf'(u)du\right\}^2=1 ,$$

"=" is obtained if and only if $f'(u)/f(u)=cu$. It follows by Theorem 3.2.2 that the limiting distribution of $\sqrt{n}(\hat{\theta}_{LS}-\theta)$ attains the bound of the asymptotic distributions if and only if f is a normal density function with mean 0 and variance σ^2. Hence it is seen by Theorem 3.2.3 that the least squares estimator is asymptotically efficient if and only if f is a normal density function with mean 0 and variance σ^2. Therefore we have now established.

Theorem 3.2.4. Under Assumptions (A.3.2.1) ~ (A.3.2.6), a necessary and sufficient condition that the least squares estimator of θ be asymptotically efficient is that f be a normal density function with mean 0 and variance σ^2.

Remark. As is immediately seen from above, Assumption (A.3.2.1) ~ (A.3.2.6) are not necessary for the proof of sufficiency.

We now consider the asymptotic efficiency of the maximum likelihood estimator in the autoregressive process. Let $\hat{\theta}_{ML}$ be the maximum likelihood estimator. By Taylor expansion we have

$$\begin{aligned} 0 &= \sum_{i=1}^{n} \frac{\partial}{\partial\theta}\log f(X_i-\hat{\theta}_{ML}X_{i-1}) \\ &= \sum_{i=1}^{n} \frac{\partial}{\partial\theta}\log f(X_i-\theta X_{i-1})+\left\{\sum_{i=1}^{n} \frac{\partial^2}{\partial\theta^2}\log f(X_i-\theta^* X_{i-1})\right\}(\hat{\theta}_{ML}-\theta) \\ &= -\sum_{i=1}^{n} X_{i-1}\psi'(X_i-\theta X_{i-1})+\left\{\sum_{i=1}^{n} X_{i-1}^2\psi''(X_i-\theta^* X_{i-1})\right\}(\hat{\theta}_{ML}-\theta) , \end{aligned}$$

where $|\theta^* - \theta| \leq |\hat{\theta}_{ML} - \theta|$ and $\psi(u) = \log u$.

Hence we have

$$\sqrt{n}(\hat{\theta}_{ML} - \theta) = \frac{\frac{1}{\sqrt{n}} \sum_{i=1}^{n} X_{i-1} \psi'(X_i - \theta X_{i-1})}{\frac{1}{n} \sum_{i=1}^{n} X_{i-1}^2 \psi''(X_i - \theta^* X_{i-1})} .$$

From Lemmas 3.2.3 and 3.2.4 it is seen that the limiting distribution of $\sqrt{n}(\hat{\theta}_{ML} - \theta)$ is normal with mean 0 and variance $(1-\theta^2)/(\sigma^2 I)$. By Theorem 3.2.3 we have established the following theorem.

Theorem 3.2.5. Under Assumptions (A.3.2.1) $\sim$ (A.3.2.5), the maximum likelihood estimator is asymptotically efficient.

3.3. Asymptotic efficiency and inefficiency of maximum probability estimators

Summary :

Suppose that X_i $(i=1,2,\ldots,n)$ are distributed according to a two-sided truncated normal distribution with unknown mean θ. It is shown that the maximum probability estimator (MPE) of Weiss and Wolfowitz is asymptotically inadmissible and has smaller concentration of probability than the asymptotically efficient estimator $\hat{\theta}_{PT} = (\min X_i + \max X_i)/2$.

Weiss and Wolfowitz ([54]) defined the concept of the maximum probability estimator (MPE) $= \hat{\theta}^r_{MP}$. It is defined as that value of d which maximizes

$$\int_{d-rc_n^{-1}}^{d+rc_n^{-1}} \prod_{i=1}^{n} f(x_i, \theta)\, d\theta$$

with appropriate choice of the sequence of constants $\{c_n\}$. They proved that $\hat{\theta}^r_{MP}$ maximizes $P_{\theta,n}\{c_n |\hat{\theta}_n - \theta| < r\}$ asymptotically ([56]). However the estimator $\hat{\theta}^r_{MP}$ depends on the choice of r, and the theorem does not say anything about

$$P_{\theta,n}\{c_n |\hat{\theta}^r_{MP} - \theta| < r'\} \quad \text{for} \quad r' \neq r$$

It can happen that $\hat{\theta}^r_{MP}$ behaves badly outside the fixed interval (-r,r). In regular cases, $\hat{\theta}^r_{MP}$ is asymptotically equivalent to the maximum likelihood estimator (MLE), and is asymptotically independent of r. But in non-regular cases $\hat{\theta}^r_{MP}$ does depend on r (which is also emphasized by Weiss and Wolfowitz themselves). The asymptotic behavior of $\hat{\theta}^r_{MP}$ is not well investigated. In [2], [4], [42] and [44] we defined the one-sided and two-sided asymptotically efficient estimator in non-regular cases, as the one which maximizes

$$\lim_{n\to\infty} P_{\theta,n}\left\{ c_n(\hat{\theta}_n - \theta) < a \right\} \quad \text{for } a > 0$$

or

$$\lim_{n\to\infty} P_{\theta,n}\left\{ c_n(\hat{\theta}_n - \theta) > -b \right\} \quad \text{for } b > 0$$

and

$$\lim_{n\to\infty} P_{\theta,n}\left\{ c_n|\hat{\theta}_n - \theta| < c \right\} \quad \text{for } c > 0$$

uniformly in θ and a or b or c among asymptotically median unbiased (AMU) estimators.

The question is : if $\hat{\theta}^r_{MP}$ is AMU, is it asymptotically efficient in the sense above ? The purpose of this section is to show that the answer is not always yes, even in the case when asymptotically efficient estimator does exist. This implies that the MPE may asymptotically inadmissible.

The case discussed below is also covered by Weiss and Wolfowitz [55]. Their Z_n is basically asymptotically equivalent to $\hat{\theta}_{PT}$ and at least as good as $\hat{\theta}^r_{MP}$ but they do not show explicitely the inadmissibility of the latter.

Let $\mathcal{X} = \textcircled{H} = R^1$. Suppose that $X_1, X_2, \ldots, X_n, \ldots$ is a sequence of independent and identically random variables with a density $f(x:\theta)$. Suppose that θ is a location parameter i.e.

$f(x:\theta)=f(x-\theta)$. Further we assume that

$$f(x) = \begin{cases} ce^{-(x^2/2)} & \text{for } |x|<1 ; \\ 0 & \text{for } |x|\geqq 1 , \end{cases}$$

where c is some constant.

It was shown for this case in section 2.4 and 2.5 that there exists a consistent estimator with order $\{n\}$, which is the largest possible order.

Let $R=(-r,r)$. The maximum probability estimator $\hat{\theta}^r_{MP}$ of θ is defined as that value of d maximizing

$$\int_{d-(r/n)}^{d+(r/n)} \prod_{i=1}^{n} f(x_i-\theta)d\theta$$

(Weiss and Wolfowitz [54], [56]), and in this case $\hat{\theta}^r_{MP}$ is shown to be

$$\hat{\theta}^r_{MP} = \begin{cases} \frac{1}{2}(\min X_i + \max X_i) & \text{for } \max X_i - \min X_i > 2 - \frac{2r}{n} ; \\ \min X_i + 1 - \frac{r}{n} & \text{for } \bar{X} > \min X_i + 1 - \frac{r}{n}, \ \max X_i - \min X_i \leqq 2 - \frac{2r}{n} ; \\ \max X_i - 1 + \frac{r}{n} & \text{for } \bar{X} < \max X_i - 1 + \frac{r}{n}, \ \max X_i - \min X_i \leqq 2 - \frac{2r}{n} ; \\ \bar{X} & \text{for } \max X_i - 1 + \frac{r}{n} \leqq \bar{X} \leqq \min X_i + 1 - \frac{r}{n} ; \end{cases}$$

where $\bar{X} = \frac{1}{n}\sum_{i=1}^{n} X_i$.

Remark : The definition of $\hat{\theta}^r_{MP}$ in the first case is not unique, but it does not affect the subsequent conclusions.

First we note that the densities of the asymptotic distributions (asymptotic densities) of $\min X_i + 1$ and $\max X_i - 1$ are given by

$$\text{(3.3.1)} \qquad g_1(u) = \begin{cases} ke^{-ku} & \text{for } u > 0 ; \\ 0 & \text{for } u \leqq 0 ; \end{cases}$$

$$\text{(3.3.2)} \qquad g_2(u) = \begin{cases} ke^{ku} & \text{for } u < 0 ; \\ 0 & \text{for } u \geqq 0 , \end{cases}$$

respectively, where $k=ce^{-1/2}$. Now we consider another estimator $\hat{\theta}_{PT}=(\min X_i + \max X_i)/2$ which is asymptotically equivalent to the Pitman (best location invariant) estimator.

Since min X_i+1 and max X_i-1 are asymptotically independent, the density of the asymptotic distribution of $\hat{\theta}_{PT}=\frac{1}{2}(\min X_i + \max X_i)$ is given by

(3.3.3) $$g(u)=ke^{-2k|u|} .$$

Next we shall obtain the asymptotic density of $\hat{\theta}^{r}_{MP}$. The asymptotic conditional density of $\frac{1}{2}(\min X_i+\max X_i)$ given $\max X_i-\min X_i > 2-\frac{2r}{n}$ is given by

$$h_1(u) = \begin{cases} \frac{k}{1-K(r)}(e^{-2k|u|}-e^{-2kr}) & \text{for } |u|<r \ ; \\ 0 & \text{for } |u|\geqq r \ ; \end{cases}$$

where $K(r)=(1+2kr)e^{-2kr}$.

This is seen as follows. The asymptotic joint density $f(t,u)$ of $T=\min X_i-\max X_i+2$ and $U=(\min X_i+\max X_i)/2$ is given by

$$f(t,u) = \begin{cases} k^2e^{-kt} & \text{for } |u|<\frac{t}{2} \ , \quad t>0 \ ; \\ 0 & \text{otherwise} \ . \end{cases}$$

The asymptotic conditional density $h_1(u)$ of U given $T<2r/n$ is also obtained by $\int_{-\infty}^{2r} f(t,u) / \int_{-\infty}^{2r} g(u)du$.

Since

$$\int_{-\infty}^{2r} g(u)du = 1-(1+2kr)e^{-2kr} = 1-K(r) \ ;$$

$$\int_{2|u|}^{2r} f(t,u)dt = \int_{2|u|}^{2r} k^2e^{-kt}dt = k(e^{-2k|u|}-e^{-2kr})$$

$$\text{for } |u|<r \ ,$$

it is seen that $h_1(u)$ is given as the above form.

It can be shown that $\bar{X}$ is asymptotically independent of min X_i and max X_i, or more precisely $\sqrt{n}\,\bar{X}$ is asymptotically normally distributed independently of $n(\min X_i+1)$ and $n(\max X_i-1)$. Hence it follows that the asymptotic conditional density of $\min X_i+1-\frac{r}{n}$ given $\bar{X}>\min X_i+1-\frac{r}{n}$ and $\max X_i-\min X_i \leqq 2-\frac{2r}{n}$ is given by

$$h_2(u) = \begin{cases} \frac{k}{K(r)} e^{-2kr} & \text{for } |u| < r \ ; \\ \frac{k}{K(r)} e^{-2k(u+r)} & \text{for } |u| \geqq r \ . \end{cases}$$

Further the asymptotic density of $\max X_i - 1 + \frac{r}{n}$ given $\overline{X} < \max X_i - 1 + \frac{r}{n}$ and $\max X_i - \min X_i \leqq 2 - \frac{2r}{n}$ is given by

$$h_3(u) = \begin{cases} \frac{k}{K(r)} e^{-2kr} & \text{for } |u| < r \ ; \\ \frac{k}{K(r)} e^{-k(r-u)} & \text{for } u \leqq -r \ . \end{cases}$$

Since the order of the probability of $\max X_i - 1 + (r/n) < \overline{X} < \min X_i + 1 - (r/n)$ is $1/n$, the probability of $\overline{X}$ given $\max X_i - 1 + (r/n) < \overline{X} < \min X_i + 1 - (r/n)$ is negligible in this case. Hence the asymptotic density of $\hat{\theta}^r_{MP}$ is given by

$$(3.3.4) \qquad g^*(u) = \begin{cases} k e^{-2k|u|} & \text{for } |u| < r \ ; \\ \frac{k}{2} e^{-k(|u|+r)} & \text{for } |u| \geqq r \ . \end{cases}$$

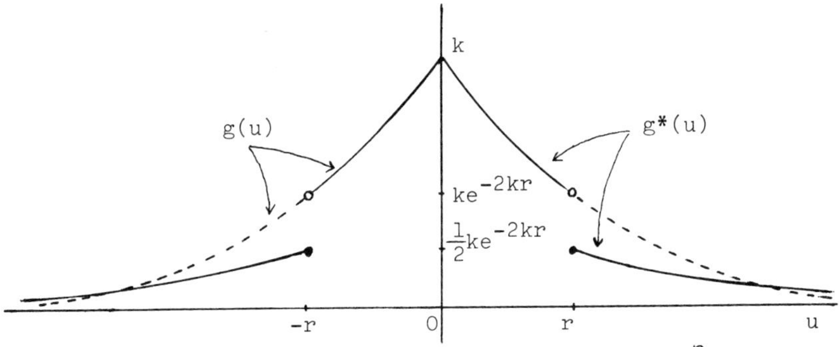

Let $P_{\theta,n}$ be probability measures with densities $\prod_{i=1}^{n} f(x_i - \theta)$. Next we shall get the lower bound of

$$(3.3.5) \qquad P_{\theta,n}\left\{ n \left| \hat{\theta}_n - \theta \right| \geqq u \right\}$$
$$= P_{\theta,n}\left\{ n(\hat{\theta}_n - \theta) < -u \right\} + P_{\theta,n}\left\{ n(\hat{\theta}_n - \theta) \geqq u \right\}$$

among AMU estimators with asymptotic distributions. Let u be an arbitrary fixed positive constant and $\hat{\theta}_n$ be AMU. Let θ_0 be any fixed

element of Ⓗ and put $\theta_1 = \theta_0 - (u/n)$ and $\theta_2 = \theta_0 + (u/n)$.
Since for each real number t $P_{\theta,n}\{\sqrt{n}(\hat{\theta}_n - \theta) \leqq t\}$ converges uniformly in a neighborhood of θ_0, it follows that for sufficiently large n

$$P_{\theta_1,n}\{\hat{\theta}_n \leqq \theta_0\} = P_{\theta_1,n}\{n(\hat{\theta}_n - \theta_1) \leqq u\} \sim P_{\theta_0,n}\{n(\hat{\theta}_n - \theta_0) \leqq u\} \ ;$$
$$P_{\theta_2,n}\{\hat{\theta}_n \geqq \theta_0\} = P_{\theta_2,n}\{n(\hat{\theta}_n - \theta_2) \leqq -u\} \sim P_{\theta_0,n}\{n(\hat{\theta}_n - \theta_0) \leqq -u\} \ .$$

In order to get the lower bound of (3.3.5) it is sufficient to find the maximum value of

$$(3.3.6) \qquad P_{\theta_2,n}\{\hat{\theta}_n \leqq \theta_0\} - P_{\theta_1,n}\{\hat{\theta}_n \leqq \theta_0\}$$

among AMU estimators $\hat{\theta}_n$. Let $B_n = \{\tilde{x}_n : \hat{\theta}_n(\tilde{x}_n) \leqq \theta_0\}$ and ϕ_n be the indicator function of B_n.
Then it follows that

$$\int_{-\infty}^{\infty} \cdots \int_{-\infty}^{\infty} \phi_n(\tilde{x}_n) \prod_{i=1}^{n} f(x_i - \theta_0) \prod_{i=1}^{n} dx_i = E_{\theta_0,n}(\phi_n) \to \frac{1}{2} (n \to \infty) \ .$$

We also have

$$(3.3.7) \qquad \int_{-\infty}^{\infty} \cdots \int_{-\infty}^{\infty} \phi_n(\tilde{x}_n) \left\{ \prod_{i=1}^{n} f(x_i, \theta_2) - \prod_{i=1}^{n} f(x_i, \theta_1) \right\} \prod_{i=1}^{n} dx_i$$
$$= E_{\theta_2,n}(\phi_n) - E_{\theta_1,n}(\phi_n) \ .$$

If the maximum value of (3.3.7) can be obtained, then it is that of (3.3.6). In a way similar to the proof of the fundamental lemma of Neyman and Pearson, it is shown that ϕ_n maximizing (3.3.7) is given by

$$\phi_n^*(\tilde{x}_n) = \begin{cases} 1, & \prod_{i=1}^{n} f(x_i, \theta_2) - \prod_{i=1}^{n} f(x_i, \theta_1) > \lambda \prod_{i=1}^{n} f(x_i, \theta_0) \ ; \\ 0, & \prod_{i=1}^{n} f(x_i, \theta_2) - \prod_{i=1}^{n} f(x_i, \theta_1) < \lambda \prod_{i=1}^{n} f(x_i, \theta_0) \ , \end{cases}$$

for some λ.
Since θ_1 and θ_2 are symmetric with respect to θ_0, it is clear that $\lambda = 0$.

Hence

$$\phi_n^*(\tilde{x}_n) = \begin{cases} 1, \text{ if } \prod_{i=1}^n f(x_i,\theta_2) > \prod_{i=1}^n f(x_i,\theta_1), \text{ that is,} \\ \quad (a) \prod_{i=1}^n f(x_i,\theta_2) > 0, \prod_{i=1}^n f(x_i,\theta_1) = 0 \quad ; \\ \text{or} \\ \quad (b) \prod_{i=1}^n f(x_i,\theta_2) > 0, \prod_{i=1}^n f(x_i,\theta_1) > 0 \text{ and } \bar{x} < c' \quad ; \\ 0, \text{ if } \prod_{i=1}^n f(x_i,\theta_2) < \prod_{i=1}^n f(x_i,\theta_1), \text{ that is,} \\ \quad (a') \prod_{i=1}^n f(x_i,\theta_2) = 0, \prod_{i=1}^n f(x_i,\theta_1) > 0 \quad ; \\ \text{or} \\ \quad (b') \prod_{i=1}^n f(x_i,\theta_2) > 0, \prod_{i=1}^n f(x_i,\theta_1) > 0 \text{ and } \bar{x} > c' \quad , \end{cases}$$

where $\bar{x} = \frac{1}{n} \sum_{i=1}^n x_i$.

Putting

$$A = \left\{ \tilde{x}_n : \max x_i - 1 > \theta_0 - \frac{u}{n} , \min x_i + 1 > \theta_0 + \frac{u}{n} \right\} ,$$

$$B = \left\{ \tilde{x}_n : \max x_i - 1 < \theta_0 - \frac{u}{n} , \min x_i + 1 < \theta_0 + \frac{u}{n} \right\} ,$$

$$C = \left\{ \tilde{x}_n : \max x_i - 1 < \theta_0 - \frac{u}{n} , \min x_i + 1 > \theta_0 + \frac{u}{n} \right\} ,$$

$$D = \left\{ \tilde{x}_n : \bar{x} < c' \right\} ,$$

$$D' = \left\{ \tilde{x}_n : \bar{x} > c' \right\} .$$

we have

$$(3.3.8) \quad \phi_n^*(\tilde{x}_n) = \begin{cases} 1, \text{ if } \tilde{x}_n \in A \cup (C \cap D) \quad ; \\ 0, \text{ if } \tilde{x}_n \in B \cup (C \cap D') . \end{cases}$$

Since $\bar{X}$, $\max X_i - 1$ and $\min X_i + 1$ are mutually asymptotically independent, it is seen that the conditional probability $P_{\theta,n}(D \mid C)$ of D given C is given by for sufficiently large n

$$P_{\theta,n}(D \mid C) = P_{\theta,n}(D)$$

In order to get

$$E_{\theta_0,n}(\phi_n^*) \to \frac{1}{2} \quad (n \to \infty)$$

we must have

$$P_{\theta_0,n}(D) \to \frac{1}{2} \quad (n \to \infty) .$$

It follows from (3.3.1), (3.3.2) and (3.3.8) that for sufficiently large n

(3.3.9) $$E_{\theta_2,n}(\phi_n^*) - E_{\theta_1,n}(\phi_n^*)$$

$$\sim P_{\theta_2,n}(A) + \frac{1}{2} P_{\theta_2,n}(C) - P_{\theta_1,n}(A) - \frac{1}{2} P_{\theta_1,n}(C)$$

$$\sim P_{\theta_2,n}(A)$$

$$\sim 1 - e^{2ku}$$

From (3.3.5), (3.3.6), (3.3.7) and (3.3.9) it is seen that

(3.3.10) $$\underline{\lim}_{n\to\infty} P_{\theta,n}\left\{ n|\hat{\theta}_n - \theta| \geqq u \right\} \geqq e^{-2ku}$$

for all AMU estimators $\hat{\theta}_n$.

Hence the right-hand side of (3.3.10) is the desired lower bound.

Therefore we have shown that the bound of the asymptotic distribution of $n|\hat{\theta}_n - \theta|$ for AMU estimators $\hat{\theta}_n$ is given by :

(3.3.11) $$\overline{\lim}_{n\to\infty} P_{\theta,n}\left\{ n|\hat{\theta}_n - \theta| < u \right\} \leqq 1 - e^{-2ku} .$$

Since both of the asymptotic densities of $\hat{\theta}_{PT}$ and $\hat{\theta}^r_{MP}$ are symmetric, it is easily seen that these estimator are AMU.

From (3.3.3) and (3.3.4) we have

(3.3.12) $$\lim_{n\to\infty} P_{\theta,n}\left\{ n|\hat{\theta}_{PT} - \theta| < u \right\} = 1 - e^{-2ku} \quad ;$$

(3.3.13) $$\lim_{n\to\infty} P_{\theta,n}\left\{ n|\hat{\theta}^r_{MP} - \theta| < u \right\} = \begin{cases} 1-e^{-2ku} & \text{for } 0 < u < r \quad ; \\ 1-e^{-k(u+r)} & \text{for } r \leqq u \quad . \end{cases}$$

Hence we see by (3.3.11), (3.3.12) and (3.3.13) that <u>$\hat{\theta}_{PT}$ is two-sided asymptotically efficient in the sense that the asymptotic distribution of $\hat{\theta}_{PT}$ attains uniformly the bound given by the right-hand side of (3.3.11) but $\hat{\theta}^r_{MP}$ is not so.</u>

Remark : In this case the MLE is asymptotically (also in small sample) uniformly (in u) inadmissible. This fact can be established in the following way. It is easily seen that

$$\hat{\theta}_{ML} = \begin{cases} \min X_i+1 & \text{if } \bar{X} > \min X_i+1 , \\ \max X_i-1 & \text{if } \bar{X} < \max X_i-1 , \\ \bar{X} & \text{if } \max X_i-1 < \bar{X} < \min X_i+1 \end{cases}$$

and asymptotically, it is equivalent to

$$\tilde{\theta}_n = \begin{cases} \min X_i+1 \text{ with probability } 1/2 . \\ \max X_i-1 \text{ with probability } 1/2 . \end{cases}$$

Since the asymptotic density of $\tilde{\theta}_n$ is symmetric, it can easily be shown that $\hat{\theta}_{ML}$ is AMU. Thus the asymptotic distribution of n $|\hat{\theta}_{ML}-\theta|$ is given by

$$\lim_{n\to\infty} P_{\theta,n}\left\{ n \,|\, \hat{\theta}_{ML} - \theta | < u \right\} = 1-e^{-ku}$$

which has twice as large a scale as the asymptotic distribution of $\hat{\theta}_{PT}$. The asymptotic efficiency of $\hat{\theta}_{ML}$ compared with $\hat{\theta}_{PT}$ is exactly 50%!

It should be also noted that $\hat{\theta}_{ML}$ is the limit of $\hat{\theta}^r_{MP}$ as $r\to 0$, and that $\hat{\theta}^r_{MP}$ is inadmissible outside the interval $-r<u<r$, and that the interval converges to the null set when r goes to zero. On the other hand when r becomes large, the interval of u in which $\hat{\theta}^r_{MP}$ is admissible extends and the probability outside the interval also concentrates to the center, and as $r\to\infty$, $\hat{\theta}^r_{MP}$ approaches to $\hat{\theta}_{PT}$ which is uniformly efficient. Thus in this case the choice of larger r is always better than the choice of smaller r, which, although remarkable enough, does not hold in the other cases.

Note also that the Pitman estimator is the generalized Bayes estimator with uniform prior and quadratic loss. $\hat{\theta}_{PT}$ is asymptotically equivalent not only to the Pitman estimator but also to any generalized Bayes estimators with symmetric loss function as given in section 5.3.

In the beginning of this chapter, we defined as AMU estimator $\hat{\theta}_n^*$ to be asymptotically efficient, if the asymptotic distribution of $n(\hat{\theta}_n^* - \theta)$ uniformly attains the bound of the asymptotic distributions of $n(\hat{\theta}_n - \theta)$ for AMU estimators $\hat{\theta}_n$ on both sides ([4], [42], [44]). It can be shown that the bound of the asymptotic distributions of $n(\hat{\theta}_n - \theta)$ for AMU estimators $\hat{\theta}_n$ is given by the following density :

$$(3.3.14) \qquad g^{**}(u) = \begin{cases} ke^{-k|u|} & \text{for } |u| \leqq \frac{1}{k}\log 2 \ ; \\ 0 & \text{for } |u| > \frac{1}{k}\log 2 \ , \end{cases}$$

(See the proof of Theorem 3.1 of [2]).
Comparing (3.3.3) and (3.3.4) we may also conclude that $\hat{\theta}_{PT}$ and $\hat{\theta}_{MP}^r$ are asymptotically equivalent for $|u| < r$ but for $|u| \geqq r$ $\hat{\theta}_{PT}$ is asymptotically better than $\hat{\theta}_{MP}^r$ in the sense that the asymptotic distribution of $\hat{\theta}_{PT}$ is uniformly nearer to the bound than that of $\hat{\theta}_{MP}^r$. Though $\hat{\theta}_{MP}^r$ and $\hat{\theta}_{PT}$ are two-sided asymptotically efficient as shown above, they do not uniformly attain the bound defined by (3.3.14). Hence they are not one-sided asymptotically efficient in the sense that their asymptotic distributions uniformly attain the above bound ([4], [42], [44]). However, it is seen that $\hat{\theta}_{PT}$ is asymptotically better than $\hat{\theta}_{MP}^r$.

It can be shown that there exists an AMU estimator $\hat{\theta}_n^*$ whose asymptotic distribution attains the bound either for $r > 0$ or $r < 0$ (but not for both). Such an estimator for positive r can be given as

$$\hat{\theta}_n^+ = \min\left\{ \min_i X_i + 1 \ , \ \max_i X_i - 1 + \frac{1}{kn}\log 2 \right\} .$$

For negative r it is also given by

$$\hat{\theta}_n^- = \max\left\{ \max_i X_i - 1 \ , \ \min_i X_i + 1 - \frac{1}{kn}\log 2 \right\} .$$

Chapter 4

Higher Order Asymptotic Efficiency

Summary :

Second order efficiency of asymptotically efficient estimators has been discussed by Fisher [20], Rao [37], [38] and others in terms of loss of information. Recently Chibisov [15], [16] has shown that a maximum likelihood estimator (MLE) is second order asymptotically efficient in this sense. Pfanzagl ([32], [33]) obtained that MLE attains the second order asymptotic efficiency in the sense adopted here. In this chapter we shall discuss second order asymptotic efficiency and proceed further to third order asymptotic efficiency. We shall show that the results can be extended to non-regular situations.

Suppose that Ⓗ is open set of R^1. For each k=1,2,..., a $\{c_n\}$-consistent estimator $\hat{\theta}_n$ is called <u>k-th order asymptotically median unbiased (or k-th order AMU)</u> estimator if for any $\vartheta \in$ Ⓗ, there exists a positive number δ such that

$$\lim_{n\to\infty} \sup_{\theta:|\theta-\vartheta|<\delta} c_n^{k-1} \left| P_{\theta,n}\{\hat{\theta}_n \leqq \theta\} - \frac{1}{2} \right| = 0 \ ;$$

$$\lim_{n\to\infty} \sup_{\theta:|\theta-\vartheta|<\delta} c_n^{k-1} \left| P_{\theta,n}\{\hat{\theta}_n \geqq \theta\} - \frac{1}{2} \right| = 0 \ .$$

For $\hat{\theta}_n$ k-th order AMU, $G_0(t,\theta)+c_n^{-1}G_1(t,\theta)+\ldots+c_n^{-(k-1)}G_{k-1}(t,\theta)$ is defined to be the <u>k-th order asymptotic distribution of $c_n(\hat{\theta}_n-\theta)$ (or $\hat{\theta}_n$ for short)</u> if

$$\lim_{n\to\infty} c_n^{k-1} \Big| P_{\theta,n}\{c_n(\hat{\theta}_n-\theta) \leqq t\} - G_0(t,\theta) - c_n^{-1}G_1(t,\theta) - \ldots - c_n^{-(k-1)}G_{k-1}(t,\theta) \Big| = 0 \ .$$

We note that $G_i(t,\theta)$ $(i=1, \ldots, k-1)$ may be general absolute continuous functions, hence the asymptotic distribution for any fixed n may not be a distribution function.

In most of the cases discussed below c_n is equal to $\sqrt{n}$, since in other cases we usually do not have any asymptotically second order efficient estimators. Hence the use of general terms for the order of consistency is superfluous here.

Let $\hat{\theta}_n$ be k-th order AMU, with k-th order asymptotic distribution $G_0(t,\theta)+c_n^{-1}G_1(t,\theta)+\ldots+c_n^{-(k-1)}G_{k-1}(t,\theta)$.
Letting $\theta_0(\in \textcircled{H})$ be arbitrary but fixed we consider the problem of testing hypothesis $H^+ : \theta = \theta_0+tc_n^{-1}(t>0)$ against $K : \theta = \theta_0$.
Put $\Phi_{1/2}=\left\{\{\phi_n\} : E_{\theta_0+tc_n^{-1},n}(\phi_n)=1/2+o(c_n^{-(k-1)}),\ 0\leqq\phi_n(\tilde{x}_n)\leqq 1\right.$ for all $\left.\tilde{x}_n\in\mathcal{X}^{(n)}\ (n=1,2,\ldots)\right\}$. Putting $A_{\hat{\theta}_n,\theta_0}=\left\{c_n(\hat{\theta}_n-\theta_0)\leqq t\right\}$, we have

$$\lim_{n\to\infty} P_{\theta_0+tc_n^{-1},n}(A_{\hat{\theta}_n,\theta_0})=\lim_{n\to\infty} P_{\theta_0+tc_n^{-1},n}\left\{\hat{\theta}_n\leqq\theta_0+tc_n^{-1}\right\} = \frac{1}{2} .$$

Hence it is seen that a sequence $\{\chi_{A_{\hat{\theta}_n,\theta_0}}\}$ of the indicators (or characteristic functions) of $A_{\hat{\theta}_n,\theta_0}$ $(n=1,2,\ldots)$ belongs to $\Phi_{1/2}$.
If

$$\sup_{\{\phi_n\}\in\Phi_{1/2}} \overline{\lim_{n\to\infty}}\, c_n^{k-1}\left\{E_{\theta_0,n}(\phi_n)-H_0^+(t,\theta_0)-c_n^{-1}H_1^+(t,\theta_0)-\ldots-c_n^{-(k-1)}\cdot H_{k-1}^+(t,\theta_0)\right\} = 0 ,$$

then we have

$$G_0(t,\theta_0)\leqq H_0^+(t,\theta_0) ;$$

and for any positive integer $j(\leqq k)$ if $G_i(t,\theta_0)=H_i^+(t,\theta_0)(i=0, \ldots, j-1)$, then

$$G_j(t,\theta_0) \leqq H_j^+(t,\theta_0) .$$

Consider next the problem of testing hypothesis $H^- : \theta = \theta_0+tc_n^{-1}$ $(t<0)$ against $K : \theta=\theta_0$.

If

$$\inf_{\{\phi_n\}\in\Phi_{1/2}} \varliminf_{n\to\infty} c_n^{k-1}\Big\{E_{\theta_0,n}(\phi_n)-H_0^-(t,\theta_0)-c_n^{-1}H_1^-(t,\theta_0)-\dots-c_n^{-(k-1)} \cdot H_{k-1}^-(t,\theta_0)\Big\} = 0 ,$$

then we have

$$G_0(t,\theta_0) \geqq H_0^-(t,\theta_0) \quad ;$$

and for any positive integer $j(\leqq k)$ if $G_i(t,\theta_0)=H_i^-(t,\theta_0)$ $(i=0, \dots , j-1)$ then $G_j(t,\theta_0)\geqq H_j^-(t,\theta_0)$.

$\hat{\theta}_n$ is said to be k-th order asymptotically efficient if its k-th order asymptotic distribution uniformly attains the bound of the k-th order asymptotic distributions of k-th order AMU estimators. That is, for each $\theta \in \Theta$

$$G_i(t,\theta) = \begin{cases} H_i^+(t,\theta) & \text{for } t>0 \quad ; \\ H_i^-(t,\theta) & \text{for } t<0 \quad , \end{cases}$$

$i=0, \dots , k-1$ ([7], [44]).

[Note that for $t=0$ we have $G_i(0,\theta)=H_i^+(0,\theta)=H_i^-(0,\theta)$ $(i=0, \dots , k-1)$ from the condition of k-th order asymptotically median unbiasedness.]

Thus if $\hat{\theta}_n^*$ is k-th order asymptotically efficeint in the above sense, for any k-th order AMU estimator $\hat{\theta}_n$, we have

$$(4.0.1) \quad \varliminf_{n\to\infty} c_n^{k-1}[P_{\theta,n}\{-a<c_n(\hat{\theta}_n^*-\theta)<b\} - P_{\theta,n}\{-a<c_n(\hat{\theta}_n-\theta)<b\}] \geqq 0$$

for all $\theta \in \Theta$ and $a>0$, $b>0$.

Note that the above definition of k-th order asymptotic efficiency gives a sufficient but not a necessary condition for (4.0.1).

4.1. Second order asymptotic efficiency

Let $X_1, X_2, \dots , X_n, \dots$ be i.i.d. random variables with a density function $f(x,\theta)$ (w.r.t. a σ-finite measure μ) satisfying

(A.3.1.1), (A.3.1.2) and (A.3.1.3). Let θ_0 be arbitrary but fixed in Ⓗ. Consider the problem of testing hypothesis $H^+ : \theta = \theta_0 + tn^{-1/2}$ $(t>0)$ against alternative $K : \theta = \theta_0$.

We put

$$Z_{ni} = \log \left\{ f(X_i, \theta_0)/f(X_i, \theta_0 + \frac{t}{\sqrt{n}}) \right\} .$$

Using Edgeworth expansion of the distribution of $\sum_{i=1}^{n} Z_{ni}$ we get the asymptotic power series of the most powerful tests.

In addition to (A.3.1.1) ∼ (A.3.1.3) we further assume the following :

(A.4.1.1) $f(x,\theta)$ is there times continuously differentiable in θ.

(A.4.1.2) There exist

$$J(\theta) = E_\theta\left[\left\{ \frac{\partial^2}{\partial\theta^2} \log f(X,\theta) \right\} \left\{ \frac{\partial}{\partial\theta} \log f(X,\theta) \right\} \right]$$

and

$$K(\theta) = E_\theta\left[\left\{ \frac{\partial}{\partial\theta} \log f(X,\theta) \right\}^3 \right]$$

and the following hold :

$$E_\theta\left[\frac{\partial^3}{\partial\theta^3} \log f(X,\theta) \right] = -3J(\theta) - K(\theta) .$$

(A.4.1.3) In the Taylor's expansion

$$\log \frac{f(x, \theta + h)}{f(x,\theta)}$$

$$= h\frac{\partial}{\partial\theta} \log f(x,\theta) + \frac{h^2}{2} \frac{\partial^2}{\partial\theta^2} \log f(x,\theta) + \frac{h^3}{6} \frac{\partial^3}{\partial\theta^3} \log f(x,\theta) + h^3 R(x,h) ,$$

$R(x,h)$ is uniformly bounded by a function $\varphi(x)$ which has moments up to the third.

We denote $(\partial/\partial\theta) f(x,\theta)$, $(\partial^2/\partial\theta^2) f(x,\theta)$ and $(\partial^3/\partial\theta^3) f(x,\theta)$ by f_θ, $f_{\theta\theta}$ and $f_{\theta\theta\theta}$, respectively.

Since

$$\frac{\partial^3}{\partial\theta^3} \log f(x,\theta) = \frac{f_{\theta\theta}(x,\theta)}{f(x,\theta)} - \frac{3f_{\theta\theta}(x,\theta) f_\theta(x,\theta)}{\{f(x,\theta)\}^2} + \frac{2\{f_\theta(x,\theta)\}^3}{\{f(x,\theta)\}^3}$$

$$= \frac{f_{\theta\theta}(x,\theta)}{f(x,\theta)} - 3\left\{\frac{\partial}{\partial\theta}\log f(x,\theta)\right\}\left\{\frac{\partial^2}{\partial\theta^2}\log f(x,\theta)\right\}$$

$$- \left\{\frac{\partial}{\partial\theta}\log f(x,\theta)\right\}^3 ,$$

it follows by the last condition of (A.4.1.2) that

$$\int f_{\theta\theta\theta}(x,\theta)d\mu(x)=0 .$$

Let $t>0$. If $\theta=\theta_0$, then

$$T_n = \sum_{i=1}^{n} Z_{ni} = -\frac{t}{\sqrt{n}}\sum_{i=1}^{n}\frac{\partial}{\partial\theta}\log (X_i,\theta_0)$$

$$-\frac{t^2}{2n}\sum_{i=1}^{n}\frac{\partial^2}{\partial\theta^2}\log f(X_i,\theta_0)$$

$$-\frac{t^3}{6n\sqrt{n}}\sum_{i=1}^{n}\frac{\partial^3}{\partial\theta^3}\log f(X_i,\theta_0) + o_p(\frac{1}{\sqrt{n}}) .$$

Hence it follows that

$$E_{\theta_0}(T_n)=\frac{t^2}{2}I(\theta_0) + \frac{t^3}{6\sqrt{n}}(3J(\theta_0)+K(\theta_0)) + o(\frac{1}{\sqrt{n}}) ,$$

$$V_{\theta_0}(T_n) = n(\frac{t^2}{n} I(\theta_0)+\frac{t^3}{n\sqrt{n}}J(\theta_0))=t^2I(\theta_0)+\frac{t^3}{\sqrt{n}} J(\theta_0)+o(\frac{1}{\sqrt{n}}) ,$$

$$E_{\theta_0}[\{T_n - E_{\theta_0}(T_n)\}^3] = -\frac{t^3}{\sqrt{n}}K(\theta_0) + o(\frac{1}{\sqrt{n}}) ,$$

Put $\theta_1=\theta_0+(t/\sqrt{n})$. If $\theta=\theta_1$, then

$$T_n = -\frac{t}{\sqrt{n}}\sum_{i=1}^{n}\frac{\partial}{\partial\theta}\log f(X_i,\theta_1)$$

$$+\frac{t^2}{2n}\sum_{i=1}^{n}\frac{\partial^2}{\partial\theta^2}\log f(X_i,\theta_1)$$

$$-\frac{t^3}{6n\sqrt{n}}\sum_{i=1}^{n}\frac{\partial^3}{\partial\theta^3}\log f(X_i,\theta_1) + o(\frac{1}{\sqrt{n}}) .$$

Hence it follows that

$$E_{\theta_1}(T_n) = -\frac{t^2}{2}I(\theta_1) +\frac{t^3}{6\sqrt{n}}(3J(\theta_1)+K(\theta_1)) + o(\frac{1}{\sqrt{n}}) ,$$

On the other hand we have

$$I(\theta_1) = I(\theta_0) + \frac{t}{\sqrt{n}}\frac{\partial}{\partial\theta}I(\theta_0) + o(\frac{1}{\sqrt{n}})$$

$$= I(\theta_0)+\frac{t}{\sqrt{n}}\frac{\partial}{\partial\theta}\int\left\{\frac{\partial}{\partial\theta}\log f(x,\theta_0)\right\}^2 f(x,\theta_0)d\mu +o(\frac{1}{\sqrt{n}})$$

$$= I(\theta_0)+\frac{t}{\sqrt{n}}\int 2\left\{\frac{\partial^2}{\partial\theta^2}\log f(x,\theta_0)\right\}\left\{\frac{\partial}{\partial\theta}\log f(x,\theta_0)\right\} f(x,\theta_0)d\mu$$

$$+\frac{t}{\sqrt{n}}\int\left\{\frac{\partial}{\partial\theta}\log f(x,\theta_0)\right\}^3 f(x,\theta_0)d\mu + o(\frac{1}{\sqrt{n}})$$

$$= I(\theta_0) + \frac{t}{\sqrt{n}}(2J(\theta_0)+K(\theta_0)) + o(\frac{1}{\sqrt{n}}) .$$

Hence we obtain

$$E_{\theta_1}(T_n) = -\frac{t^2}{2}I(\theta_0) - \frac{t^3}{6\sqrt{n}}(3J(\theta_0)+2K(\theta_0)) + o(\frac{1}{\sqrt{n}})$$

Since

$$J(\theta_1) = J(\theta_0) + \frac{t}{\sqrt{n}}\frac{\partial}{\partial\theta}J(\theta_0) + o(\frac{1}{\sqrt{n}}) ;$$

$$K(\theta_1) = K(\theta_0) + \frac{t}{\sqrt{n}}\frac{\partial}{\partial\theta}K(\theta_0) + o(\frac{1}{\sqrt{n}}) ,$$

it follows by a similar way as above that

$$V_{\theta_1}(T_n) = t^2 I(\theta_0) + \frac{t^3}{\sqrt{n}}(J(\theta_0)+K(\theta_0)) + o(\frac{1}{\sqrt{n}})$$

$$E_{\theta_1}[\{T_n - E_{\theta_1}(T_n)\}^3] = -\frac{t^3}{\sqrt{n}}K(\theta_0) + o(\frac{1}{\sqrt{n}})$$

Letting a_n be the rejection bound, we have

$$P_{\theta_1,n}\{T_n < a_n\} = P_{\theta_1,n}\left\{\frac{T_n+(t^2 I(\theta_0)/2)}{t\sqrt{I(\theta_0)}} < \frac{a_n+(t^2 I(\theta_0)/2)}{t\sqrt{I(\theta_0)}}\right\} .$$

Putting $c_n = \{a_n+(t^2 I(\theta_0)/2)\}/(t\sqrt{I(\theta_0)})$, we obtain

$$P_{\theta_1,n}\{T_n < a_n\}$$

$$= \Phi(c_n)-\phi(c_n)\left\{-\frac{t^2}{6\sqrt{n}\,I(\theta_0)}(3J(\theta_0)+2K(\theta_0))+\frac{t}{2\sqrt{n}\,I(\theta_0)}(J(\theta_0)+K(\theta_0))c_n\right.$$

$$\left. -\frac{1}{6\sqrt{nI(\theta_0)}\,I(\theta_0)}K(\theta_0)(c_n^2-1)\right\} + o(\frac{1}{\sqrt{n}}) ,$$

where $\Phi(u)=\int_{-\infty}^{u}\phi(x)dx$ with $\phi(x)=\frac{1}{\sqrt{2\pi}}e^{-\frac{x^2}{2}}$

If $P_{\theta_1,n}\{T_n < a_n\}=1/2$, then it follows that $c_n=O(1/\sqrt{n})$ and $\Phi(c_n)=\frac{1}{2} + c_n\phi(c_n)$.

Since

$$c_n = -\frac{t^2}{6\sqrt{n}\,I}(3J+2K) + \frac{K}{6\sqrt{nI}\;I} + o(\frac{1}{\sqrt{n}}) ,$$

where I, J and K denote $I(\theta_0)$, $J(\theta_0)$ and $K(\theta_0)$, respectively, we have

$$a_n = -\frac{t^2 I}{2} - \frac{t^3}{6\sqrt{n}\;I}(3J+2K) + \frac{tK}{6\sqrt{n}\;I} + o_p(\frac{1}{\sqrt{n}}) .$$

Then we have

$$P_{\theta_0,n}\left\{ T_n \geqq a_n \right\}$$

$$= 1 - P_{\theta_0,n}\left\{ T_n < a_n \right\}$$

$$= 1 - P_{\theta_0,n}\left\{ \frac{T_n-(t^2I/2)}{t\sqrt{I}} - c_n < -t\sqrt{I} \right\}$$

$$= 1 - \Phi(-t\sqrt{I}) + \phi(-t\sqrt{I})\left\{ \frac{t^2}{6\sqrt{nI}}(3J+K) - c_n + \frac{t}{2\sqrt{n}\;I}J(-t\sqrt{I}) - \frac{1}{6\sqrt{nI}\;I}K(t^2I-1)\right\} + o(\frac{1}{\sqrt{n}})$$

$$= \Phi(t\sqrt{I}) + \frac{t^2}{6\sqrt{nI}}(3J+2K)\,\phi(t\sqrt{I}) + o(\frac{1}{\sqrt{n}}) .$$

Since θ_0 is arbitrary, it follows that for each $t > 0$

$$G_\theta(t) = \Phi(t\sqrt{I(\theta)}) ,$$

$$\gamma_\theta(t) = \frac{t^2}{6\sqrt{I(\theta)}}(3J(\theta)+2K(\theta))\,\phi(t\sqrt{I(\theta)}) .$$

In a similar way as the case $t > 0$, we have for each $t < 0$,

$$G_\theta(t) = \Phi(-t\sqrt{I(\theta)}) ,$$

$$\gamma_\theta(t) = -\frac{t^2}{6\sqrt{I(\theta)}}(3J(\theta)+2K(\theta))\,\phi(t\sqrt{I(\theta)}) .$$

Therefore we have now established the following :

<u>Theorem 4.1.1.</u> Under conditions (A.3.1.1)$\sim$(A.3.1.3) and (A.4.1.1)$\sim$(A.4.1.3), if

(4.1.1) $$P_{\theta,n}\left\{ \sqrt{nI(\theta)}\,(\hat{\theta}_n - \theta) \leqq z \right\} = \Phi(z) + \frac{3J(\theta)+2K(\theta)}{6\sqrt{n}\;I(\theta)^{3/2}}z^2\phi(z) + o(\frac{1}{\sqrt{n}}) ,$$

then $\hat{\theta}_n$ is second order asymptotically efficient.

(4.1.1) means that $\sqrt{nI(\theta)}\ (\hat{\theta}_n - \theta)$ has an asymptotic distribution with mean $-\left\{(3J(\theta)+2K(\theta))/(6\sqrt{nI(\theta)})\right\} + o(1/\sqrt{n})$ and with variance $1+o(1/\sqrt{n})$ and third moment $-(3J(\theta)+2K(\theta))/(\sqrt{n}\ I(\theta)^{3/2})$.

Example : Let $X_1, X_2, \ldots, X_n, \ldots$ be independently and identically distributed random variables with an exponential distribution having the following density function $f(x,\theta)$:

(4.1.2)
$$f(x,\theta) = \begin{cases} \frac{1}{\theta} e^{-\frac{x}{\theta}}, & x > 0, \\ 0, & x \leqq 0. \end{cases}$$

Since

$$\log f(X,\theta) = -\frac{X}{\theta} - \log \theta ,$$

$$\frac{\partial}{\partial\theta}\log f(X,\theta) = \frac{X}{\theta^2} - \frac{1}{\theta} ,$$

$$\frac{\partial^2}{\partial\theta^2}\log f(X,\theta) = -\frac{2X}{\theta^3} + \frac{1}{\theta^2} ,$$

it follows that

$$I(\theta) = \frac{1}{\theta^4}E_\theta[\ (X-\theta)^2] = \frac{1}{\theta^2} ,$$

$$J(\theta) = -\frac{1}{\theta^5}E_\theta[\ (X-\theta)(2X-\theta)] = -\frac{2}{\theta^5}E\ [(X-\theta)^2] = -\frac{2}{\theta^3},$$

$$K(\theta) = \frac{1}{\theta^6}E_\theta[\ (X-\theta)^3] = \frac{2}{\theta^3} .$$

If $\sqrt{n}(\hat{\theta}_n - \theta)/\theta$ has an asymptotic distribution with mean $1/(3\sqrt{n})$ and variance $1+o(1/\sqrt{n})$ and third moment $2/\sqrt{n}$, then $\hat{\theta}_n$ is second order asymptotically efficient.

The maximum likelihood estimator of θ is given by $\hat{\theta}_{ML}=\bar{X}=\sum_{i=1}^{n}X_i/n$. Putting

$$\hat{\theta}^*_{ML} = (1+\frac{1}{3n})\bar{X} ,$$

we have

$$E_\theta\left[\frac{\sqrt{n}(\hat{\theta}^*_{ML}-\theta)}{\theta}\right] = \frac{1}{3\sqrt{n}}$$

$$V_\theta\left[\frac{\sqrt{n}\,\hat{\theta}^*_{ML}}{\theta}\right] = \left(1+\frac{1}{3n}\right)^2 = 1 + o\left(\frac{1}{\sqrt{n}}\right) \quad ,$$

$$E_\theta\left[\left\{\frac{\sqrt{n}(\hat{\theta}^*_{ML}-\theta)}{\theta}\right\}^3\right] = \frac{2}{\sqrt{n}}\left(1+\frac{1}{3n}\right)^3 = \frac{2}{\sqrt{n}} + o\left(\frac{1}{\sqrt{n}}\right) \quad .$$

<u>Hence $\hat{\theta}^*_{ML}$ is second order asymptotically efficient.</u>
In this case $\bar{X}$ is a sufficient statistic and the distribution has monotone likelihood ratio. Consider the problem of testing hypothesis $\theta = \theta_1$ against alternative $\theta = \theta_0$, where $\theta_1 = \theta_0 + (t/\sqrt{n})(t>0)$. Then the rejection region of the most powerful test is given by the following form :

$$\bar{X} < c \ .$$

On the other hand since the cumulants of $\bar{X}$ are given as follows :

$$V_\theta(\bar{X}) = \frac{\theta^2}{n} \quad ,$$

$$E_\theta[(\bar{X}-\theta)^3] = \frac{2\theta^3}{n^2} \quad .$$

$\sqrt{n}(\bar{X}-\theta)/\theta$ has an asymptotic distribution with mean 0 and variance 1 and third cumulant $2/\sqrt{n}$.

Using Edgeworth expansion we have

$$(4.1.3) \quad P_{\theta,n}\left\{\frac{\sqrt{n}(\bar{X}-\theta)}{\theta} < c\right\}$$

$$= \Phi(c) - \frac{1}{3\sqrt{n}}\,\phi(c)(c^2-1) + o\left(\frac{1}{\sqrt{n}}\right)$$

Let $\theta = \theta_1 (= \theta_0 + t/\sqrt{n})(t>0)$. If (4.1.3) agrees with 1/2 up to the order of $1/\sqrt{n}$, then the following must hold :

$$c\,\phi(c) - \frac{1}{3\sqrt{n}}\,\phi(c)(c^2-1) \fallingdotseq 0$$

Hence it follows that $c=-1/(3\sqrt{n})$. The rejection region of the level 1/2 test of hypothesis $\theta=\theta_1$ is given by

$$\bar{X} < (1-\frac{1}{3n})\theta_1 \quad .$$

This agrees asymptotically with

$$\hat{\theta}^*_{ML} = (1+\frac{1}{3n})\bar{X}(<\theta_1)$$

up to the order of $1/\sqrt{n}$.

In this case, since $\bar{X}$ is sufficient and complete it shall be asymptotically efficient up to any order, if the meaning of asymptotic efficiency is properly defined. This completes the example.

Assume that $\sqrt{n}(\hat{\theta}_n-\theta)$ has the asymptotic normal distribution with mean 0 and variance $1/I(\theta)$, and that the cumulants of the asymptotic distribution are given as follows :

$$\sqrt{n}\, E_\theta(\hat{\theta}_n-\theta) = \frac{1}{\sqrt{n}}c_1(\theta) + o(\frac{1}{\sqrt{n}}) \ ;$$

$$nV_\theta(\hat{\theta}_n) = \frac{1}{I(\theta)} + \frac{1}{\sqrt{n}}c_2(\theta) + o(\frac{1}{\sqrt{n}}) \ ;$$

$$E_\theta[\{\sqrt{n}(\hat{\theta}_n-E_\theta(\hat{\theta}_n))\}^3] = \frac{1}{\sqrt{n}}c_3(\theta) + o(\frac{1}{\sqrt{n}}) \ .$$

Further the m-th ($m \geqq 4$) cumulants of $\sqrt{n}\,\hat{\theta}_n$ is assumed to be less than order $1/\sqrt{n}$. We assume that $c_1(\theta)$, $c_2(\theta)$ and $c_3(\theta)$ are continuous in θ. Put $\hat{\theta}^*_n = \hat{\theta}_n-(1/n)u(\hat{\theta}_n)$. Since $\hat{\theta}^*_n$ agrees with $\hat{\theta}_n-(1/n)u(\hat{\theta}_n)$ up to the order of $1/n$, the cumulants of the asymptotic distribution of $\sqrt{nI(\theta)}\,(\hat{\theta}^*_n-\theta)$ are given as follows :

$$\sqrt{nI(\theta)}\, E_\theta(\hat{\theta}^*_n-\theta) = \sqrt{\frac{I(\theta)}{n}}\left\{c_1(\theta)-u(\theta)\right\} + o(\frac{1}{\sqrt{n}}) \ ;$$

$$nI(\theta)V_\theta(\hat{\theta}^*_n) = 1 + \frac{I(\theta)c_2(\theta)}{\sqrt{n}} + o(\frac{1}{\sqrt{n}}) \ ;$$

$$E_\theta[\{\sqrt{nI(\theta)}\,(\hat{\theta}^*_n-E_\theta(\hat{\theta}^*_n))\}^3] = \frac{\{I(\theta)\}^{3/2}c_3(\theta)}{\sqrt{n}} + o(\frac{1}{\sqrt{n}}) \ .$$

If

$$u(\theta) = c_1(\theta) - \frac{1}{6}I(\theta)c_3(\theta) ,$$

then in the Edgeworth expansion the following holds :

$$P_{\theta,n}\left\{\hat{\theta}_n^* \leqq \theta\right\} = \frac{1}{2} + o\left(\frac{1}{\sqrt{n}}\right) .$$

It follows that

(4.1.4) $$P_{\theta,n}\left\{ \sqrt{nI(\theta)}\,(\hat{\theta}_n^* - \theta) < t \right\}$$

$$= \Phi(t) - \frac{1}{\sqrt{n}}\,\phi(t)\,\left[\frac{1}{2}I(\theta)c_2(\theta)t + \frac{1}{6}\,I(\theta)^{\frac{3}{2}}c_3(\theta)t^2\right] + o\left(\frac{1}{\sqrt{n}}\right) .$$

Hence it is seen that (4.1.4) agrees with the bound (4.1.1) if and only if $c_2(\theta)=0$ and $c_3(\theta)=-(3J(\theta)+2K(\theta))/I(\theta)^3$.

If $c_2(\theta)=0$, then $c_3(\theta)$ must automatically be equal to $-(3J(\theta)+2K(\theta))/I(\theta)^3$.

Indeed since $\phi(t)t^2 > 0$ for all $t \neq 0$, it follows that (4.1.1) $\gtrless$ (4.1.4) uniformly in θ if $c_3(\theta) \gtrless -(3J(\theta)+2K(\theta))/I(\theta)^3$. This means that the bound (4.1.1) fails to hold at either positive or negative t. Hence if $c_2(\theta)=0$, we must have $c_3(\theta)=-(3J(\theta)+2K(\theta))/I(\theta)^3$. This apparently seems mysterious. But it will be naturally seen from different examples that if the second moment is decided up to order $1/\sqrt{n}$, then the third moment is done.

The second order asymptotic efficiency of the maximum likelihood estimator will be verified using the above fact.
The continuous differentiability of the likelihood function is assumed up to the necessary order. Let $\hat{\theta}_{ML}$ be the maximum likelihood estimator.

We have

$$0 = \sum_{i=1}^{n} \frac{\partial}{\partial\theta} \log f(X_i, \hat{\theta}_{ML})$$

$$= \sum_{i=1}^{n} \frac{\partial}{\partial\theta} \log f(X_i, \theta) + \sum_{i=1}^{n} \left\{ \frac{\partial^2}{\partial\theta^2} \log f(X_i, \theta) \right\} (\hat{\theta}_{ML} - \theta)$$

$$+ \frac{1}{2} \sum_{i=1}^{n} \left\{ \frac{\partial^3}{\partial\theta^3} \log f(X_i, \theta^*) \right\} (\hat{\theta}_{ML} - \theta)^2 ,$$

where $|\theta^* - \theta| \leqq |\hat{\theta}_{ML} - \theta|$.

Putting $T_n = \sqrt{n}(\hat{\theta}_{ML} - \theta)$ we have

$$0 = \frac{1}{\sqrt{n}} \sum_{i=1}^{n} \frac{\partial}{\partial\theta} \log f(X_i, \theta) + \frac{1}{n} \left\{ \sum_{i=1}^{n} \frac{\partial^2}{\partial\theta^2} \log f(X_i, \theta) \right\} T_n$$

$$+ \frac{1}{2n\sqrt{n}} \left\{ \sum_{i=1}^{n} \frac{\partial^3}{\partial\theta^3} \log f(X_i, \theta^*) \right\} T_n^2 + o_p\left(\frac{1}{\sqrt{n}}\right)$$

Put

$$Z_1(\theta) = \frac{1}{n} \sum_{i=1}^{n} \frac{\partial}{\partial\theta} \log f(X_i, \theta) \quad ;$$

$$Z_2(\theta) = \frac{1}{\sqrt{n}} \sum_{i=1}^{n} \left\{ \frac{\partial^2}{\partial\theta^2} \log f(X_i, \theta) + I(\theta) \right\} \quad ;$$

$$W(\theta) = \frac{1}{n} \sum_{i=1}^{n} \frac{\partial^3}{\partial\theta^3} \log f(X_i, \theta^*) \ .$$

Then $Z_1(\theta)$ and $Z_2(\theta)$ have the asymptotic normal distributions with mean 0 and variance $I(\theta)$ and $L(\theta)$ $(=E_\theta[\{(\partial^2/\partial\theta^2)\log f(X,\theta)+I(\theta)\}^2])$ respectively, and covariance $J(\theta)$. Also $W(\theta)$ converges in probability to $-3J(\theta)-K(\theta)$. Hence it follows that

$$Z_1(\theta) + \left\{ -I(\theta) + \frac{1}{n} Z_2(\theta) \right\} T_n - \frac{3J(\theta)+K(\theta)}{2\sqrt{n}} T_n^2 = o_p\left(\frac{1}{\sqrt{n}}\right) .$$

Therefore we have the following theorem.

Theorem 4.1.2. Let $\hat{\theta}_{ML}$ be the maximum likelihood estimator. Then

$$\sqrt{n}(\hat{\theta}_{ML} - \theta) = \frac{1}{I(\theta)} Z_1(\theta) + \frac{1}{\sqrt{n}\, I(\theta)^2} Z_1(\theta) Z_2(\theta) - \frac{3J(\theta)+K(\theta)}{2\sqrt{n}\, I(\theta)^3} Z_1^2(\theta) + o_p\left(\frac{1}{\sqrt{n}}\right).$$

Since covariance of $Z_1(\theta)$ and the term of order $1/\sqrt{n}$ is zero, it follows that

$$V_\theta(T_n) = \frac{1}{I(\theta)} + o\left(\frac{1}{\sqrt{n}}\right) .$$

Hence it is seen that <u>the maximum likelihood estimator (properly adjusted to be second order AMU) is second order asymptotically efficient.</u> If indeed we calculate the asymptotic cumulants, they are as follows :

$$E_\theta(T_n) = -\frac{J(\theta)+K(\theta)}{2\sqrt{n}\,I(\theta)^2} + o(\frac{1}{\sqrt{n}}) ;$$

$$V_\theta(T_n) = \frac{1}{I(\theta)} + o(\frac{1}{\sqrt{n}}) ;$$

$$E_\theta[\{T_n - E_\theta(T_n)\}^3] = -\frac{3J(\theta)+2K(\theta)}{\sqrt{n}\,I(\theta)^3} + o(\frac{1}{\sqrt{n}}) .$$

<u>Hence we have shown that the third order moment is actually equal to the quantity derived above.</u>

Thus we have established the following :

<u>Theorem 4.1.3.</u> Let $\hat{\theta}_{ML}$ be the maximum likelihood estimator. Then

$$\hat{\theta}^*_{ML} = \hat{\theta}_{ML} + \frac{K(\hat{\theta}_{ML})}{6nI(\hat{\theta}_{ML})^2}$$

is second order asymptotically efficient.

As another example, we consider the location parameter case. Let $X_1, X_2, \dots, X_n, \dots$ be a sequence of independently and identically distributed random variables with a density function $f(x-\theta)$ w.r.t. a Lebesgue measure. It is well known that under appropriate regularity conditions the best linear estimator $\hat{\theta}_n = \sum_{i=1}^{n} c_{in}X_{(i|n)} + c_{on}$ is asymptotically efficient, where $X_{(1|n)} < \dots < X_{(n|n)}$ and c_{in} are the optimal constants established by Blom [14] and others.

Then the estimator $\hat{\theta}_n$ is second asymptotically efficient. Indeed, if $U_{(i \mid n)}$ are order statistics from the uniform distribution, then it follows that

$$F^{-1}(X_{(i \mid n)}) = \theta + F^{-1}(\frac{i}{n+1})$$
$$+ \left\{ F^{-1}(\frac{i}{n+1}) \right\}' (U_{(i \mid n)} - \frac{i}{n+1})$$
$$+ \frac{1}{2} \left\{ F^{-1}(\frac{i}{n+1}) \right\}'' (U_{(i \mid n)} - \frac{i}{n+1})^2 + R .$$

Let $c_{on} = -\sum_{i=1}^{n} c_{in} F^{-1}(i/(n+1))$ and $\sum_{i=1}^{n} c_{in} = 1$.

$$a_{in} = \left\{ F^{-1}(\frac{i}{n+1}) \right\}' ,$$

$$b_{in} = \left\{ F^{-1}(\frac{i}{n+1}) \right\}'' ,$$

we have

$$\sqrt{n}\,(\hat{\theta}_n - \theta) = \sqrt{n} \sum_{i=1}^{n} c_{in}\, a_{in}\, (U_{(i \mid n)} - \frac{i}{n+1})$$
$$+ \frac{\sqrt{n}}{2} \sum_{i=1}^{n} c_{in}\, b_{in} (U_{(i \mid n)} - \frac{i}{n+1})^2 + \sqrt{n}\, R_{in} .$$

Then the asymptotic variance is given by

$$V_\theta(\sqrt{n}(\hat{\theta}_n - \theta))$$
$$= n \sum_{i=1}^{n} c_{in}^2\, a_{in}^2 \frac{i(n+1-i)}{(n+1)(n+2)} + 2n \sum_{i<j}\sum c_{in}\, c_{jn}\, a_{in}\, a_{jn} \frac{i(n+1-j)}{(n+1)^2(n+2)}$$
$$+ n \sum_{i \leqq j}\sum c_{in}\, c_{jn}\, a_{in}\, a_{jn}\, \frac{i(n+1-j)(n+1-2J)}{(n+1)^3(n+2)(n+3)}$$
$$+ n \sum_{i>j}\sum c_{in}\, c_{jn}\, a_{in}\, b_{jn} \frac{j(n+1-i)(n+1-2j)}{(n+1)^3(n+2)(n+3)} + o(\frac{1}{n})$$

Let $J(u)$ be a function sufficiently smooth and put $c_{in}=(1/n)J(i/(n+1))$. Putting $h(u)=f(F^{-1}(u))$, we have

$$(F^{-1}(u))'=h(u)^{-1}$$

$$(F^{-1}(u))''=-h'(u)h(u)^{-2} .$$

Since

$$V(\sqrt{n}(\hat{\theta}_n-\theta))$$
$$= 2\iint_{u<v} u(1-v)J(u)J(v)h(u)^{-1}h(v)^{-1}dudv$$
$$+ \frac{1}{n}\iint_{u<v} u(1-v)(1-2v)J(u)J(v)h(u)^{-1}h'(v)h(v)^{-2}dudv$$
$$+ \frac{1}{n}\iint_{u>v} v(1-u)(1-2v)J(u)J(v)h(u)^{-1}h'(v)h(v)^{-2}dudv$$
$$+ o(\frac{1}{n}) ,$$

it follows that if

$$J(u) = -h(u)h''(u)/I ,$$

where $I = \int \{f'(x)\}^2/f(x)dx = \int \{h'(u)\}^2 du$,

then we have

$$V(\sqrt{n}(\hat{\theta}_n-\theta)) = \frac{1}{I} + o(\frac{1}{\sqrt{n}}) .$$

Let $\hat{\theta}^*_n$ be the modification of $\hat{\theta}_n$ which is second order AMU. Then it will be shown that $\hat{\theta}^*_n$ is second order asymptotically efficient.

Since

$$J = -\int\left\{\frac{d^2}{dx^2}\log f(x)\right\}\left\{\frac{d}{dx}\log f(x)\right\} f(x)dx$$
$$= \frac{1}{2}\int\left\{\frac{d}{dx}\log f(x)\right\}^2 f'(x)dx$$
$$= \frac{1}{2}\int\left\{\frac{d}{dx}\log f(x)\right\}^3 f(x)dx$$
$$= \frac{1}{2}\int\{h'(u)\}^3 du$$
$$= -\frac{K}{2} ,$$

the third asymptotic moment of the second order asymptotically efficient estimator must be $-K/(2n^2I^3)$.

Indeed, since

$$\mathrm{cov}[\ (U_{(i\mid n)}-\frac{i}{n+1})(U_{(j\mid n)}-\frac{j}{n+1})\ ,\ (U_{(k\mid n)}-\frac{k}{n+1})^2\ ]$$

$= 2 \text{ cov } (U_{(i|n)}, U_{(k|n)})\text{cov}(U_{(j|n)}, U_{(k|n)}) + o(\frac{1}{n})$;

$E[(U_{(i|n)} - \frac{i}{n+1})(U_{(j|n)} - \frac{j}{n+1})(U_{(k|n)} - \frac{k}{n+1}]$

$= \frac{i(n+1-j)(n+1-k)}{(n+1)^2(n+2)(n+3)}$, $i \leqq j \leqq k$,

it follows that

$$E[\{\hat{\theta}_n - E(\hat{\theta}_n)\}^3]$$

$$= \frac{6}{n^2}\iiint_{u<v<w} u(1-2v)(1-w)J(u)J(v)J(w)h(u)^{-1}h(v)^{-1}h(w)^{-1}dudvdw$$

$$+ \frac{3}{n^2}\int\left\{\int(\min(u,v)-uv)J(u)h(u)du\right\}^2 J(v)h(v)^{-2}h'(v)dv+o(\frac{1}{n^2}) .$$

Putting $J(u)=-h(u)h''(u)/I$, we have

$$E[\{\hat{\theta}_n - E(\hat{\theta}_n)\}^3]$$

$$= -\frac{1}{n^2I^3}\int\{h'(u)\}^3du - \frac{1}{n^2I^3}\int h(u)h'(u)h''(u)du + o(\frac{1}{n^2})$$

$$= -\frac{1}{n^2I^3}\int\{h'(u)\}^3du + \frac{1}{2n^2I^3}\int\{h'(u)\}^3du + o(\frac{1}{n^2})$$

$$= -\frac{K}{2n^2I^3} + o(\frac{1}{n^2}) .$$

Hence we have shown that the third asymptotic moment is $-K/(2n^2I^3)$.

It should be noted that the second order asymptotic efficiency of the MLE depends on the assumption of differentiability of the density function.

The following example makes this point clear.

Suppose that X_i's are independently and identically distributed according to the double (bilateral) exponential distribution with the density function

$$f(x-\theta) = \frac{1}{2}e^{-|x-\theta|} ,$$

where θ is unknown location parameter. In this case it can be shown that the first order asymptotic bound is obtained with I=1, and it can be easily shown that the MLE

$$\hat{\theta}_{ML} = \text{med } X_i$$

is asymptotically efficient.

The second order asymptotic bound is obtained after some algebraic manipulations as follows :

$$(4.1.5) \qquad \underline{\Phi(t) - \frac{t^2}{6\sqrt{n}}\,\phi(t)\,\text{sgn}\,t} \ .$$

But the asymptotic distribution of $\hat{\theta}_{ML}$ is given by

$$(4.1.6) \quad \underline{P_{\theta,n}\left\{\sqrt{n}(\hat{\theta}_{ML}-\theta) \leqq t\right\} = \Phi(t) - \frac{t^2}{2\sqrt{n}}\,\phi(t)\,\text{sgn}\,t + o\left(\frac{1}{\sqrt{n}}\right)} \ ,$$

which differs in the second term from the above bound.

In the remainder of this section 4.1 we shall prove (4.1.5) and (4.1.6) ([42]).

First we shall obtain the bound of the second order asymptotic distributions of second order AMU estimators.

We consider the likelihood ratio test for testing the hypothesis $\theta = \theta_0 + tn^{-1/2}$ $(t > 0)$ against the alternative $\theta = \theta_0$.

Put $Z_{ni} = \log f(X_i, \theta_0)/f(X_i, \theta_1)$ with $\theta_1 = \theta_0 + tn^{-1/2}$. Setting $\Delta = tn^{-1/2}$, we have

$$Z_{ni} = \begin{cases} \Delta & \text{for } X_i \leqq \theta_0 \ ; \\ \theta_0 + \theta_1 - 2X_i & \text{for } \theta_0 \leqq X_i \leqq \theta_1 \ ; \\ -\Delta & \text{for } X_i > \theta_1 \ . \end{cases}$$

Then we obtain

$$E_{\theta_0}(Z_{ni}) = \int_{-\infty}^{0} \frac{1}{2}\,\Delta\, e^{u} du + \int_0^{\Delta} \frac{1}{2}(\Delta - 2u)e^{-u} du - \int_{\Delta}^{\infty} \frac{1}{2}\,\Delta\, e^{-u} du$$

$$= \frac{1}{2}\Delta + \frac{1}{2}\Delta(1-e^{-\Delta}) - \left\{1-(1+\Delta)e^{-\Delta}\right\} - \frac{\Delta}{2}e^{-\Delta}$$

$$= \frac{\Delta^2}{2} - \frac{\Delta^3}{6} + o(\Delta^3) \ ;$$

$$E_{\theta_1}(Z_{ni}) = \int_{-\infty}^{-\Delta} \frac{1}{2}\Delta e^u du + \int_{-\Delta}^{0} \frac{1}{2}(-\Delta - 2u)e^u du - \int_0^{\infty} \frac{1}{2}\Delta e^{-u} du$$

$$= \frac{\Delta}{2} e^{-\Delta} - \frac{\Delta}{2}(1-e^{-\Delta}) + \left\{1-(1+\Delta)e^{-\Delta}\right\} - \frac{\Delta}{2}$$

$$= -\frac{\Delta^2}{2} + \frac{\Delta^3}{6} + o(\Delta^3) .$$

We also have

$$E_{\theta_0}(Z_{ni}^2) = \int_{-\infty}^{0} \frac{1}{2}\Delta^2 e^u du + \int_0^{\Delta} \frac{1}{2}(\Delta-2u)^2 e^{-u} du + \int_{\Delta}^{\infty} \frac{1}{2}\Delta^2 e^{-u} du$$

$$= \frac{1}{2}\Delta^2 + \frac{1}{2}\Delta^2(1-e^{-\Delta}) - 2\Delta\left\{1-(1+\Delta)e^{-\Delta}\right\}$$

$$+ 2\left\{2 - (\Delta^2+2\Delta+2)e^{-\Delta}\right\} + \frac{1}{2}\Delta^2 e^{-\Delta}$$

$$= \Delta^2 - \frac{1}{3}\Delta^3 + o(\Delta^3) \quad ;$$

$$V_{\theta_0}(Z_{ni}) = \Delta^2 - \frac{1}{3}\Delta^3 + o(\Delta^3) \quad ;$$

$$E_{\theta_1}(Z_{ni}^2) = E_{\theta_0}(Z_{ni}^2) \quad ;$$

$$V_{\theta_1}(Z_{ni}) = V_{\theta_0}(Z_{ni}) = \Delta^2 - \frac{1}{3}\Delta^3 + o(\Delta^3) \quad ;$$

$$E_{\theta_0}(Z_{ni}^3) = \int_{-\infty}^{0} \frac{1}{2}\Delta^3 e^u du + \int_0^{\Delta} \frac{1}{2}(\Delta - 2u)^3 e^{-u} du - \int_{\Delta}^{\infty} \frac{1}{2}\Delta^3 e^{-u} du$$

$$= \frac{1}{2}\Delta^3 - \frac{1}{2}\Delta^3 e^{-\Delta} + o(\Delta^3)$$

$$= o(\Delta^3) \quad ;$$

$$E_{\theta_1}(Z_{ni}^3) = -E_{\theta_0}(Z_{ni}^3) = o(\Delta^3) \quad ;$$

$$E_{\theta_0}[\{Z_{ni} - E_{\theta_0}(Z_{ni})\}^3] = -E_{\theta_1}[\{Z_{ni} - E_{\theta_1}(Z_{ni})\}^3] = o(\Delta^3) .$$

Hence we have

$$E_{\theta_0}(\sum_{i=1}^{n} Z_{ni}) = \frac{1}{2}n\Delta^2 - \frac{1}{6}n\Delta^3 + o(n\Delta^3)$$

$$= \frac{t^2}{2} - \frac{t^3}{6\sqrt{n}} + o(\frac{1}{\sqrt{n}}) \quad ;$$

$$E_{\theta_1}(\sum_{i=1}^{n} Z_{ni}) = -\frac{1}{2}n\Delta^2 + \frac{1}{6}n\Delta^3 + o(n\Delta^3)$$

$$= \frac{t^2}{2} - \frac{t^3}{6\sqrt{n}} + o(\frac{1}{\sqrt{n}}) \quad ;$$

$$V_{\theta_0}(\sum_{i=1}^{n} Z_{ni}) = V_{\theta_1}(\sum_{i=1}^{n} Z_{ni})$$
$$= n\Delta^2 - \frac{1}{3}n\Delta^3 + o(n\Delta^3)$$
$$= t^2 - \frac{t^3}{3\sqrt{n}} + o(\frac{1}{\sqrt{n}}) ;$$

$$E_{\theta_1}[\{\sum_{i=1}^{n} Z_{ni} - E_{\theta_1}(\sum_{i=1}^{n} Z_{ni})\}^3] = -E_{\theta_0}[\{\sum_{i=1}^{n} Z_{ni} - E_{\theta_0}(\sum_{i=1}^{n} Z_{ni})\}^3] = o(\frac{1}{\sqrt{n}}) .$$

Since

$$P_{\theta_1,n}\{\sum_{i=1}^{n} Z_{ni} \geqq -\frac{t^2}{2} + \frac{t^3}{6\sqrt{n}}\} = \frac{1}{2} + o(\frac{1}{\sqrt{n}}) ;$$

$$P_{\theta_0,n}\{\sum_{i=1}^{n} Z_{ni} \geqq -\frac{t^2}{2} + \frac{t^3}{6\sqrt{n}}\}$$

$$= P_{\theta_0,n}\{\sum_{i=1}^{n} Z_{ni} + \frac{t^2}{2} - \frac{t^3}{6\sqrt{n}} \geqq -t^2 + \frac{t^3}{3\sqrt{n}}\}$$

$$= \Phi(t\sqrt{1 - \frac{t}{3\sqrt{n}}}) + o(\frac{1}{\sqrt{n}})$$

$$= \Phi(t) - \frac{t^2}{6\sqrt{n}}\phi(t) + o(\frac{1}{\sqrt{n}}) ,$$

it follows that for every AMU estimator $\hat{\theta}_n$

$$\overline{\lim_{n\to\infty}} \sqrt{n}\,[P_{\theta,n}\{\sqrt{n}(\hat{\theta}_n - \theta) \leqq t\} - \Phi(t) + \frac{t^2}{6\sqrt{n}}\phi(t)] \leqq 0$$

for $t>0$.

By a similar way as the case $t>0$ it may be shown that for every AMU estimator $\hat{\theta}_n$

$$\underline{\lim}_{n\to\infty} \sqrt{n}[P_{\theta,n}\{\sqrt{n}(\hat{\theta}_n - \theta) \leqq t\} - \Phi(t) - \frac{t^2}{6\sqrt{n}}\phi(t)] \geqq 0$$

for $t<0$.

Hence we conclude that the second order asymptotic bound is given by (4.1.5).

Secondly we shall obtain the second order asymptotic distribution of the maximum likelihood estimator $\hat{\theta}_{ML}$.

Let $X_1, X_2, \ldots , X_n$ be rearranged in order from least to greatest and let the ordered random variables be $X_{(1)}, X_{(2)}, \ldots, X_{(n)}$, where $X_{(1)} \leqq X_{(2)} \leqq \cdots \leqq X_{(n)}$. These new random variables are called the order statistics of $X_1, X_2, \ldots, X_n$, If n is an odd number, i.e. n=2m+1, then the MLE $\hat{\theta}_{ML}$ is given by $\hat{\theta}_{ML}=X_{(m+1)}$. We shall obtain the value of $P_{\theta_0,n}\left\{X_{(m+1)} \leqq \theta_0+tn^{-1/2}\right\}$ for $t>0$. In $X_1, X_2, \ldots, X_n$ let N be the number of X_i satisfying $X_i > \theta_0+tn^{-1/2}$. Since $X_{(m+1)} \leqq \theta_0+tn^{-1/2}$ is equivalent to $N<m+1$, it follows that

(4.1.7) $$P_{\theta_0,n}\left\{X_{(m+1)} \leqq \theta_0+tn^{-1/2}\right\} = P\left\{N<m+1\right\} .$$

The number N is also distributed according to the binomial distribution B(n,p) with

$$p=\frac{1}{2}\int_{\Delta}^{\infty} e^{-u}du = \frac{1}{2}e^{-\Delta}=\frac{1}{2}\left\{1 - \Delta + \frac{\Delta^2}{2} + o(\Delta^2)\right\} .$$

Then we have

$$E(N)=np \quad ;$$

$$V(N)=np(1-p) \quad ;$$

$$E[(N-np)^3]=np(1-p)(1-2p) \quad ;$$

and

(4.1.8) $$P\left\{N<m+1\right\} =P\left\{N<m+\frac{1}{2}\right\} =P\left\{N\leqq\frac{n}{2}\right\} .$$

Putting

$$Z = \frac{N-np}{\sqrt{n}\ /2} ,$$

we have

$$E(Z)=0 \ ;$$

$$V(Z)=4p(1-p)=4\left\{\frac{1}{4} -(p-\frac{1}{2})^2\right\} =1-\Delta^2=1-o(\frac{1}{n}) \ ;$$

$$E(Z^3)= \frac{8}{\sqrt{n}}p(1-p)(1-2p)=\frac{8}{\sqrt{n}}(1-2p)\left\{\frac{1}{4} - (p- \frac{1}{2})^2\right\} =o(\frac{1}{n}) .$$

Using Edgeworth expansion we obtain

$$P\left\{N\leqq\frac{n}{2}\right\}$$

$$= P\left\{Z \leqq \sqrt{n}(1-2p)\right\}$$

$$= P\left\{Z \leqq \sqrt{n}\,\Delta - \frac{\sqrt{n}}{2}\,\Delta^2\right\} + o\left(\frac{1}{\sqrt{n}}\right)$$

$$= P\left\{Z \leqq t - \frac{t^2}{2\sqrt{n}}\right\} + o\left(\frac{1}{\sqrt{n}}\right)$$

$$= \Phi\left(t - \frac{t^2}{2\sqrt{n}}\right) + o\left(\frac{1}{\sqrt{n}}\right)$$

$$= \Phi(t) - \frac{t^2}{2\sqrt{n}}\,\phi(t) + o\left(\frac{1}{\sqrt{n}}\right) .$$

From (4.1.7) and (4.1.8) we have

$$(4.1.9) \quad \lim_{n\to\infty} \sqrt{n}\left| P_{\theta,n}\left\{\sqrt{n}(\hat{\theta}_{ML} - \theta) \leqq t\right\} - \Phi(t) + \frac{t^2}{2\sqrt{n}}\,\phi(t)\right| = 0$$

for $t > 0$.

By a similar way as the case $t > 0$ we obtain

$$(4.1.10) \quad \lim_{n\to\infty} \sqrt{n}\left| P_{\theta,n}\left\{\sqrt{n}(\hat{\theta}_{ML} - \theta) \leqq t\right\} - \Phi(t) - \frac{t^2}{2\sqrt{n}}\,\phi(t)\right| = 0$$

for $t < 0$.

If n is an even number, i.e. $n=2m$, then the MLE $\hat{\theta}_{ML}$ is given by $\left\{X_{(m)}+X_{(m+1)}\right\}/2$. In a similar way as the case of the odd number we have

$$P_{\theta,n}\left\{X_{(m)} \leqq \theta + tn^{-1/2}\right\}$$

$$= P\left\{N < m+1\right\}$$

$$= P\left\{N \leqq m + \frac{1}{2}\right\}$$

$$= P\left\{N \leqq \frac{n+1}{2}\right\}$$

$$= P\left\{Z \leqq \sqrt{n}(1-2p) + \frac{1}{\sqrt{n}}\right\}$$

$$= P\left\{Z \leqq t - \frac{t^2-1}{2\sqrt{n}}\right\} + o\left(\frac{1}{n}\right) .$$

Put

$$X_{(m+1)} - X_{(m)} = \frac{2}{n}U + o\left(\frac{1}{n}\right) .$$

Then it is seen that U and $X_{(m)}$ are independent and U is a random variables with an exponential distribution.

Indeed, if X_i's have an identical continuous distribution, then $V_i = -\log(1-F(X_i))$ $(i=1,2,\dots,n)$ are independent random variables with an exponential distribution. Let $V_{(1)}, V_{(2)}, \dots, V_{(n)}$ be the order statistics of $V_1, V_2, \dots, V_n$. Put $U_i = (n-i+1)(V_{(i)} - V_{(i-1)})$ $(i=1,2,\dots,n)$. Then U_i's are independent random variables with an identical exponential distribution. Hence we have

$$X_{(m)} = F^{-1}(1-\exp(-V_{(m)}));$$

$$X_{(m+1)} = F^{-1}(1-\exp(-V_{(m+1)})).$$

Since the density function $f(x)$ of $F(x)$ is continuous, it follows that

$$X_{(m+1)} - X_{(m)} = \frac{m}{nf(F^{-1}(\frac{m}{n}))}(V_{(m+1)} - V_{(m)}) + R$$

$$= \frac{1}{nf(F^{-1}(\frac{m}{n}))} U_{m+1} + R .$$

Here R has smaller order then n^{-1} because the order of R is smaller than that of $V_{(m+1)} - V_{(m)}$. Hence it is easily seen that $U_{(m+1)}$ is independent of $V_{(m)}$ and also $X_{(m)}$.

Since

$$P_{\theta,n}\left\{\hat{\theta}_{ML} \leqq \theta + tn^{-1/2}\right\}$$

$$= P_{\theta,n}\left\{X_{(m)} + \frac{X_{(m+1)} - X_{(m)}}{2} < \theta + tn^{-1/2}\right\}$$

$$= P_{\theta,n}\left\{X_{(m)} - \theta < tn^{-1/2} - n^{-1}U + o(\frac{1}{n})\right\}$$

$$= E^U[P_{\theta,n}\left\{X_{(m)} - \theta < tn^{-1/2} - n^{-1}U + o(\frac{1}{n}) \mid U\right\}]$$

$$= E[P\left\{Z \leqq t - \frac{U}{\sqrt{n}} - \frac{t^2-2}{2\sqrt{n}}\right\}] + o(\frac{1}{\sqrt{n}})$$

$$= E[P\left\{Z \leqq t - \frac{t^2}{2\sqrt{n}} - \frac{U-1}{\sqrt{n}}\right\}] + o(\frac{1}{\sqrt{n}}),$$

and E(U-1)=0, it follows that

$$(4.1.11) \quad P_{\theta,n}\{\hat{\theta}_{ML} \leqq \theta + tn^{-1/2}\}$$

$$= P\left\{Z \leqq t - \frac{t^2}{2\sqrt{n}}\right\} + o\left(\frac{1}{\sqrt{n}}\right)$$

$$= \Phi(t) - \frac{t^2}{2\sqrt{n}}\phi(t) + o\left(\frac{1}{\sqrt{n}}\right) .$$

For the case of n=2m we have the same second order asymptotic distribution as the case n=2m+1. Hence it is concluded by (4.1.9), (4.1.10) and (4.1.11) that (4.1.5) and (4.1.6) hold.

4.2. Third order asymptotic efficiency

Now we can proceed to the discussion of the third order asymptotic efficiency in a way completely analogous to in the second order.

The continuous differentiability of the likelihood function is assumed up to the fourth order. Putting $\psi' = (\partial/\partial\theta)\log f(x,\theta)$, $\psi'' = (\partial^2/\partial\theta^2)\log f(x,\theta)$ and $\psi^{(i)} = (\partial^i/\partial\theta^i)\log f(x,\theta)$ (i=3,4), we have

$$T_n = \sum_{i=1}^{n} Z_{ni}$$

$$= -\frac{t}{n}\sum_{i=1}^{n}\psi'(X_i) - \frac{t^2}{2n}\sum_{i=1}^{n}\psi''(X_i) - \frac{t^3}{6n\sqrt{n}}\sum_{i=1}^{n}\psi^{(3)}(X_i)$$

$$-\frac{t^4}{24n^2}\sum_{i=1}^{n}\psi^{(4)}(X_i) + o\left(\frac{1}{n}\right) ,$$

where $Z_{ni} = \log\{f(X_i,\theta_0)/f(X_i,\ \theta_0 + (t/\sqrt{n}))\}$.

Define

$$L = E_{\theta_0}[\psi^{(3)}(X)\psi'(X)] \quad ,$$

$$M = E_{\theta_0}[\psi''(X)^2] \quad ,$$

$$N = E_{\theta_0}[\psi''(X)\psi'(X)^2] \quad ,$$

$$H = E_{\theta_0}[\psi'(X)^4]$$

Since under appropriate conditions the following holds :

$$E_{\theta_0}\{\psi^{(4)}(X)\} = -4L - 3M - 6N - H \quad ,$$

the asymptotic cumulants are given as follows (also under some set of regularity conditions which are almost obvious but tedious to give explicitely) :

$$E_{\theta_0}(T_n)= \frac{t^2 I}{2} + \frac{t^3}{6\sqrt{n}}(3J+K) + \frac{t^4}{24n}(4L+3M+6N+H) + o(\frac{1}{n})$$

$$V_{\theta_0}(T_n)=t^2 I + \frac{t^3}{\sqrt{n}} J + \frac{t^4}{4n}(M-I^2) + \frac{t^4}{3n} L + o(\frac{1}{n})$$

$$E_{\theta_0}[\{T_n - E_{\theta_0}(T_n)\}^3] \equiv \gamma_{\theta_0}(T_n) = -\frac{t^3}{n} K - \frac{3t^4}{2n}(N+I^2) + o(\frac{1}{n})$$

$$E_{\theta_0}[\{T_n - E_{\theta_0}(T_n)\}^4] - 3\{V_{\theta_0}(T_n)\}^2 \equiv \delta_{\theta_0}(T_n) = \frac{t^4}{n}(H-3I^2) + o(\frac{1}{n})$$

If $\theta = \theta_1 (= \theta_0 + (t/\sqrt{n}))$, then for any measurable function g

$$E_{\theta_1}\{g(X)\}$$

$$= E_{\theta_0}[g(X)\{1 + \frac{t}{\sqrt{n}} \frac{f_{\theta_0}'(X)}{f_{\theta_0}(X)} + \frac{t^2}{2n} \frac{f_{\theta_0}''(X)}{f_{\theta_0}(X)} + \frac{t^3}{6n\sqrt{n}} \frac{f_{\theta_0}^{(3)}(X)}{f_{\theta_0}(X)}\}] + o(\frac{1}{n})$$

$$= E_{\theta_0}[g(X)\{1 + \frac{t}{\sqrt{n}} \psi'(X) + \frac{t^2}{2n}(\psi''(X) + \psi'(X)^2)$$

$$+ \frac{t^3}{6n\sqrt{n}}(\psi^{(3)}(X) + 3\psi''(X)\psi'(X) + \psi'(X)^3)\}] + o(\frac{1}{n})$$

Hence we have

$$E_{\theta_1}(T_n) = E_{\theta_0}(T_n) - E_{\theta_0}[\{\frac{t}{\sqrt{n}} \sum_{i=1}^{n} \psi'(X_i) + \frac{t^2}{2n} \sum_{i=1}^{n} \psi''(X_i) + \frac{t^3}{6n\sqrt{n}} \sum_{i=1}^{n} \psi^{(3)}(X_i)\}$$

$$\cdot \{\frac{t}{\sqrt{n}} \sum_{i=1}^{n} \psi'(X_i) + \frac{t^2}{2n} \sum_{i=1}^{n} (\psi''(X_i) + \psi'(X_i)^2)$$

$$+ \frac{t^3}{6n\sqrt{n}} \sum_{i=1}^{n} (\psi^{(3)}(X_i) + 3\psi''(X_i)\psi'(X_i) + \psi'(X_i)^3)\}]$$

$$+ o(\frac{1}{n})$$

$$= -\frac{t^2}{2} I - \frac{t^2}{6\sqrt{n}}(3J+2K) - \frac{t^4}{24n}(4L+3M+12N+3H) \quad .$$

Similarly we obtain

$$V_{\theta_1}(T_n)=t^2I+\frac{t^3}{\sqrt{n}}(J+K)+\frac{t^4}{12n}(4L+3M+18N+6H-3I^2)+o(\frac{1}{n}) \ ,$$

$$\gamma_{\theta_1}(T_n)= -\frac{t^3}{\sqrt{n}} K - \frac{t^4}{2n}(3N+2H-3I^2) + o(\frac{1}{n})$$

$$\delta_{\theta_1}(T_n) = \frac{t^4}{n}(H-3I^2) + o(\frac{1}{n}) \ .$$

Put $V_n=\left\{ T_n - E_{\theta_1}(T_n) \right\} / (t\sqrt{I}\ )$.

Denote that

$$V_{\theta_1}(V_n) = \frac{1}{t^2I}V_{\theta_1}(T_n)$$

$$= 1 + \frac{1}{\sqrt{n}}\beta_1 + \frac{1}{n}\beta_2 + o(\frac{1}{n}) \ ,$$

$$\gamma_{\theta_1}(V_n) = \frac{1}{t^3I\sqrt{I}}\ \gamma_{\theta_1}(T_n)$$

$$= \frac{1}{\sqrt{n}}\gamma_1 + \frac{1}{n}\gamma_2 + o(\frac{1}{n}) \ ,$$

$$\delta_{\theta_1}(V_n) = \frac{1}{t^4I}\ \delta_{\theta_1}(T_n)$$

$$= \frac{1}{n}\delta + o(\frac{1}{n}) \ .$$

Now assuming that the distribution of the likelihood ratio V_n is absolutely continuous with respect to the Lebesgue measure, then the Edgeworth expansion can be applied and we have

$$P_{\theta_1,n}\left\{ V_n \leqq a \right\}$$

$$= \Phi(a) - \phi(a)\left\{ \frac{1}{2}(\frac{\beta_1}{\sqrt{n}} + \frac{\beta_2}{n})a + \frac{1}{6}(\frac{\gamma_1}{\sqrt{n}} + \frac{\gamma_2}{n})(a^2-1) \right.$$

$$+ (\frac{\delta}{24n} + \frac{\beta_1^2}{8n})(a^3-3a) + \frac{\beta_1\gamma_1}{12n}(a^4+6a^2+3)$$

$$\left. + \frac{\gamma_1^2}{72n}(a^5-10a^3-15a) \right\} + o(\frac{1}{n}) \ .$$

Choose a such that

$$P_{\theta_1,n}\left\{ V_n \leqq a \right\} = \frac{1}{2} + o(\frac{1}{n}) \ .$$

Since

$$\Phi(a) = \frac{1}{2} + a\,\phi(a) - \frac{1}{2}a^3\,\phi'(a) + \dots ,$$

it follows that

$$a = \frac{\beta_1}{2\sqrt{n}}\,a - \left(\frac{\gamma_1}{6\sqrt{n}} + \frac{\gamma_2}{n}\right) + \frac{1}{4\,n}\,\beta_1\gamma_1 + o\left(\frac{1}{n}\right)$$

$$= -\frac{\gamma_1}{6\sqrt{n}} - \frac{\gamma_2}{n} + \frac{\beta_1\gamma_1}{6\,n} + o\left(\frac{1}{n}\right) .$$

If for $\theta = \theta_0$ we denote

$$E_{\theta_1}(V_n) - E_{\theta_0}(V_n) - a = -t\sqrt{I} + \frac{1}{\sqrt{n}}\alpha_1' + \frac{1}{n}\alpha_2' + o\left(\frac{1}{n}\right) ,$$

$$V_{\theta_0}(V_n) = \frac{1}{\sqrt{n}}\beta_1' + \frac{1}{n}\beta_2' + o\left(\frac{1}{n}\right) ,$$

$$\gamma_{\theta_0}(V_n) = \frac{1}{\sqrt{n}}\gamma_1' + \frac{1}{n}\gamma_2' + o\left(\frac{1}{n}\right) ,$$

$$\delta_{\theta_0}(V_n) = \frac{1}{n}\delta_1' + o\left(\frac{1}{n}\right) ,$$

then we have

(4.2.1) $$P_{\theta_0, n}\left\{V_n \geqq a\right\}$$

$$= 1 - P_{\theta_0, n}\left\{V_n < a\right\}$$

$$= \Phi(t') + \phi(t')\Big\{\frac{1}{\sqrt{n}}\alpha_1' + \frac{1}{n}\alpha_2' - \left(\frac{\beta_1'}{2\sqrt{n}} + \frac{\beta_2'}{2\,n} + \frac{\alpha_1'^2}{2\,n}\right)t'$$

$$+ \left(\frac{\gamma_1'}{6\sqrt{n}} + \frac{\gamma_2'}{2\,n} + \frac{\alpha_1'\beta_1'}{2\,n}\right)(t'^2 - 1)$$

$$- \left(\frac{\delta_1'}{24n} + \frac{\alpha_1'\gamma_1'}{6n} + \frac{\beta_1'^2}{8n}\right)(t'^3 - 3t')$$

$$+ \frac{\beta_1'\gamma_1'}{12n}(t'^4 - 6t'^2 + 3)$$

$$- \frac{\gamma_1'^2}{72n}(t'^5 - 10t'^3 + 15t')\Big\} + o\left(\frac{1}{n}\right) ,$$

where $t' = t\sqrt{I}$.

Substituting in (4.2.1)

$$\alpha_1' = \frac{t'^2}{2I^{3/2}}(2J+K) - \frac{K}{6I^{3/2}} ,$$

$$\alpha_2' = \frac{t'^3}{12I^2}(4L+3M+6N+3I^2) + \frac{t'}{6I^3}K(J+K) ,$$

$$\beta_1' = \frac{t'J}{I^{3/2}} ,$$

$$\beta_2' = \frac{t'^2}{4I^2}(M-I^2) + \frac{t'^2}{3I^2} L ,$$

$$\gamma_1' = - \frac{K}{I^{3/2}} ,$$

$$\gamma_2' = - \frac{3t'}{2I^2} (N+I^2) ,$$

$$\delta = \frac{1}{I}(H-3I^2) ,$$

we obtain

$$P_{\theta_0,n}\left\{ V_n \geqq a \right\}$$

$$= \Phi (t')+\phi(t')[\frac{t'^2}{6\sqrt{n}\, I^{3/2}} (3J+2K) - \frac{t'^5}{72I^{3/2}} (3J+2K)^2$$

$$- \frac{t'^3}{72nI^3} \left\{ (3J+2K)^2-3I(4L+3M+6N-H+6I^2) \right\}$$

$$+ \frac{t^3}{72nI^3} \left\{ 2K^2+9I(2N+H-I^2) \right\}]$$

$$+ o(\frac{1}{n}) .$$

Hence it follows that for $t > 0$

$$P_{\theta,n} \left\{ \sqrt{nI} (\hat{\theta}_n - \theta) \leqq t \right\}$$

$$\leqq \Phi(t)- \phi(t)\left\{ \frac{\beta_3''}{6\sqrt{n}} + \frac{\beta_3''}{6\sqrt{n}} (t^2-1) + \frac{\beta_3''^2}{72n}(t^5-10t^3+15t) \right.$$

$$+ (\frac{\beta_4''}{24n} + \frac{\beta_3''^2}{36n})(t^3-3t)$$

$$\left. + (\frac{\beta_2''}{2n} + \frac{\beta_3''^2}{72n})t \right\} + o(\frac{1}{n}) ,$$

where $\beta_3''=-(3J+2K)/I^{3/2}$,

$$\beta_4''= \frac{1}{I^3} \left\{ 3(3J+2K)^2 - I(4L+3M+6N-H+6I^2) \right\} ,$$

$$\beta_2'' = \frac{1}{36I^3}\left\{17(3J+2K)^2 - 2K^2 - 9I(4L+3M+8N+5I^2)\right\} .$$

Further the moment of the third order asymptotically efficient estimator must be as follows :

(4.2.2) $$E_\theta(\hat{\theta}_n) = \theta + \frac{\beta_3''}{6\sqrt{n}\, I^{3/2}} + o\left(\frac{1}{n\sqrt{n}}\right) ,$$

(4.2.3) $$V_\theta(\sqrt{n}\,\hat{\theta}_n) = I + \frac{\beta_2''}{nI} + o\left(\frac{1}{n}\right) ,$$

(4.2.4) $$\gamma_\theta(\sqrt{n}\,\hat{\theta}_n) = \frac{\beta_3''}{\sqrt{n}\, I^{3/2}} + o\left(\frac{1}{n}\right) ,$$

(4.2.5) $$\delta_\theta(\sqrt{n}\,\hat{\theta}_n) = \frac{\beta_4''}{nI^2} + o\left(\frac{1}{n}\right) :$$

Next we shall consider the example of the exponential distribution with the density (4.1.2). Since

$$\hat{\theta}_n = \left(1+\frac{1}{3n}\right)\bar{X} ,$$

it follows that

(4.2.6) $$E_\theta(\hat{\theta}_n) = \theta + \frac{\theta}{3n} ,$$

(4.2.7) $$V_\theta(\sqrt{n}\,\hat{\theta}_n) = \left(1+\frac{2}{3n}\right)\theta^2 ,$$

(4.2.8) $$\gamma_\theta(\sqrt{n}\,\hat{\theta}_n) = \frac{2}{n}\,\theta^3 + o\left(\frac{1}{n}\right) ,$$

(4.2.9) $$\delta_\theta(\sqrt{n}\,\hat{\theta}_n) = \frac{6}{n}\,\theta^4 + o\left(\frac{1}{n}\right) .$$

Since $I=1/\theta^2$, $J=-2/\theta^3$, $K=2/\theta^3$, $L=6/\theta^4$, $M=5/\theta^4$, $N=-5/\theta^4$ and $H=9/\theta^4$, it follows from (4.2.2) ~ (4.2.5) that $\beta_3''=2$, $\beta_4''=6$ and $\beta_2''=2/3$. Then the moments of (4.2.2) ~ (4.2.5) are equal to those of (4.2.6) ~ (4.2.9), respectively. This fact should hold because $\bar{X}$ is a sufficient statistic.

We next consider the maximum likelihood estimator $\hat{\theta}_{ML}$.

Putting $T_n = \sqrt{n}(\hat{\theta}_{ML} - \theta)$, we have

$$Z_1 + (-1 + \frac{1}{\sqrt{n}} Z_2) T_n + \frac{1}{2\sqrt{n}} \left\{ -(3J+K) + \frac{1}{\sqrt{n}} Z_3 \right\} T^2 - \frac{4L+3M+6N+H}{6n} T_n^3 = o_p(\frac{1}{n}) ,$$

where $Z_3 = \frac{1}{\sqrt{n}} \sum_{i=1}^{n} \left\{ \psi^{(3)}(X_i) + 3J + K \right\}$.

We also obtain

$$T_n = \frac{1}{I} Z_1 + \frac{1}{\sqrt{n}\, I^2} (Z_1 Z_2 - \frac{3J+K}{2I} Z_1^2)$$

$$+ \frac{1}{nI^3} \left\{ Z_1 Z_2^2 + \frac{1}{2} Z_1^2 Z_3 - \frac{3(3J+K)}{2I} Z_1^2 Z_2 \right.$$

$$\left. + \frac{(3J+K)^2}{2I^2} Z_1^3 - \frac{4L+3M+6N+H}{6I} Z_1^3 \right\} + o_p(\frac{1}{n}) .$$

Since

$$E(Z_1^2 Z_2) = \frac{1}{\sqrt{n}} (N+I^2) + o(\frac{1}{\sqrt{n}}) ,$$

$$E(Z_1^3 Z_2) = 3\, IJ + o(1) ,$$

$$E(Z_1^4 Z_2^2) = 3\,(M-I^2) + 12J^2 + o(1) , \quad \text{etc.,}$$

it follows that

$$E(T_n) = - \frac{J+K}{2\sqrt{n}\, I^2} + o(\frac{1}{n}) ,$$

$$V(T_n) = \frac{1}{I} + \frac{7J^2+14JK+5K^2}{2nI^4} - \frac{L+4N+H+I^2}{nI^3} + o(\frac{1}{n}) ,$$

$$\gamma(T_n) = - \frac{3J+2K}{nI} + o(\frac{1}{n}) ,$$

$$\delta(T_n) = \frac{12(J+K)(2J+K)}{nI^5} - \frac{4L+12N+3H+3I^2}{nI^4} + o(\frac{1}{n}) .$$

We define a modified maximum likelihood estimator $\hat{\theta}^*_{ML}$ as follows :

$$\hat{\theta}^*_{ML} = \hat{\theta}_{ML} + \frac{K(\hat{\theta}_{ML})}{6nI^2(\hat{\theta}_{ML})} .$$

Then $\hat{\theta}^*_{ML}$ is third order AMU, that is,

$$P_{\theta,n} \left\{ \hat{\theta}^*_{ML} \leqq \theta \right\} = \frac{1}{2} + o(\frac{1}{n}) .$$

Further it follows that

$$nV(\hat{\theta}^*_{ML})=n(1+\frac{1}{6n}\frac{\partial}{\partial\theta}\frac{K(\theta)}{I(\theta)^2})^2V(\hat{\theta}_{ML})$$

$$= V(T_n)+\frac{(3N+H)I-2K(2J+K)}{3nI^4}+o(\frac{1}{n}) ,$$

where

$$\frac{\partial}{\partial\theta}K(\theta) = \frac{\partial}{\partial\theta}\int\{\psi'(x)\}^3 f(x)dx$$

$$= 3\int\psi''(x)\,\psi'(x)^2 f(x)dx + \int\{\psi'(x)\}^4 f(x)dx$$

$$= 3N+H ,$$

$$\frac{\partial}{\partial\theta}I(\theta) = 2J+K .$$

In general the maximum likelihood estimator is seen not to be third order asymptotically efficient.

It is enough to consider a special case, and we take the Cauchy distribution as an example, which has the following density function $f(x,\theta)$:

$$f(x,\theta) = \frac{1}{\pi(1+(x-\theta)^2)} .$$

Let $X_1, X_2, \dots, X_n, \dots$ be independent identically distributed random variables with above density.

Since

$$\frac{\partial}{\partial\theta}\log f(x,0) = -\frac{\partial}{\partial x}\log f(x,0) = \frac{2x}{1+x^2} ,$$

$$\frac{\partial^2}{\partial\theta^2}\log f(x,0) = \frac{\partial^2}{\partial x^2}\log f(x,0) = -\frac{2(1-x^2)}{(1+x^2)^2} ,$$

$$\frac{\partial^3}{\partial\theta^3}\log f(x,0) = -\frac{\partial^3}{\partial x^3}\log f(x,0) = -\frac{4x(3-x^2)}{(1+x^2)^3} ,$$

it follows that $I=1/2$, $J=K=0$, $L=-3/4$, $M=7/8$, $N=-1/8$ and $H=3/8$.

Since the density is symmetric, the maximum likelihood estimator $\hat{\theta}_{ML}$ is third order AMU. The second and fourth cumulants are given as follows :

(4.2.10) $$V(\sqrt{n}\,\hat{\theta}_{ML}) = 2+\frac{5}{n} + o(\frac{1}{n}) ,$$

(4.2.11) $$\gamma(\sqrt{n}\,\hat{\theta}_{ML}) = \frac{66}{n} + o(\frac{1}{n}) .$$

On the other hand since the values of the bounds of the asymptotic distribution are given by

$$\beta_2'' = 1/4 , \quad \beta_4'' = 0 ,$$

they are larger than the above moments (4.2.10) and (4.2.11). Hence $\hat{\theta}_{ML}$ is not third order asymptotically efficient.

4.3. Second order asymptotic efficiency of estimators when X_i's are not i.i.d.

Let X_i's be independent but not identically distributed random variables with density functions $f_i(x_i, \theta)$. Let $\hat{\theta}_n$ be a second order AMU estimator. When $c_n(\hat{\theta}_n - \theta)$ has an asymptotic distribution, we obtain the bound of the distributions of $\hat{\theta}_n$.

Let θ_0 be arbitrary but fixed in Θ. Consider the most powerful test of the problem of testing hypothesis $\theta = \theta_1 (= \theta_0 + tc_n^{-1} a,\ a > 0)$ against alternative $\theta = \theta_0$. Then the test statistic is given by

(4.3.1) $$T_n = \sum_{i=1}^{n} \log \frac{f_i(X_i, \theta_1)}{f_i(X_i, \theta_0)}$$

$$= -\sum_{i=1}^{n} \left\{ \frac{\partial}{\partial\theta} \log f_i(X_i, \theta_0) \right\} \cdot (c_n^{-1} a)$$

$$- \frac{1}{2} \sum_{i=1}^{n} \left\{ \frac{\partial^2}{\partial\theta^2} \log f_i(X_i, \theta_0) \right\} (c_n^{-1} a)^2$$

$$- \frac{1}{6} \sum_{i=1}^{n} \left\{ \frac{\partial^3}{\partial\theta^3} \log f_i(X_i, \theta_0) \right\} (c_n^{-1} a)^3$$

$$+ \sum_{i=1}^{n} R_{in}$$

Put

$$I_i = E_{\theta_0}[\{ \frac{\partial}{\partial\theta} \log f_i(X_i, \theta_0) \}^2] = - E_{\theta_0}[\frac{\partial^2}{\partial\theta^2} \log f_i(X_i, \theta)] ,$$

It follows that the variance of the first term of the right-hand side of (4.3.1) is given by $(c_n^{-1}\ a)^2 \sum_{i=1}^{n} I_i$. When $c_n = \sqrt{\tilde{I}_n}$, where $I_n = \sum_{i=1}^{n} I_i$, it's variance converges to some finite value.
Suppose that the following "Lindeberg type" condition holds : for every $\varepsilon > 0$,

$$\lim_{n\to\infty} \frac{1}{\tilde{I}} \sum_{i=1}^{n} E_{\theta_0}\left[\left\{ \frac{\partial}{\partial\theta}\log f_i(X_i,\theta_0) \right\}^2 \chi_{in}(\varepsilon) \right] = 0 \quad ,$$

where

$$\chi_{in}(\varepsilon) = \begin{cases} 1, \text{ if } \frac{\partial}{\partial\theta}\log f_i(X_i,\theta_0) > \varepsilon\sqrt{\tilde{I}_n} \quad , \\ 0, \text{ otherwise} . \end{cases}$$

Then the first term of the right-hand side of (4.3.1) is asymptotically normal with mean 0 and variance a^2. A necessary condition for this to hold is that $\lim_{n\to\infty} \max_{1\leq i\leq n} I_i/\tilde{I}_n = 0$.
If a similar condition as above holds for the second derivative : for every $\varepsilon > 0$,

$$\lim_{n\to\infty} \frac{1}{\tilde{I}_n} \sum_{i=1}^{n} E_{\theta_0}\left[\left\{ -\frac{\partial^2}{\partial\theta^2}\log f_i(X_i,\theta_0) \right\} \chi^*_{in}(\varepsilon) \right] = 0 \ ,$$

where

$$\chi^*_{in}(\varepsilon) = \begin{cases} 1, \quad \text{if } -\frac{\partial^2}{\partial\theta^2}\log f_i(X_i,\theta_0) > \varepsilon\ \tilde{I}_n \quad , \\ 0, \quad \text{otherwise} \ , \end{cases}$$

then the second term of the right-hand side of (4.3.1) converges in probability to $a^2/2$.

If under alternative $\theta = \theta_0$, similar regularity conditions as above hold, then the test statistic T_n is asymptotically normal with mean $-a^2/2$ and variance a^2. Put $c_n = \sqrt{\tilde{I}_n}$. For $a > 0$, the supremum of the power functions is given by $\Phi(a)$. Hence <u>if $\sqrt{\tilde{I}_n}(\hat{\theta}_n - \theta)$ is asymptotically normal with mean 0 and variance 1, then $\hat{\theta}_n$ is an asymptotically efficient estimator.</u>

Let $\hat{\theta}_{ML}$ be a maximum likelihood estimator.
Then

$$\begin{aligned} 0 &= \sum_{i=1}^{n} \left\{ \frac{\partial}{\partial\theta}\log f_i(X_i,\theta) \right\} (\hat{\theta}_{ML} - \theta) \\ &+ \sum_{i=1}^{n} \left\{ \frac{\partial^2}{\partial\theta^2}\log f_i(X_i,\theta) \right\} (\hat{\theta}_{ML} - \theta)^2 \\ &+ \sum_{i=1}^{n} R'_{in} \end{aligned}$$

Under some regularity conditions $\sqrt{\tilde{I}_n}(\hat{\theta}_{ML}-\theta)$ is asymptotically normal with mean 0 and variance 1. Hence the maximum likelihood estimator $\hat{\theta}_{ML}$ is asymptotically efficient.

In order to consider the higher order terms of the bound of the power functions, we must decide the orders of

$$\tilde{K}_n=\sum_{i=1}^{n}K_i=\sum_{i=1}^{n}E_{\theta_0}\left[\left\{\frac{\partial}{\partial\theta}\log f_i(X_i,\theta_0)\right\}^3\right]$$

and

$$\tilde{J}_n=\sum_{i=1}^{n}J_i=\sum_{i=1}^{n}E_{\theta_0}\left[\left\{\frac{\partial^2}{\partial\theta^2}\log f_i(X_i,\theta_0)\right\}\left\{\frac{\partial}{\partial\theta}\log f_i(X_i,\theta_0)\right\}\right].$$

The asymptotic moments of the test statistics T_n are formally obtained as follows :

$$E_{\theta_0}(T_n) = \frac{a^2}{2} - \frac{3\tilde{J}_n+\tilde{K}_n}{6\tilde{I}_n^{3/2}}a^3 + o\left(\frac{3\tilde{J}_n+\tilde{K}_n}{\tilde{I}_n^{3/2}}\right)$$

$$V_{\theta_0}(T_n) = a^2 + \frac{\tilde{J}_n}{\tilde{I}_n^{3/2}}a^3 + o\left(\frac{\tilde{J}_n}{\tilde{I}_n^{3/2}}\right),$$

$$E_{\theta_0}[\{T_n - E_{\theta_0}(Y_n)\}^3] = -\frac{\tilde{K}_n}{\tilde{I}_n^{3/2}}a^3 + o\left(\frac{\tilde{K}_n}{\tilde{I}_n^{3/2}}\right).$$

Similarly we have for $\theta = \theta_1$

$$E_{\theta_1}(T_n) = -\frac{a^2}{2} - \frac{3\tilde{J}_n+2\tilde{K}_n}{6\tilde{I}_n^{3/2}}a^3 + o\left(\frac{3\tilde{J}_n+2\tilde{K}_n}{\tilde{I}_n^{3/2}}\right),$$

$$V_{\theta_1}(T_n) = a^2 + \frac{\tilde{J}_n+\tilde{K}_n}{\tilde{I}_n^{3/2}}a^3 + o\left(\frac{\tilde{J}_n+\tilde{K}_n}{\tilde{I}_n^{3/2}}\right),$$

$$E_{\theta_1}[\{T_n - E_{\theta_1}(T_n)\}^3] = -(\tilde{K}_n/\tilde{I}_n^{\frac{3}{2}})a^3 + o\left(\frac{\tilde{K}_n}{\tilde{I}_n^{3/2}}\right).$$

If $\tilde{J}_n$ and $\tilde{K}_n$ have the same order which is less than $\tilde{I}_n^{\frac{3}{2}}$ and the residual terms are negligible, we can apply Edgeworth expansion to the asymptotic distribution of the statistic T_n. Put $d_n^{-1}=\tilde{K}_n/\tilde{I}_n^{\frac{3}{2}}$.

If

$$\lim_{n\to\infty} d_n \left| P_{\theta,n}\{\hat{\theta}_n \leqq \theta\} - 1/2 \right| = 0$$

uniformly in any neighborhood of θ_0 , then for $a > 0$

$$(4.3.2) \quad \overline{\lim_{n\to\infty}} d_n[P_{\theta_0,n} \{\sqrt{\tilde{I}_n}(\hat{\theta}_n - \theta) \leq a\} - \Phi(a) - \frac{3\tilde{J}_n + 2\tilde{K}_n}{6\tilde{I}_n^{3/2}} a^2 \phi(a)] \leq 0$$

For $a < 0$, a similar inequality as above holds.

If the equality "=" of (4.3.2) holds, then $\hat{\theta}_n$ may be called second order asymptotically efficient. The necessary condition for second order asymptotic efficiency is that $\hat{\theta}_n$ has the following asymptotic moments :

$$E_\theta\{\sqrt{\tilde{I}_n}(\hat{\theta}_n - \theta)\} = -\frac{3\tilde{J}_n + 2\tilde{K}_n}{3\tilde{I}_n^{3/2}} + o(\frac{1}{d_n}) \ ;$$

$$V_\theta\{\sqrt{\tilde{I}_n}(\hat{\theta}_n - \theta)\} = 1 + o(\frac{1}{d_n}) \ ;$$

$$E_\theta\{\sqrt{\tilde{I}_n}(\hat{\theta}_n - \theta)\}^3 = -\frac{3\tilde{J}_n + 2\tilde{K}_n}{\tilde{I}_n^{3/2}} + o(\frac{1}{d_n}) \ .$$

If $\hat{\theta}_n$ has a smooth asymptotic distribution and $V_\theta\{\sqrt{\tilde{I}_n}(\hat{\theta}_n - \theta)\} = 1+o(1/d_n)$, then the adjusted estimator $\hat{\theta}_n^*$ which is second order asymptotically median unbiased is second order asymptotically efficient.

Let

$$\hat{\theta}_{ML}^* = \hat{\theta}_{ML} + \frac{\tilde{K}_n(\hat{\theta}_{ML})}{6n\tilde{I}_n(\hat{\theta}_{ML})^2} \ .$$

Under regularity conditions $\hat{\theta}_{ML}^*$ is second order asymptotically efficient.

In the above we did not give any explicit relation between n and $\tilde{I}_n$ and d_n, hence no direct relation between $c_n = \sqrt{\tilde{I}_n}$ and d_n. However if the following generally hold :

$$\tilde{I}_n / n \to \bar{I} \quad (n \to \infty) \ ,$$

$$\tilde{J}_n / n \to \bar{J} \quad (n \to \infty) \ ,$$

$$\tilde{K}_n / n \to \bar{K} \quad (n \to \infty) \ ,$$

then it follows that $d_n = \sqrt{n}$. Since the second term of the asymptotic distribution has order of $1/\sqrt{n}$, a similar results as the i.i.d. case holds. But these facts do not always hold. For the purpose of illustration we consider the following examples.

Let $X_i = \theta z_i + U_i$, i=1,2,..., where z_i's are constants and U_i's are independent identically distributed random variables with a density f(u) w.r.t. a Lebesgue measure. Putting

$$I = \int \left\{ \frac{d}{du} \log f(u) \right\}^2 f(u)du = \int \frac{\{f'(u)\}^2}{f(u)} du ,$$

$$J = \int \left\{ \frac{d^2}{du^2} \log f(u) \right\} \left\{ \frac{d}{du} \log f(u) \right\} f(u)du$$

$$= \int \left[\frac{f''(u)f'(u)}{f(u)} - \frac{\{f'(u)\}^3}{\{f(u)\}^2} \right] du ,$$

$$K = \int \left\{ \frac{d}{du} \log f(u) \right\}^3 f(u)du$$

$$= \int \frac{\{f'(u)\}^3}{\{f(u)\}^2} du ,$$

we have for each i

$$I_i = z_i^2 I , \quad J_i = -z_i^3 J , \quad K_i = -z_i^3 K .$$

Then it follows that

$$\tilde{I}_n = \left(\sum_{i=1}^n z_i^2\right) I , \quad \tilde{J} = -\left(\sum_{i=1}^n z_i^3\right) J , \quad \tilde{K} = -\left(\sum_{i=1}^n z_i^3\right) K .$$

By Lindeberg's condition we obtain

$$\max_{1 \leq i \leq n} z_i^2 \Big/ \sum_{i=1}^n z_i^2 \to 0 \quad (n \to \infty) .$$

Since

$$\frac{\left|\sum_{i=1}^n z_i^3\right|}{\left(\sum_{i=1}^n z_i^2\right)^{3/2}} \leqq \frac{\max |z_i|}{\left(\sum_{i=1}^n z_i^2\right)^{1/2}} \to 0 \quad (n \to \infty) ,$$

it follows that

$$\tilde{J}_n \big/ \tilde{I}_n^{3/2} \to 0 \quad (n \to \infty) ,$$

$$\tilde{K}_n \big/ \tilde{I}_n^{3/2} \to 0 \quad (n \to \infty) .$$

Then we can take $d_n = (\sum_{i=1}^{n} z_i^2)^{3/2} / (\sum_{i=1}^{n} z_i^3)$.

If $\sum_{i=1}^{n} z_i^2/n \to m_2 (n \to \infty)$ and $\sum_{i=1}^{n} z_i^3/n \to m_3 (\neq 0)$, then it follows that $d_n = O(\sqrt{n})$. Then the second term of the asymptotic distribution has order of 1.

But letting $z_i = i$, we have for sufficiently large n

$$\tilde{I}_n \fallingdotseq n(n+1)(2n+1)I/6 \sim n^3 I/3 .$$

Hence <u>if $\sqrt{nI^3/3}\ (\hat{\theta}_n - \theta)$ is asymptotically normal with mean 0 and variance 1, then $\hat{\theta}_n$ is an asymptotically efficient estimator.</u>

Since for sufficiently large n

$$\tilde{J}_n = -\frac{n^2(n+1)^2}{4} J \sim -\frac{n^4}{4} J ,$$

$$\tilde{K}_n = -\frac{n^2(n+1)^2}{4} K \sim -\frac{n^4}{4} K ,$$

it follows that $d_n = O(\sqrt{n})$. Hence in this case the second term of the asymptotic distribution also has order $1/\sqrt{n}$.

If $z_i \to 0\ (i \to \infty)$, then it follows by Lindeberg's conditions that the following must hold :

$$\tilde{I}_n = (\sum_{i=1}^{n} z_i^2) I \to \infty \quad (n \to \infty) .$$

But it is possible that $\sum_{i=1}^{n} z_i^2 \to \infty\ (n \to \infty)$ and $\sum_{i=1}^{n} z_i^3 < \infty$.

Since $d_n = O(\tilde{I}_n^{\frac{1}{2}})$, letting $z_i = 1/\sqrt{i}$ we obtain $\tilde{I}_n = O(\log n)$.

Hence <u>$\sqrt{\log n}\ (\hat{\theta}_n - \theta)$ is asymptotically normal and the second term of the asymptotic distribution has order of $(\log n)^{1/2}$</u> .

Let X_i's be independently distributed random variables with the following density :

$$f(x_i, \theta) = \frac{1}{\theta^{p_i} \Gamma(p_i)} x^{p_i - 1} e^{-x/\theta} , \quad x > 0 ,$$

where p_i's are known positive integers and θ is an unknown parameter.

Since

$$\log f_i(X_i, \theta) = (p_i - 1) \log X_i - X_i/\theta - p_i \log \theta - \log \Gamma(p_i) ,$$

$$\frac{\partial}{\partial\theta}\log f_i(X_i,\theta) = \frac{X_i}{\theta^2} - \frac{p_i}{\theta} ,$$

$$\frac{\partial^2}{\partial\theta^2}\log f_i(X_i,\theta) = -\frac{2X_i}{\theta^3} + \frac{p_i}{\theta^2} ,$$

it follows that

$$I_i = E_\theta\left(\frac{X_i}{\theta^2} - \frac{p_i}{\theta}\right)^2 = \frac{p_i}{\theta^2} ,$$

$$J_i = -E_\theta\left(\frac{X_i}{\theta^2} - \frac{p_i}{\theta}\right)\left(\frac{2X_i}{\theta^3} - \frac{p_i}{\theta^2}\right) = -\frac{2p_i}{\theta^3} ,$$

$$K_i = E_\theta\left(\frac{X_i}{\theta^2} - \frac{p_i}{\theta}\right)^3 = \frac{2p_i}{\theta^3} .$$

Then we have

$$\tilde{I}_n = \sum_{i=1}^n p_i/\theta^2 , \quad \tilde{J}_n = -2\sum_{i=1}^n p_i/\theta^3 , \quad \tilde{K}_n = 2\sum_{i=1}^n p_i/\theta^3 .$$

Since $d_n = (\sum_{i=1}^n p_i)^{1/2}$, the asymptotic distribution of the second order asymptotically efficient estimator $\hat{\theta}_n$ is given by

$$P_{\theta,n}\left\{\sqrt{\sum_{i=1}^n p_i}\,(\hat{\theta}_n - \theta)/\theta < a\right\} = \Phi(a) + \frac{1}{3\sqrt{\sum_{i=1}^n p_i}}\, a^2\phi(a) + o\left(\frac{1}{\sqrt{\sum_{i=1}^n p_i}}\right) .$$

The maximum likelihood estimator $\hat{\theta}_{ML}$ of θ is also given by $\sum_{i=1}^n X_i / \sum_{i=1}^n p_i$. Let

$$\hat{\theta}^*_{ML} = \left(1 + \frac{1}{3\sum_{i=1}^n p_i}\right)\hat{\theta}_{ML} .$$

Then $\hat{\theta}^*_{ML}$ is a second order asymptotically efficient estimator.

Since in this case $\sum_{i=1}^n X_i$ is a sufficient statistic, the second order AMU estimator based on it must be second order asymptotically efficient. Indeed it is easily shown that the asymptotic distribution of $\hat{\theta}^*_{ML}$ agrees with the above bound of the asymptotic distributions.

4.4. Second order asymptotic efficiency of estimators in multiparameter cases

In the previous section second order asymptotic efficiency of estimators is discussed for one-dimensional case. In this section we adopt a similar approach to multiparameter cases and obtain a straight-forward generalization of the one-parameter case.

Let $X_1, X_2, \ldots, X_n, \ldots$ be a sequence of independent identically distributed random variables with the density $f(x, \theta, \xi)$, where θ is a real valued parameter and ξ is a real (vector) valued parameter. We assume that ξ is a nuisance parameter. We shall define an estimator of θ to be k-th order asymptotically efficient if its k-th order asymptotic distribution attains the bound of the k-th order asymptotic distributions of k-th order asymptotically median unbiased (AMU) estimators of θ. We shall obtain the bound of the second order asymptotic distributions of second order AMU estimators and show that a modified maximum likelihood estimator is second order asymptotically efficient.

Let $\mathcal{X}$ be an abstract sample space whose generic point is denoted by x, $\mathcal{B}$ a σ-field of subsets of $\mathcal{X}$ and let Θ and Ξ be parameter spaces which are assumed to be open sets in R^1 and R^p respectively. (We denote by R^p a Euclidean p-space with a norm $\|\cdot\|$.) We assume that $\xi=(\xi_1, \ldots, \xi_p)$ $(\in \Xi)$ is a nuisance parameter. We consider a sequence of classes of probability measures $\{P_{\theta,\xi,i} : (\theta,\xi) \in \Theta \times \Xi\}$ $(i=1,2,\ldots)$ each defined over $(\mathcal{X},\mathcal{B})$. We shall denote by $(\mathcal{X}^{(n)},\mathcal{B}^{(n)})$ the n-fold direct products of $(\mathcal{X},\mathcal{B})$ and the corresponding product measures by $P^{(n)}_{\theta,\xi} = P_{\theta,\xi,1} \times \cdots \times P_{\theta,\xi,n}$. Let $\hat{\theta}_n$ be a $\{c_n\}$-consistent estimator.

Definition 4.4.1. For each k=1,2,..., $\hat{\theta}_n$ is k-th order asymptotically median unbiased (or k-th order AMU) estimator if for any $\vartheta_0=(\theta_0,\xi_0) \in \Theta \times \Xi$, there exists a positive number δ such that

$$\lim_{n\to\infty} \sup_{\vartheta \in \Theta\times\Xi : \|\vartheta_0-\vartheta\|<\delta} c_n^{k-1} \left| P^{(n)}_{\vartheta,\xi}\{\hat{\theta}_n \leq \theta\} - \frac{1}{2} \right| = 0 ;$$

$$\lim_{n\to\infty} \sup_{\vartheta \in \Theta\times\Xi : \|\vartheta_0-\vartheta\|<\delta} c_n^{k-1} \left| P^{(n)}_{\vartheta,\xi}\{\hat{\theta}_n \geq \theta\} - \frac{1}{2} \right| = 0 .$$

Definition 4.4.2. Suppose that $\hat{\theta}_n$ is k-th order asymptotically median unbiased, $G_0(t,\theta,\xi)+c_n^{-1}G_1(t,\theta,\xi)+\dots+c_n^{-(k-1)}G_{k-1}(t,\theta,\xi)$ is defined to be its k-th order asymptotic distribution if

$$\lim_{n\to\infty} c_n^{k-1}\left| P^{(n)}_{\theta,\xi}\{c_n(\hat{\theta}_n-\theta)\leqq t\} - G_0(t,\theta,\xi) - c_n^{-1} G_1(t,\theta,\xi) - \dots - c_n^{-(k-1)}G_{k-1}(t,\theta,\xi)\right| = 0 .$$

Consider the problem of testing hypothesis $H^+:\theta=\theta_0+tn^{-1/2}(t>0)$, against $K:\theta=\theta_0,\ \xi=\xi$.

Put $\Phi_{1/2}=\left\{\{\phi_n\} : E^{(n)}_{\theta_0+tc_n^{-1},\xi}(\phi_n)=1/2+o(c_n^{-(k-1)}),\ 0\leqq\phi_n(\tilde{x}_n)\leqq 1\right.$ for all $\left.\tilde{x}_n\in\mathcal{X}^{(n)}\ (n=1,2,\dots)\right\}$. Putting $A_{\hat{\theta}_n,\theta}=\left\{c_n(\hat{\theta}_n-\theta)\leqq t\right\}$, we have

$$\lim_{n\to\infty} P^{(n)}_{\theta_0+tc_n^{-1},\xi}(A_{\hat{\theta}_n,\theta_0})=\lim_{n\to\infty} P^{(n)}_{\theta_0+tc_n^{-1},\xi}\left\{\hat{\theta}_n\leqq\theta_0+tc_n^{-1}\right\} = \frac{1}{2}.$$

Hence it is seen that a sequence $\left\{\chi_{A_{\hat{\theta}_n,\theta_0}}\right\}$ of the indicators of $A_{\hat{\theta}_n,\theta_0}$ $(n=1,2,\dots)$ belongs to $\Phi_{1/2}$. If for $\xi=\xi_0$ and for each $t>0$

$$\sup_{\{\phi_n\}\in\Phi_{1/2}} \overline{\lim_{n\to\infty}}\, c_n^{k-1}\Big\{E^{(n)}_{\theta_0,\xi_0}(\phi_n)-H_0^+(t,\theta_0,\xi_0)-c_n^{-1}H_1^+(t,\theta_0,\xi_0) - \dots - c_n^{-(k-1)}H^+_{k-1}(t,\theta_0,\xi_0)\Big\} = 0$$

then we have

$$G_0(t,\theta_0,\xi_0)\leqq H_0^+(t,\theta_0,\xi_0)\quad ;$$

and for any positive integer $j(\leqq k)$ if

$$G_i(t,\theta_0,\xi_0) = H_i^+(t,\theta_0,\xi_0)\ (i=0,\dots,j-1)\ ,$$

then $G_j(t,\theta_0,\xi_0)\leqq H_j^+(t,\theta_0,\xi_0)$.

Consider next the problem of testing hypothesis $H^-:\theta=\theta_0+tn^{-\frac{1}{2}}$ $(t<0)$, against $K:\theta=\theta_0,\ \xi=\xi_0$. If for $\xi=\xi_0$ and for each $t<0$

$$\inf_{\{\phi_n\}\in\Phi_{1/2}} \underline{\lim_{n\to\infty}}\, c_n^{k-1}\Big\{E^{(n)}_{\theta_0,\xi_0}(\phi_n)-H_0^-(t,\theta_0,\xi_0)-c_n^{-1}H_1^-(t,\theta_0,\xi_0) - \dots\ c_n^{-(k-1)}H^-_{k-1}(t,\theta_0,\xi_0)\Big\} = 0 ,$$

then we have

$$G_0(t,\theta_0,\xi_0) \geqq H_0^-(t,\theta_0,\xi_0) \ ;$$

and for any positive integer $j(\leqq k)$ if

$$G_i(t,\theta_0,\xi_0)=H_i^-(t,\theta_0,\xi_0) \quad (i=0,\ldots,j-1) \ ,$$

then

$$G_j(t,\theta_0,\xi_0) \geqq H_j^-(t,\theta_0,\xi_0) \ .$$

Definition 4.4.3. Suppose that $\hat{\theta}_n$ is k-th order asymptotically median unbiased. It is called k-th order asymptotically efficient if for each $\theta \in \Theta$ and each $\xi \in \Xi$

$$G_i(t,\theta,\xi) = \begin{cases} H_i^+(t,\theta,\xi) & \text{for } t>0 \ , \\ H_i^-(t,\theta,\xi) & \text{for } t<0 \ , \end{cases}$$

$i=0, \ldots , k-1$.

In the subsequent discussion of this section we shall deal only with the second order asymptotic efficiency, in the case when $c_n=n^{1/2}$ but the line of discussion can be extended to more general cases.

We assume Ξ is an open set of R^1 because in the subsequent discussion there arises no substantial difference between R^1 and R^p $(p \geqq 2)$. Let $X_1, X_2, \ldots , X_n, \ldots$ be a sequence of i.i.d. random variables having a density function $f(x,\theta,\xi)$.

First we shall obtain the bound of the power functions. Consider the problem of testing hypothesis $H^+: \theta = \theta_0+tn^{-\frac{1}{2}}$ $(t>0)$ $\xi = \xi_0+un^{-\frac{1}{2}}$, against $K: \theta = \theta_0, \xi = \xi_0$, where u is an arbitrary but fixed constant. Putting $\theta_1 = \theta_0+tn^{-\frac{1}{2}}$ and $\xi_1=\xi_0+un^{-\frac{1}{2}}$, we define Z_{ni} as follows:

$$Z_{ni} = \log \frac{f(X_i,\theta_0,\xi_0)}{f(X_i,\theta_1,\xi_1)} \ .$$

The most powerful test is given by the following rejection region

$$\sum_{i=1}^{n} Z_{ni} > c \ ,$$

where c is some constant.

For each $i=1,2, \ldots$, expand Z_{ni} as follows :

$$Z_{ni}=-\frac{\partial}{\partial\theta}\log f(X_i,\theta_0,\xi_0)(tn^{-\frac{1}{2}})-\frac{\partial}{\partial\xi}\log f(X_i,\theta_0,\xi_0)(un^{-\frac{1}{2}})$$

$$-\frac{1}{2}\frac{\partial^2}{\partial\theta^2}\log f(X_i,\theta_0,\xi_0)(t^2n^{-1})-\frac{1}{2}\frac{\partial^2}{\partial\xi^2}\log f(X_i,\theta_0,\xi_0)(u^2n^{-1})$$

$$-\frac{\partial^2}{\partial\theta\partial\xi}\log f(X_i,\theta_0,\xi_0)(tun^{-1})$$

$$-\frac{1}{6}\frac{\partial^3}{\partial\theta^3}\log f(X_i,\theta_0,\xi_0)(t^3n^{-\frac{3}{2}})-\frac{1}{6}\frac{\partial^3}{\partial\xi^3}\log f(X_i,\theta_0,\xi_0)(u^3n^{-\frac{3}{2}})$$

$$-\frac{1}{2}\frac{\partial^3}{\partial\theta^2\partial\xi}\log f(X_i,\theta_0,\xi_0)(t^2un^{-\frac{3}{2}})$$

$$-\frac{1}{2}\frac{\partial^3}{\partial\theta\partial\xi^2}\log f(X_i,\theta_0,\xi_0)(tu^2n^{-\frac{3}{2}})+o_p(n^{-\frac{3}{2}}) .$$

Hence $\sum_{i=1}^{n} Z_{ni}$ is asymptotically normal. If $\theta=\theta_0$ and $\xi=\xi_0$, then the asymptotic mean μ_0 and the asymptoic variance σ_0^2 of $\sum_{i=1}^{n} Z_{ni}$ are given by

$$\mu_0=\frac{1}{2}(I_{00}t^2+I_{11}u^2+2I_{01}tu)+o(n^{-\frac{1}{2}}) ;$$

$$=I_{00}t^2+I_{11}u^2+2I_{01}tu+o(n^{-\frac{1}{2}}) ,$$

where

$$I_{00}=E_{\theta_0,\xi_0}[\{\frac{\partial}{\partial\theta}\log f(X_i,\theta_0,\xi_0)\}^2]=-E_{\theta_0,\xi_0}[\frac{\partial^2}{\partial\theta^2}\log f(X_i,\theta_0,\xi_0)] :$$

$$I_{01}=E_{\theta_0,\xi_0}[\frac{\partial}{\partial\theta}\log f(X_i,\theta_0,\xi_0)\frac{\partial}{\partial\xi}\log f(X_i,\theta_0,\xi_0)]$$

$$=-E_{\theta_0,\xi_0}[\frac{\partial^2}{\partial\theta\partial\xi}\log f(X_i,\theta_0,\xi_0)] ;$$

$$I_{11}=E_{\theta_0,\xi_0}[\{\frac{\partial}{\partial\xi}\log f(X_i,\theta_0,\xi_0)\}^2]=-E_{\theta_0,\xi_0}[\frac{\partial^2}{\partial\xi^2}\log f(X_i,\theta_0,\xi_0)] .$$

If $\theta=\theta_1$, $\xi=\xi_1$, then it follows that

$$E_{\theta_1,\xi_1}[\frac{\partial}{\partial\theta}\log f(X_i,\theta_0,\xi_0)]$$

$$=E_{\theta_0,\xi_0}[\frac{\partial}{\partial\theta}\log f(X_i,\theta_0,\xi_0)\frac{f(X_i,\theta_1,\xi_1)}{f(X_i,\theta_0,\xi_0)}]$$

$$=E_{\theta_0,\xi_0}[\{\frac{\partial}{\partial\theta}\log f(X_i,\theta_0,\xi_0)\}^2](tn^{-1/2})$$

$$+ E_{\theta_0,\xi_0}[\frac{\partial}{\partial\theta}\log f(X_i,\theta_0,\xi_0)\frac{\partial}{\partial\xi}\log f(X_i,\theta_0,\xi_0)](un^{-1/2})+o(n^{-1/2}) ;$$

$$E_{\theta_1,\xi_1}[\frac{\partial}{\partial\xi}\log f(X_i,\theta_0,\xi_0)]$$

$$= E_{\theta_0,\xi_0}[\frac{\partial}{\partial\theta}\log f(X_i,\theta_0,\xi_0)\frac{\partial}{\partial\xi}\log f(X_i,\theta_0,\xi_0)](tn^{-1/2})$$

$$+ E_{\theta_0,\xi_0}[\left\{\frac{\partial}{\partial\xi}\log f(X_i,\theta_0,\xi_0)\right\}^2] (un^{-1/2}) + o(n^{-1/2}) .$$

For $\theta=\theta_1$ and $\xi=\xi_1$, the asymptotic mean μ_1 and the asymptotic variance σ_1^2 of $\sum_{i=1}^{n} Z_{ni}$ are given by

$$\mu_1 = -\frac{1}{2}(I_{00}t^2 + I_{11}u^2 + 2I_{10}tu) + o(n^{-1/2}) ;$$

$$\sigma_1^2 = \sigma_0^2 + o(n^{-1/2}) .$$

Hence the asymptotic power of the most powerful test of $\sum_{i=1}^{n} Z_{ni}$ is obtained as follows :

$$(4.4.1) \quad \Phi(\frac{\mu_0+\mu_1}{\sigma_0}) = \Phi(\sqrt{I_{00}t^2+I_{11}u^2+2I_{01}tu})$$

where $\Phi(x) = \int_{-\infty}^{x} \frac{1}{\sqrt{2\pi}} e^{-u^2/2}du$.

Since u can take arbitrary values, then the power function of the tests of (composite) hypothesis is not larger than the infimum of (4.4.1) with respect to u. A u minimizing $I_{00}t^2 + I_{11}u^2 + 2I_{01}tu$ is given by $u_0=-(I_{01}/I_{11})t$, and it follows that

$$\Phi(\sqrt{I_{00}t^2+I_{11}u_0^2+2I_{01}tu_0}) = \Phi(\sqrt{I^*}\, t) ,$$

where $I^*=I_{00}-(I_{01}^2/I_{11})$.

In order to obtain the expansion of $\sum_{i=1}^{n} Z_{ni}$ up to the order of n we put

$$J_{000}=E_{\theta_0,\xi_0}[\frac{\partial^2}{\partial\theta^2}\log f(X_i,\theta_0,\xi_0)\frac{\partial}{\partial\theta}\log f(X_i,\theta_0,\xi_0)] ;$$

$$J_{001}=E_{\theta_0,\xi_0}[\frac{\partial^2}{\partial\theta^2}\log f(X_i,\theta_0,\xi_0)\frac{\partial}{\partial\xi}\log f(X_i,\theta_0,\xi_0)] ;$$

$$J_{010}=E_{\theta_0,\xi_0}[\frac{\partial^2}{\partial\theta\partial\xi}\log f(X_i,\theta_0,\xi_0)\frac{\partial}{\partial\theta}\log f(X_i,\theta_0,\xi_0)] ;$$

$$J_{011}=E_{\theta_0,\xi_0}[\frac{\partial^2}{\partial\theta\partial\xi}\log f(X_i,\theta_0,\xi_0)\frac{\partial}{\partial\xi}\log f(X_i,\theta_0,\xi_0)] ;$$

$J_{110}=E_{\theta_0,\xi_0}[\frac{\partial^2}{\partial\xi^2}\log f(X_i,\theta_0,\xi_0)\frac{\partial}{\partial\theta}\log f(X_i,\theta_0,\xi_0)]$;

$J_{111}=E_{\theta_0,\xi_0}[\frac{\partial^2}{\partial\xi^2}\log f(X_i,\theta_0,\xi_0)\frac{\partial}{\partial\xi}\log f(X_i,\theta_0,\xi_0)]$;

$K_{000}=E_{\theta_0,\xi_0}[\{\frac{\partial}{\partial\theta}\log f(X_i,\theta_0,\xi_0)\}^3]$;

$K_{001}=E_{\theta_0,\xi_0}[\{\frac{\partial}{\partial\theta}\log f(X_i,\theta_0,\xi_0)\}^2\{\frac{\partial}{\partial\xi}\log f(X_i,\theta_0,\xi_0)\}]$;

$K_{011}=E_{\theta_0,\xi_0}[\{\frac{\partial}{\partial\theta}\log f(X_i,\theta_0,\xi_0)\}\{\frac{\partial}{\partial\xi}\log f(X_i,\theta_0,\xi_0)\}^2]$;

$K_{111}=E_{\theta_0,\xi_0}[\{\frac{\partial}{\partial\xi}\log f(X_i,\theta_0,\xi_0)\}^3]$.

Under suitable regularity conditions the following hold :

$E_{\theta_0,\xi_0}[\frac{\partial^3}{\partial\theta^3}\log f(X_i,\theta_0,\xi_0)] = -3J_{000} - K_{000}$;

$E_{\theta_0,\xi_0}[\frac{\partial^3}{\partial\theta^2\partial\xi}\log f(X_i,\theta_0,\xi_0)] = -J_{001}-2J_{010}-K_{001}$;

$E_{\theta_0,\xi_0}[\frac{\partial^3}{\partial\theta\partial\xi^2}\log f(X_i,\theta_0,\xi_0)] = -J_{110}-2J_{011}-K_{011}$;

$E_{\theta_0,\xi_0}[\frac{\partial^3}{\partial\xi^3}\log f(X_i,\theta_0,\xi_0)] = -3J_{111}-K_{111}$.

If $\theta=\theta_0$ and $\xi=\xi_0$, then the asymptotic mean μ_0 of $\sum_{i=1}^{n} Z_{ni}$ up to order $n^{-1/2}$ is given by :

$$\mu_0= \frac{1}{2}(I_{00}t^2+I_{11}u^2+2I_{01}tu)$$
$$+\frac{1}{2}n^{-\frac{1}{2}}\{(J_{000}t^3+(J_{001}+2J_{010})t^2u+(J_{110}+2J_{011})tu^2+J_{111}u^3\}$$
$$+\frac{1}{6}n^{-1/2}(K_{000}t^3+3K_{001}t^2u+3K_{011}t^2u+K_{111}u^3)+o(n^{-1/2}) .$$

When $u=-(I_{01}/I_{11})t$, μ_0 can be written as

$$\mu_0 = \frac{1}{2}I^*t^2 + \frac{1}{6}n^{-1/2}(3J^*+K^*)t^3 + o(n^{-1/2}) .$$

Similarly the asymptotic variance σ_0^2 and the asymptotic third moment γ_0 of $\sum_{i=1}^{n} Z_{ni}$ are given by

$$\sigma_0^2 = I^*t^2 + J^*t^3n^{-1/2} + o(n^{-1/2}) \quad ;$$

$$\gamma_0 = E_{\theta_0,\xi_0}\left(\sum_{i=1}^{n} Z_{ni} - \mu_0\right)^3 = K^*t^3n^{-1/2} + o(n^{-1/2}) \ .$$

If $\theta = \theta_1$ and $\xi = \xi_1$, then it follows that

$$\log \frac{f(X_i,\theta_1,\xi_1)}{f(X_i,\theta_0,\xi_0)}$$

$$= 1 + \frac{\partial}{\partial\theta}\log f(X_i,\theta_0,\xi_0)(tn^{-1/2}) + \frac{\partial}{\partial\xi}\log f(X_i,\theta_0,\xi_0)(un^{-1/2})$$

$$+ \frac{1}{2}\left\{\frac{\partial^2}{\partial\theta^2}\log f(X_i,\theta_0,\xi_0)(t^2n^{-1}) + \frac{\partial^2}{\partial\xi^2}\log f(X_i,\theta_0,\xi_0)(u^2n^{-1})\right.$$

$$\left. + 2\frac{\partial^2}{\partial\theta\partial\xi}\log f(X_i,\theta_0,\xi_0)(tun^{-1})\right\}$$

$$+ \frac{1}{2}\left[\left\{\frac{\partial}{\partial\theta}\log f(X_i,\theta_0,\xi_0)\right\}^2 t^2n^{-1} + \left\{\frac{\partial}{\partial\xi}\log f(X_i,\theta_0,\xi_0)\right\}^2 u^2n^{-1}\right.$$

$$+ 2\left\{\frac{\partial}{\partial\theta}\log f(X_i,\theta_0,\xi_0)\right\}\left\{\frac{\partial}{\partial\xi}\log f(X_i,\theta_0,\xi_0)\right\}(tun^{-1}) + o_p(n^{-1}) \ .$$

Hence we have

$$\mu_1 = \mu_0 + (I_{00}t^2 + I_{11}u^2 + 2I_{01}tu)$$

$$- n^{-1/2}\left\{J_{000}t^3 + (J_{001}+2J_{010})t^2u + (J_{110}+2J_{011})tu^2 + J_{111}u^3\right\}$$

$$- \frac{1}{2}n^{-1/2}(K_{000}t^3 + 3K_{001}t^2u + 3K_{011}tu^2 + K_{111}u^3) + o(n^{-1/2}) \ .$$

If $u = -(I_{01}/I_{00})t$, then it follows that

$$\mu_1 = -\frac{1}{2}I^*t^2 - \frac{1}{6}n^{-1/2}(3J^*+2K^*)t^3 + o(n^{-1/2}) \ .$$

Similarly we obtain

$$\sigma_1^2 = \sigma_0^2 + n^{-1/2}(K_{000}t^3 + 3K_{001}t^2u + 3K_{011}tu^2 + K_{111}u^3) + o(n^{-1/2})$$

$$= I^*t^2 + n^{-1/2}(J^*+K^*)t^3 + o(n^{-1/2}) \ ;$$

$$\gamma_1 = E_{\theta_1,\xi_1}\left(\sum_{i=1}^{n} Z_{ni} - \mu_1\right)^3 = \gamma_0 + o(n^{-1/2}) = -K^*n^{-1/2}t^3 + o(n^{-1/2}) \ .$$

If we choose a c such that

$$P_{\theta_1,\xi_1}^{(n)}\left\{\sum_{i=1}^{n} Z_{ni} > c\right\} = \frac{1}{2} + o(n^{-1/2}) \ ,$$

then $c(=c_1)$ is given by

$$c_1 = -\frac{I^*}{2}t^2 - \frac{3J^*+2K^*}{6}n^{-1/2} + \frac{K^*}{6I^*}n^{-1}$$

Hence we have

(4.4.2) $P^{(n)}_{\theta_0,\xi_0}\left\{\sum_{i=1}^{n} Z_{ni} > c_1\right\} = \Phi(t\sqrt{I^*}) + \phi(t\sqrt{I^*})\frac{3J^*+2K^*}{6\sqrt{I^*}}t^2 n^{-\frac{1}{2}} + o(n^{-\frac{1}{2}})$,

where $\Phi(x) = \int_{-\infty}^{x}\phi(u)du$ with $\phi(u) = \frac{1}{\sqrt{2\pi}}e^{-u^2/2}$.

The bound of the power functions of the problem of testing hypothesis is obtained from (4.4.2), that is,

(4.4.3) $H_0^+(t, \theta_0, \xi_0) = \Phi(t\sqrt{I^*})$

(4.4.4) $H_1^+(t, \theta_0, \xi_0) = \phi(t\sqrt{I^*})\frac{(3J^*+2K^*)}{6\sqrt{I^*}}t^2$

By a similar way as the case $t > 0$ it may be shown that in the case $t < 0$ the bound of the power functions is also given by (4.4.3) and (4.4.4). Thus we have established the following :

Theorem 4.4.1. The bound for the distributions of second order asymptotically median unbiased estimators is given by

$$G_0(t, \theta_0, \xi_0) = \Phi(t\sqrt{I^*}) \quad ;$$

$$G_1(t, \theta_0, \xi_0) = \phi(t\sqrt{I^*})\left\{\frac{(3J^*+2K^*)}{6\sqrt{I^*}}t^2\right\},$$

where I^*, J^* and K^* are defined in context.

Let $\hat{\theta}(=\{\hat{\theta}_n\})$ and $\hat{\xi}(=\{\hat{\xi}_n\})$ be maximum likelihood estimators of θ and ξ, respectively.

Since

$$\sum_{i=1}^{n}\frac{\partial}{\partial\theta}\log f(X_i, \hat{\theta}, \hat{\xi}) = 0 \ ;$$

$$\sum_{i=1}^{n}\frac{\partial}{\partial\xi}\log f(X_i, \hat{\theta}, \hat{\xi}) = 0 \ ,$$

expanding them in the neighborhood of $\theta = \theta_0$ and $\xi = \xi_0$ respectively, we get

$$\sum_{i=1}^{n}\frac{\partial}{\partial\theta}\log f(X_i,\theta_0,\xi_0)$$
$$+\sum_{i=1}^{n}\frac{\partial^2}{\partial\theta^2}\log f(X_i,\theta_0,\xi_0)(\hat{\theta}-\theta_0)+\sum_{i=1}^{n}\frac{\partial^2}{\partial\theta\partial\xi}\log f(X_i,\theta_0,\xi_0)(\hat{\xi}-\xi_0)$$
$$+\frac{1}{2}\left\{\sum_{i=1}^{n}\frac{\partial^3}{\partial\theta^3}\log f(X_i,\theta_0,\xi_0)(\hat{\theta}-\theta_0)^2+\sum_{i=1}^{n}\frac{\partial^3}{\partial\theta\partial\xi^2}\log f(X_i,\theta_0,\xi_0)(\hat{\xi}-\xi_0)^2\right.$$
$$\left.+2\sum_{i=1}^{n}\frac{\partial^3}{\partial\theta^2\partial\xi}\log f(X_i,\theta_0,\xi_0)(\hat{\theta}-\theta_0)(\hat{\xi}-\xi_0)\right\}$$
$$=o_p(1)$$

$$\sum_{i=1}^{n}\frac{\partial}{\partial\xi}\log f(X_i,\theta_0,\xi_0)$$
$$+\sum_{i=1}^{n}\frac{\partial^2}{\partial\theta\partial\xi}\log f(X_i,\theta_0,\xi_0)(\hat{\theta}-\theta_0)+\sum_{i=1}^{n}\frac{\partial^2}{\partial\xi^2}\log f(X_i,\theta_0,\xi_0)(\hat{\xi}-\xi_0)$$
$$+\frac{1}{2}\left\{\sum_{i=1}^{n}\frac{\partial^3}{\partial\theta^2\partial\xi}\log f(X_i,\theta_0,\xi_0)(\hat{\theta}-\theta_0)^2+\sum_{i=1}^{n}\frac{\partial^3}{\partial\xi^3}\log f(X_i,\theta_0,\xi_0)(\hat{\xi}-\xi_0)^2\right.$$
$$\left.+2\sum_{i=1}^{n}\frac{\partial^3}{\partial\theta\partial\xi^2}\log f(X_i,\theta_0,\xi_0)(\hat{\theta}-\theta_0)(\hat{\xi}-\xi_0)\right\}$$
$$=o_p(1)$$

Putting

$$Z_0 = n^{-1/2}\sum_{i=1}^{n}\frac{\partial}{\partial\theta}\log f(X_i,\theta_0,\xi_0)\ ;$$
$$Z_1 = n^{-1/2}\sum_{i=1}^{n}\frac{\partial}{\partial\xi}\log f(X_i,\theta_0,\xi_0)\ ;$$
$$Z_{00} = n^{-1/2}\sum_{i=1}^{n}\left\{\frac{\partial^2}{\partial\theta^2}\log f(X_i,\theta_0,\xi_0)+I_{00}\right\}\ ;$$
$$Z_{11} = n^{-1/2}\sum_{i=1}^{n}\left\{\frac{\partial^2}{\partial\xi^2}\log f(X_i,\theta_0,\xi_0)+I_{11}\right\}\ ;$$
$$Z_{01} = n^{-1/2}\sum_{i=1}^{n}\left\{\frac{\partial^2}{\partial\theta\partial\xi}\log f(X_i,\theta_0,\xi_0)+I_{01}\right\}\ ,$$

we see that they are asymptotically normal with mean 0.

Since

$$0=Z_0+(-I_{00}+n^{-1/2}Z_{00})n^{1/2}(\hat{\theta}-\theta_0)+(-I_{01}+n^{-1/2}Z_{01})n^{1/2}(\hat{\xi}-\xi_0)$$
$$+\frac{1}{2}(-3J_{000}-K_{000})n^{1/2}(\hat{\theta}-\theta_0)^2+\frac{1}{2}(-J_{110}-2J_{011}-K_{011})n^{1/2}(\hat{\xi}-\xi_0)^2$$
$$+(-J_{001}-2J_{010}-K_{001})n^{1/2}(\hat{\theta}-\theta_0)(\hat{\xi}-\xi_0)+o_p(n^{-1/2})\ ;$$

$$0=Z_1+(-I_{01}+n^{-1/2}Z_{01})n^{1/2}(\hat{\theta}-\theta_0)+(-I_{11}+n^{-1/2}Z_{11})n^{1/2}(\hat{\xi}-\xi_0)$$
$$+\frac{1}{2}(-J_{001}-2J_{010}-K_{001})n^{1/2}(\hat{\theta}-\theta_0)^2+\frac{1}{2}(-3J_{111}-K_{111})n^{1/2}(\hat{\xi}-\xi_0)^2$$
$$+(-J_{110}-2J_{011}-K_{011})n^{1/2}(\hat{\theta}-\theta_0)(\hat{\xi}-\xi_0)+o_p(n^{-1/2}) ,$$

it follows that

$$(I_{00}I_{11}-I_{01}{}^2)n^{1/2}(\hat{\theta}-\theta_0)$$
$$= I_{11}Z_0-I_{01}Z_1+n^{-1/2}(I_{11}Z_{00}-I_{01}Z_{01})n^{1/2}(\hat{\theta}-\theta_0)$$
$$+n^{-1/2}(I_{11}Z_{01}-I_{01}Z_{11})n^{1/2}(\hat{\xi}-\xi_0)$$
$$-\frac{1}{2}n^{-1/2}(3J_{000}I_{11}-J_{001}I_{01}-2J_{010}I_{01}+K_{000}I_{11}-K_{001}I_{01})\left\{n^{1/2}(\hat{\theta}-\theta_0)\right\}^2$$
$$-\frac{1}{2}n^{-1/2}(J_{110}I_{11}+2J_{011}I_{11}-3J_{111}I_{01}+K_{011}I_{11}-K_{111}I_{01})\left\{n^{1/2}(\hat{\xi}-\xi_0)\right\}^2$$
$$-n^{-1/2}(J_{001}I_{11}-J_{110}I_{01}+2J_{010}I_{11}-2J_{011}I_{01}+K_{001}I_{11}$$
$$-K_{011}I_{01})\left\{n^{1/2}(\hat{\theta}-\theta_0)(\hat{\xi}-\xi_0)\right\}+o_p(n^{-1/2})$$

Putting $D=I_{00}I_{11}-I_{01}{}^2$, $a=-I_{01}/I_{11}$, $b=-I_{01}/I_{00}$,

we have

$$n^{1/2}(\hat{\theta}-\theta_0)$$
$$=\frac{I_{11}}{D}(Z_0+aZ_1)+\frac{I_{11}{}^2}{D^2}n^{-1/2}(Z_{00}+aZ_{01})(Z_0+aZ_1)$$
$$+\frac{I_{11}I_{00}}{D^2}n^{-1/2}(Z_{01}+aZ_{11})(Z_1+bZ_0)$$
$$-\frac{I_{11}{}^3}{2D^3}n^{-1/2}(3J_{000}+J_{001}a+2J_{010}a+K_{000}+K_{001}a)(Z_0+aZ_1)^2$$
$$-\frac{I_{11}I_{00}{}^2}{2D^3}n^{-1/2}(J_{110}+2J_{011}+3J_{111}a+K_{011}+K_{111}a)(Z_1+bZ_0)^2$$
$$-\frac{I_{11}{}^2I_{00}}{D^3}n^{-1/2}(J_{001}+J_{110}a+2J_{010}+2J_{011}a+K_{001}+K_{001}a)(Z_0+aZ_1)(Z_1+bZ_0)$$
$$+o_p(n^{-1/2}) .$$

Since

$$E_{\theta_0,\xi_0}(Z_0{}^2)=I_{00},\ E_{\theta_0,\xi_0}(Z_1{}^2)=I_{11},\ E_{\theta_0,\xi_0}(Z_0Z_1)=I_{01} ,$$

$$E_{\theta_0,\xi_0}(Z_{00}Z_0) = J_{000} \ , \quad E_{\theta_0,\xi_0}(Z_{00}Z_1) = J_{001} \ ,$$

$$E_{\theta_0,\xi_0}(Z_{01}Z_0) = J_{010} \ , \quad E_{\theta_0,\xi_0}(Z_{01}Z_1) = J_{011} \ ,$$

$$E_{\theta_0,\xi_0}(Z_{11}Z_0) = J_{110} \ , \quad E_{\theta_0,\xi_0}(Z_{11}Z_1) = J_{111} \ ,$$

it follows that

$$E_{\theta_0,\xi_0}[\, n^{1/2}(\hat{\theta} - \theta_0)] = -\frac{J^*+K^*}{2I^{*2}} n^{-1/2} + o(n^{-1/2}) \ ;$$

$$V_{\theta_0,\xi_0}(n^{1/2}(\hat{\theta} - \theta_0)) = \frac{1}{I^*} + o(n^{-1/2}) \ ,$$

$$E_{\theta_0,\xi_0}[\left\{n^{1/2}(\hat{\theta} - \theta_0)\right\}^3] = -\frac{3J^*+2K^*}{I^{*3}} n^{-1/2} + o(n^{-1/2}) \ .$$

If we define $\hat{\theta}^*$ by

$$\hat{\theta}^* = \hat{\theta} - \frac{K^*}{6nI^*}$$

then $\hat{\theta}^*$ is second order AMU and second order asymptotically efficient at $\theta = \theta_0$ and $\xi = \xi_0$. Since K* and I* depend on θ_0 and ξ_0, we may change $\hat{\theta}^*$ into the following form $\hat{\theta}^{**}$:

$$\hat{\theta}^{**} = \hat{\theta} - \frac{K^*(\hat{\theta}, \hat{\xi})}{6nI^*(\hat{\theta}, \hat{\xi})} \ .$$

Thus we have established the following :

<u>Theorem 4.4.2.</u> $\hat{\theta}^{**}$ is second order AMU and second order asymptotically efficient.

For any consistent estimator $\hat{\theta}_n^0$ if the asymptotic variance of $\hat{\theta}_n^0$ is given by

$$V_{\theta,\xi}(n^{1/2}(\hat{\theta}_n^0 - \theta)) = \frac{1}{I^*} + o(n^{-1/2}) \ ,$$

then we can transform $\hat{\theta}_n^0$ into $\hat{\theta}_n^{0*}$ such that the latter is second order AMU. <u>In a similar way as section 4.1 if a second order asymptotic distribution exists, it can be shown that $\hat{\theta}_n^{0*}$ must be second order asymptotic efficient.</u>

We assume that $\hat{\theta}_n^0$ has a second order asymptotic distribution and

$$E_{\theta,\xi}[\, n^{1/2}(\hat{\theta}_n^0 - \theta)\,] = c_1(\theta,\xi)n^{-1/2} + o(n^{-1/2}) \; ;$$

$$V_{\theta,\xi}(\, n^{1/2}(\hat{\theta}_n^0 - \theta)) = \frac{1}{I^*(\theta,\xi)} + o(n^{-1/2})$$

$$E_{\theta,\xi}[\{n^{1/2}(\hat{\theta}_n^0 - \theta)\}^3] = c_3(\theta,\xi)n^{-1/2} + o(n^{-1/2}) \; .$$

Let $\hat{\theta}^0(=\{\hat{\theta}_n^0\})$ and $\hat{\xi}^0$ $(=\{\xi_n^0\})$ be consistent estimators of θ and ξ, respectively. Putting

$$\hat{\theta}_n^{0*} = \hat{\theta}_n - n^{-1}\left\{ c_1(\hat{\theta}^0, \hat{\xi}^0) - \frac{1}{6} I^*(\hat{\theta}^0, \hat{\xi}^0) c_3(\hat{\theta}^0, \hat{\xi}^0) \right\} ,$$

we see that $\hat{\theta}_n^{0*}$ is second order asymptotically efficient.

4.5. Second order asymptotic efficiency in an autoregressive process

We consider second order asymptotic efficiency in a simple autoregressive process as a typical case of non-independent samples. Let $X_i = \theta X_{i-1} + U_i, i=1,2, \ldots$, where $X_0=0$ and $\{ U_i : i = 1,2, \ldots \}$ is a sequence of independent identically distributed real random variables having a density f with mean 0 and variance σ^2 .

Throughout this section 4.5 we assume that $\mathfrak{X} = R^1$ and Ⓗ is an open interval (-1,1) and consider an autoregressive process $\{X_i\}$ given above.

In this section it will be shown that the bound of the second order asymptotic distributions of second order AMU estimators of θ is obtained using the best test statistics and that a modified least squares estimator of θ is second order asymptotically efficient if f is a normal density with mean 0 and variance σ^2.

We assume the following :

(A.4.5.1) f is continuously differentiable four times and $f(u)>0$ for all $u\in R^1$.

Further it is assumed that

$$I=\int_{-\infty}^{\infty}\{\psi'(u)\}^2 f(u)du \ ; \ J=\int_{-\infty}^{\infty}\psi'(u)\psi''(u)f(u)du \ ; \ K=\int_{-\infty}^{\infty}\{\psi'(u)\}^3 f(u)du$$

exist and are well defined.

Let θ_0 be arbitrary but fixed in Ⓗ. Consider the problem of testing hypothesis $H:\theta=\theta_0+(t/\sqrt{n})(t>0)$ against alternative $K:\theta=\theta_0$. Puttint $\theta_1=\theta_0+\Delta$ with $\Delta=t/\sqrt{n}$, we define Z_{ni} as follows :

$$Z_{ni}=\log\frac{f(X_i-\theta_0X_{i-1})}{f(X_i-\theta_1X_{i-1})}$$

If $\theta=\theta_0$, then we have

$$(4.5.1)\qquad Z_{ni}=\log\frac{f(U_i)}{f(U_i-\Delta X_{i-1})}$$

$$=\Delta\psi'(U_i)X_{i-1}-\frac{\Delta^2}{2}\psi''(U_i)X_{i-1}^2+\frac{\Delta^3}{6}\psi'''(U_i)X_{i-1}^3-\frac{\Delta^4}{24}\psi^{(4)}(U_i^*)X_{i-1}^4 ,$$

where $\psi(u)=\log f(u)$ and for each i U_i^* lies between U_i and $U_i-\Delta X_{i-1}$.
If $\theta=\theta_1$, then we have

$$(4.5.2)\qquad Z_{ni}=\log\frac{f(U_i+\Delta X_{i-1})}{f(U_i)}$$

$$=\Delta\psi'(U_i)X_{i-1}+\frac{\Delta^2}{2}\psi''(U_i)X_{i-1}^2+\frac{\Delta^3}{6}\psi'''(U_i)X_{i-1}^3+\frac{\Delta^4}{24}\psi^{(4)}(U_i^{**})X_{i-1}^4 ,$$

where for each i U_i^{**} lies between U_i and $U_i+\Delta X_{i-1}$.

For the subsequence discussion of this section we assume the following :

(A.4.5.2) $d^4\log f(u)/du^4$ $(=\psi^{(4)}(u))$ is a bounded function and $\lim_{u\to\pm\infty}f(u)=\lim_{u\to\pm\infty}f'(u)=\lim_{u\to\pm\infty}f''(u)=0$ and $E[|U_i|^4]<\infty$.

Then we have

$$\int_{-\infty}^{\infty}\psi'''(u)f(u)du=-3\int_{-\infty}^{\infty}\psi''(u)\psi'(u)f(u)du-\int_{-\infty}^{\infty}\{\psi'(u)\}^3f(u)du .$$

Put $\mu_3=\int_{-\infty}^{\infty}u^3f(u)du$.

Since $E(\psi''(U_i))=-I$, $E_\theta(X_{i-1}^2)=\sigma^2 \frac{1-\theta^{2(i-1)}}{1-\theta^2}$ and $E_\theta(X_{i-1}^3)=\mu_3\frac{1-\theta^{3(i-1)}}{1-\theta^3}$,

it follows from (4.5.1) that

$$(4.5.3)\quad E_{\theta_0}(\sum_{i=1}^n Z_{ni}) = \frac{n\Delta^2\sigma^2 I}{2(1-\theta_0^2)} - \frac{\Delta^3 n(3J+K)}{6(1-\theta_0^3)} + o(n\Delta^3) .$$

similarly we have

$$(4.5.4)\quad V_{\theta_0}(\sum_{i=1}^n Z_{ni}) = E_{\theta_0}[\ \{\sum_{i=1}^n (Z_{ni}-E_{\theta_0}(Z_{ni}))\}^2\ ]$$

$$= \frac{n\Delta^2\sigma^2 I}{1-\theta_0^2} - \frac{\Delta^3 nJ\mu_3}{1-\theta_0^3} + o(n\Delta^3)$$

Further we have

$$(4.5.5)\quad E_{\theta_0}[\{\sum_{i=1}^n Z_{ni} - E_{\theta_0}(\sum_{i=1}^n Z_{ni})\}^3] = \frac{\Delta^3 nK\mu_3}{1-\theta_0^3} + o(n\Delta^3)$$

Since $\Delta=t/\sqrt{n}$, it follows from (4.5.3), (4.5.4) and (4.5.5) that

$$(4.5.6)\quad E_{\theta_0}(\sum_{i=1}^n Z_{ni}) = \frac{t^2\sigma^2 I}{2(1-\theta_0^2)} - \frac{t^3\mu_3(3J+K)}{6\sqrt{n}(1-\theta_0^3)} + o(\frac{1}{\sqrt{n}}) ;$$

$$(4.5.7)\quad V_{\theta_0}(\sum_{i=1}^n Z_{ni}) = \frac{t^2\sigma^2 I}{1-\theta_0^2} - \frac{t^3\mu_3 J}{\sqrt{n}(1-\theta_0^3)} + o(\frac{1}{\sqrt{n}}) ;$$

$$(4.5.8)\quad E_{\theta_0}[\{\sum_{i=1}^n Z_{ni} - E_{\theta_0}(\sum_{i=1}^n Z_{ni})\}^3] = \frac{t^3 K\mu_3}{\sqrt{n}(1-\theta_0^3)} + o(\frac{1}{\sqrt{n}}) .$$

Since $\theta_1=\theta_0+(t/\sqrt{n})$, it follows that for sufficiency large n

$$(4.5.9)\quad \frac{1}{1-\theta_1^2} = \frac{1}{1-\theta_0^2} + \frac{2t\theta_0}{\sqrt{n}(1-\theta_0^2)^2} + o(\frac{1}{\sqrt{n}}) ;$$

$$(4.5.10)\quad \frac{1}{1-\theta_1^3} = \frac{1}{1-\theta_0^3} + \frac{3t\theta_0^2}{\sqrt{n}(1-\theta_0^2)^2} + o(\frac{1}{\sqrt{n}}) .$$

In a similar way as the case $\theta=\theta_0$ we have from (4.5.2), (4.5.9) and (4.5.10)

$$(4.5.11)\quad E_{\theta_1}\left(\sum_{i=1}^{n} Z_{ni}\right) = -\frac{t^2\sigma^2 I}{2(1-\theta_0^2)} - \frac{t^3\sigma^2 I\theta_0}{\sqrt{n}(1-\theta_0^2)^2} - \frac{t^3\mu_3(3J+K)}{6\sqrt{n}(1-\theta_0^3)} + o\left(\frac{1}{\sqrt{n}}\right);$$

$$(4.5.12)\quad V_{\theta_1}\left(\sum_{i=1}^{n} Z_{ni}\right) = \frac{t^2\sigma^2 I}{1-\theta_0^2} + \frac{2t^3\sigma^2 I\theta_0}{\sqrt{n}(1-\theta_0^2)^2} + \frac{\mu_3 J}{\sqrt{n}(1-\theta_0^3)} + o\left(\frac{1}{\sqrt{n}}\right);$$

$$(4.5.13)\quad E_{\theta_1}\left[\left\{\sum_{i=1}^{n} Z_{ni} - E_{\theta_1}\left(\sum_{i=1}^{n} Z_{ni}\right)\right\}^3\right] = \frac{t^3 K\mu_3}{\sqrt{n}(1-\theta_0^3)} + o\left(\frac{1}{\sqrt{n}}\right).$$

$$(4.5.14)\quad E_{\theta}\left(\sum_{i=1}^{n} Z_{ni}\right) = \mu_1 + \frac{1}{\sqrt{n}} c_1 + o\left(\frac{1}{\sqrt{n}}\right);$$

$$(4.5.15)\quad V_{\theta}\left(\sum_{i=1}^{n} Z_{ni}\right) = v^2 + \frac{1}{\sqrt{n}} c_2 + o\left(\frac{1}{\sqrt{n}}\right);$$

$$(4.5.16)\quad E_{\theta}\left[\left\{\sum_{i=1}^{n} Z_{ni} - E_{\theta}\left(\sum_{i=1}^{n} Z_{ni}\right)\right\}^3\right] = \frac{1}{\sqrt{n}} c_3 + o\left(\frac{1}{\sqrt{n}}\right),$$

then using Edgeworth expansion(*) we have

$$(4.5.17)\quad P_{\theta,n}\left\{\sum_{i=1}^{n} Z_{ni} \leq a\right\}$$

$$= \Phi\left(\frac{a-\mu}{v}\right) - \phi\left(\frac{a-\mu}{v}\right)\left[\frac{c_1}{v\sqrt{n}} + \frac{c_2}{2v^2\sqrt{n}}\left(\frac{a-\mu}{v}\right) + \frac{c_3}{6v^3\sqrt{n}}\left\{\left(\frac{a-\mu}{v}\right)^2 - 1\right\}\right] + o\left(\frac{1}{\sqrt{n}}\right),$$

where $\Phi(u) = \int_{-\infty}^{u} \phi(x)dx$ with $\phi(x) = \frac{1}{\sqrt{2\pi}} e^{-x^2/2}$.

Next we shall choose a such that

$$P_{\theta_1,n}\left\{\sum_{i=1}^{n} Z_{ni} \leq a\right\} = \frac{1}{2} + o\left(\frac{1}{\sqrt{n}}\right)$$

(*) Since Z_{ni} are not independent identically distributed, the usual form of Edgeworth expansion may not be applicable but in this particular situation we can expand $\sum_{i=1}^{n} Z_{ni}$ in a polynomial in terms of the independent U_i's and then apply the argument of Bhattacharya and Ghosh [13].

For this purpose putting

$$\Phi\left(\frac{a-\mu}{v}\right)-\phi\left(\frac{a-\mu}{v}\right)\left[\frac{c_1}{v\sqrt{n}}+\frac{c_2}{2v^2\sqrt{n}}\left(\frac{a-\mu}{v}\right)+\frac{c_3}{6v^3\sqrt{n}}\left\{\left(\frac{a-\mu}{v}\right)^2-1\right\}\right]$$

$$=\frac{1}{2}+o\left(\frac{1}{\sqrt{n}}\right)$$

$$=\Phi(0)+o\left(\frac{1}{\sqrt{n}}\right) \quad ,$$

we have

$$\frac{a-\mu}{v}=O\left(\frac{1}{\sqrt{n}}\right) \quad .$$

Since

$$\Phi\left(\frac{a-\mu}{v}\right)-\Phi(0)=\phi(\xi)\,\frac{a-\mu}{v}=\phi\left(\frac{a-\mu}{v}\right)\left(\frac{c_1}{v\sqrt{n}}-\frac{c_3}{6v^3\sqrt{n}}\right)+o\left(\frac{1}{\sqrt{n}}\right)$$

where ξ lies between 0 and $(a-\mu)/v$,

we obtain

$$(4.5.18) \qquad a=\mu+\frac{c_1}{\sqrt{n}}-\frac{c_3}{6v^2\sqrt{n}}+o\left(\frac{1}{\sqrt{n}}\right) \quad .$$

From (4.5.12)$\sim$(4.5.17) and (4.5.18) we have

$$(4.5.19) \quad a=-\frac{t^2\sigma^2 I}{2(1-\theta_0^2)}-\frac{t^3\mu_3(3J+K)}{6\sqrt{n}(1-\theta_0^3)}-\frac{t^3\sigma^2 I\theta_0}{\sqrt{n}(1-\theta_0^2)^2}-\frac{t^3K\mu_3(1-\theta_0^2)}{6\sqrt{n}t^2\sigma^2 I(1-\theta_0^3)}+o\left(\frac{1}{\sqrt{n}}\right) \quad .$$

On the other hand since

$$P_{\theta_0,n}\left\{\sum_{i=1}^n Z_{ni}\geqq a\right\}=P_{\theta_0,n}\left\{-\left(\sum_{i=1}^n Z_{ni}-a-\frac{t^2\sigma^2 I}{1-\theta_0^2}\right)\leqq\frac{t^2\sigma^2 I}{1-\theta_0^2}\right\} ,$$

Putting

$$W_n=-\left(\sum_{i=1}^n Z_{ni}-a-\frac{t^2\sigma^2 I}{1-\theta_0^2}\right)$$

we have from (4.5.6), (4.5.7), (4.5.8) and (4.5.19)

$$(4.5.20) \qquad E_{\theta_0}(W_n)=-\frac{t^3\sigma^2 I\theta_0}{\sqrt{n}(1-\theta_0^2)^2}-\frac{tK\mu_3(1-\theta_0^2)}{6\sqrt{n}\sigma^2 I(1-\theta_0^3)}+o\left(\frac{1}{\sqrt{n}}\right) \quad ;$$

$$(4.5.21) \qquad V_{\theta_0}(W_n)=\frac{t^2\sigma^2 I}{1-\theta_0^2}-\frac{t^3\mu_3 J}{\sqrt{n}(1-\theta_0^3)}+o\left(\frac{1}{\sqrt{n}}\right) \quad ;$$

$$(4.5.22) \qquad E_{\theta_0}\left[\left\{W_n-E_{\theta_0}(W_n)\right\}^3\right]=-\frac{t^3K\mu_3}{\sqrt{n}(1-\theta_0^3)}+o\left(\frac{1}{\sqrt{n}}\right) \quad .$$

From (4.5.17), (4.5.20), (4.5.21) and (4.5.22) we obtain

$$P_{\theta_0,n}\left(\left\{W_n \leqq \frac{t^2\sigma^2 I}{1-\theta_0^2}\right\}\right)$$

$$=\Phi\left(\frac{t\sigma\sqrt{I}}{\sqrt{1-\theta_0^2}}\right)+\phi\left(\frac{t\sigma\sqrt{I}}{\sqrt{1-\theta_0^2}}\right)\left\{\frac{t^2\sigma\sqrt{I}\,\theta_0}{\sqrt{n}(1-\theta_0^2)^{3/2}}+\frac{t^2\sqrt{1-\theta_0^3}\,\mu_3}{6\sigma\sqrt{I}\sqrt{n}(1-\theta_0^3)}(3J+K)\right\}+o\left(\frac{1}{\sqrt{n}}\right)$$

If $\hat{\theta}_n$ is second order asymptotically median unbiased, then it follows by the fundamental lemma of Neyman and Pearson that for all $t>0$

$$\overline{\lim_{n\to\infty}}\sqrt{n}\left[P_{\theta_0,n}\left(\left\{\sqrt{n}(\hat{\theta}_n-\theta_0)\leqq t\right\}\right)-\Phi\left(\frac{t\sigma\sqrt{I}}{\sqrt{1-\theta_0^2}}\right)\right.$$
$$\left.-\phi\left(\frac{t\sigma\sqrt{I}}{\sqrt{1-\theta_0^2}}\right)\left\{\frac{t^2\sigma\sqrt{I}\,\theta_0}{\sqrt{n}(1-\theta_0^2)^{3/2}}+\frac{t^2\sqrt{1-\theta_0^2}\,\mu_3}{6\sigma\sqrt{I}\sqrt{n}(1-\theta_0^3)}(3J+K)\right\}\right]\leqq 0\ .$$

In a similar way as the case $t>0$, we have by the fundamental lemma of Neyman and Pearson

$$\underline{\lim_{n\to\infty}}\sqrt{n}\left[P_{\theta_0,n}\left(\left\{\sqrt{n}(\hat{\theta}_n-\theta_0)\leqq t\right\}\right)-\Phi\left(\frac{t\sigma\sqrt{I}}{\sqrt{1-\theta_0^2}}\right)\right.$$
$$\left.-\phi\left(\frac{t\sigma\sqrt{I}}{\sqrt{1-\theta_0^2}}\right)\left\{\frac{t^2\sigma\sqrt{I}\,\theta_0}{\sqrt{n}(1-\theta_0^2)^{3/2}}+\frac{t^2\sqrt{1-\theta_0^2}\,\mu_3}{6\sigma\sqrt{I}\sqrt{n}(1-\theta_0^3)}(3J+K)\right\}\right]\geqq 0\ .$$

for all $t<0$.

Since θ_0 is arbitrary, we have now established the following :

<u>Theorem 4.5.1.</u> Under Assumptions (A.4.5.1) and (A.4.5.2), the bound of the second order asymptotic distribution of second order AMU estimators is given by

$$\Phi\left(\frac{t\sigma\sqrt{I}}{\sqrt{1-\theta^2}}\right)+\phi\left(\frac{t\sigma\sqrt{I}}{\sqrt{1-\theta^2}}\right)\left\{\frac{t^2\sigma\sqrt{I}\,\theta}{\sqrt{n}(1-\theta^2)^{3/2}}+\frac{t^2\sqrt{1-\theta^2}\,\mu_3}{6(1-\theta^3)\sqrt{In}}(3J+K)\right\}.$$

The least squares estimator $\hat{\theta}_{LS}$ of θ is $(\sum_{i=2}^{n}X_iX_{i-1})/\sum_{i=2}^{n}X_{i-1}^2$. Under Assumptions(A 4.5.1) and(A4.5.2) $\hat{\theta}_{LS}$ is a $\{\sqrt{n}\}$ -consistent estimator. We assume that f is a normal density. Then Assumptions (A.4.5.1) and(A.4.5.2) hold.

Since
$$\sqrt{n}\,(\hat{\theta}_{LS}-\theta) = \frac{(1/\sqrt{n})\sum_{i=2}^{n} U_i X_{i-1}}{(1/n)\sum_{i=2}^{n} X_{i-1}^2}$$
it follows that $\sqrt{n}(\hat{\theta}_{LS}-\theta) \leqq t$ if and only if
$$\frac{1}{\sqrt{n}} \sum_{i=2}^{n} U_i X_{i-1} - \frac{t}{n} \sum_{i=2}^{n} X_{i-1}^2 \leqq 0 .$$
Put
$$Z_n = \frac{1}{\sqrt{n}} \sum_{i=2}^{n} U_i X_{i-1} - \frac{t}{n} \sum_{i=2}^{n} X_{i-1}^2 .$$
Since
$$E_\theta(Z_n) = -\frac{t\sigma^2}{1-\theta^2} + o\left(\frac{1}{\sqrt{n}}\right) ;$$
$$V_\theta(Z_n) = \frac{\sigma^4}{1-\theta^2} - \frac{4\sigma^4\theta t}{\sqrt{n}(1-\theta^2)^2} + o\left(\frac{1}{\sqrt{n}}\right) ;$$
$$E_\theta\left[\left\{Z_n - E_\theta(Z_n)\right\}^3\right] = \frac{6\sigma^6\theta}{\sqrt{n}(1-\theta^2)^2} + o\left(\frac{1}{\sqrt{n}}\right) ,$$
Using Edgeworth expansion we obtain
$$P_{\theta,n}\left\{\sqrt{n}\,(\hat{\theta}_{LS}-\theta) \leqq t\right\}$$
$$= P_{\theta,n}\left\{ Z_n \leqq 0 \right\}$$
$$= \Phi\left(\frac{t}{\sqrt{1-\theta^2}}\right) + \phi\left(\frac{t}{\sqrt{1-\theta^2}}\right)\left\{\frac{\theta t^2}{\sqrt{n}(1-\theta^2)^{3/2}} + \frac{\theta}{\sqrt{n}\sqrt{1-\theta^2}}\right\} + o\left(\frac{1}{\sqrt{n}}\right) .$$
As is immediately seen from above, $\hat{\theta}_{LS}$ is not second order AMU. Hence we define a modified least squares estimator $\hat{\theta}^*_{LS}$ as follows :
$$\hat{\theta}^*_{LS} = \left(1 + \frac{1}{n}\right) \hat{\theta}_{LS} .$$
Then $\hat{\theta}^*_{LS}$ is second order AMU and

(4.5.24) $$P_{\theta,n}\left\{\sqrt{n}(\hat{\theta}^*_{LS}-\theta)\leqq t\right\} = \Phi\left(\frac{t}{\sqrt{1-\theta^2}}\right) + \phi\left(\frac{t}{\sqrt{1-\theta^2}}\right)\frac{\theta t^2}{\sqrt{n}(1-\theta^2)^{3/2}} + o\left(\frac{1}{\sqrt{n}}\right) .$$

Since $\sigma^2 I=1$ and $\mu_3=0$, it follows from Theorem 4.5.1 and (4.5.24) that the second order asymptotic distribution of $\hat{\theta}^*_{LS}$ attains the bound of the second order asymptotic distributions of second order AMU estimators. Therefore we have now established the following :

Theorem 4.5.2. If f is a normal density with mean 0 and variance σ^2, then $\hat{\theta}^*_{LS}$ is second order asymptotically efficient.

Chapter 5

Second Order and Third Order Asymptotic Efficiency of The Maximum Likelihood Estimator and Other Estimators

5.1 Third order asymptotic efficiency of the maximum likelihood estimator for the multiparameter exponential case

Summary :

Third order asymptotic efficiency of the maximum likelihood estimator (MLE) has been discussed by J.Pfanzagl and W.Wefelmeyer [34], (who adopted the terminology) and also by J.K.Ghosh and K.Subramanyam [21], for cases where sufficient statistics exist. In this section we shall establish more general results for the multiparameter exponential family, introducing a differential operator, and show that (modified) MLE is always optimal up to the order n^{-1} among a similar modified class of estimators.

The method is not resticted to the exponential family and will be applied to more general cases in a subsequent discussion.

Suppose that $X_1, X_2, \dots, X_n$ are independently and identically distributed according to an exponential type distribution, i.e. with the density (w.r.t. σ-finite measure μ)

$$f(x,\theta)=h(x)c(\theta)\exp\left\{\sum_{i=1}^{m} s_i(\theta)t_i(x)\right\},$$

where θ is a p-dimensional real vector-valued parameter which belongs to an open set $\Theta \subset R^p$, and $s_i(\theta)$'s are continuous real valued functions and $t_i(x)$'s are real valued measurable functions. Define

$$c^*(s_1, \dots, s_m) = \int h(x)\exp\left\{\sum_{i=1}^{m} s_i t_i(x)\right\} d\mu(x)$$

as a real valued function of $s_1, \dots, s_m$.

Let $S^* \subset R^m$ be the set of $(s_1, \dots, s_m)$ for which c^* is finite.

We assume that for every $\theta \in \text{Ⓗ}$, $(s_1(\theta), \dots, s_m(\theta))$ is an inner point of S^*.

We have, obviously that $c = c^{*-1}$.

We assume that $s_1(\theta), \dots, s_m(\theta)$ are linearly independent as θ varies in Ⓗ, and also that $t_1(x), \dots, t_m(x)$ are linear independent functions of x. (The domain of x need not be specified)

For the sample $(X_1, \dots, X_n)$, a sufficient satistic is given by

$$T = (T_1, \dots, T_m) ;$$
$$T_i = \sum_{j=1}^{m} t_i(X_j)/n .$$

From the assumptions T has moment generating function given by

$$E(\exp \sum_{i=1}^{n} \xi_i T_i)$$
$$= c^*(s_1(\theta)+\xi_1, \dots, s_m(\theta)+\xi_m)/c^*(s_1(\theta), \dots, s_m(\theta)) .$$

And the joint cumulants of T_i's are given by

$$\kappa_{ijk\dots} = \frac{\partial^r}{\partial s_i \partial s_j \partial s_k \cdots} \log c^* \Big|_{s_1=s_1(\theta), \dots, s_m=s_m(\theta)} .$$

We assume that the T_i's are continuously (w.r.t. the Lebesgue measure) distributed and that their joint density admits an Edgeworth expansion.

We also assume that $s_i(\theta)$'s are continuously differentiable four times.

Now we shall define a class of estimators called <u>extended regular (ER) estimators</u> as those which can be expressed as

$$\hat{\theta} = g(T_1, \dots, T_m) + \frac{1}{n} h(T_1, \dots, T_m) ,$$

where g is three times continuously differentiable and h is continuously differentiable, and both are independent of n.

An ER estimator is consistent iff

(5.1.1) $$g(\mu_1(\theta), \dots, \mu_m(\theta)) \equiv \theta ,$$

where $\mu_i(\theta)=E_\theta(T_i)$, $(i=1, \ldots, m)$

and if $\hat{\theta}$ is a consistent ER estimator, $\sqrt{n}(\hat{\theta}-\theta)$ is asymptotically distributed according to the normal distribution with mean vector 0 and the variance-covariance matrix

$$\Sigma = \sum_i \sum_j \sigma_{ij} a_i a_j' \quad , \quad \sigma_{ij}=V_\theta(T_i, T_j)$$

where a_i's are vectors defined by

$$(5.1.2) \qquad a_i = \frac{\partial g}{\partial T_i}\bigg|_{T_i=\mu_i(\theta)}$$

It is well known that

$$\Sigma - I^{-1} \geqq 0 \text{ (positive semi-definite)}$$

where I is the Fisher Information matrix given by

$$\begin{aligned} I &= -E_\theta\left[\frac{\partial^2}{\partial\theta\partial\theta'}\log f(X,\theta)\right] \\ &= E_\theta\left[\frac{\partial}{\partial\theta}\log f(X,\theta)\frac{\partial}{\partial\theta'}\log f(X,\theta)\right] \\ &= E_\theta\left[\sum_i \frac{\partial s_i}{\partial\theta}(T_i-\theta) \cdot \sum_j \frac{\partial s_j}{\partial\theta'}(T_i-\theta)\right] \\ &= \sum_i \sum_j \sigma_{ij} \frac{\partial s_i}{\partial\theta} \frac{\partial s_j}{\partial\theta} \quad . \end{aligned}$$

Also it is easily seen that the equality is attained iff

$$(5.1.3) \qquad a_i = I^{-1} \frac{\partial s_i}{\partial\theta}$$

We call an ER estimator which satisfies (5.1.1) and (5.1.3) an <u>ER best asymptotically normal (BAN) estimator.</u>

Now we expand each component of $\hat{\theta}$ in a Taylor series as follows

$$\begin{aligned} (5.1.4) \quad \hat{\theta}_\alpha = \theta_\alpha &+ \sum_i a_i^\alpha (T_i-\mu_i) + \frac{1}{2}\sum_i\sum_j b_{ij}^\alpha (T_i-\mu_i)(T_j-\mu_j) \\ &+ \frac{1}{6}\sum_i\sum_j\sum_k c_{ijk}^\alpha (T_i-\mu_i)(T_j-\mu_j)(T_k-\mu_k) \\ &+ \frac{1}{\sqrt{n}}\Delta_\alpha + \frac{1}{n}\sum_i d_i^\alpha (T_i-\mu_i) + R_\alpha \quad , \end{aligned}$$

where $a_i^\alpha = \dfrac{\partial g_\alpha}{\partial T_i}$, $b_{ij}^\alpha = \dfrac{\partial^2 g_\alpha}{\partial T_i \partial T_j}$, $c_{ijk}^\alpha = \dfrac{\partial^3 g_\alpha}{\partial T_i \partial T_j \partial T_k}$,

$$\Delta_\alpha = h(\mu_1, \ldots, \mu_m) \quad , \quad d_i^\alpha = \frac{\partial h}{\partial T_i}$$

and all derivatives are evaluated at $T_i \equiv \mu_i(\theta)$.

By putting $T_i' = \sqrt{n}(T_i - \mu_i)$, (5.1.1) can be rewritten as

$$\sqrt{n}(\hat{\theta}_\alpha - \theta_\alpha) = \sum_i a_i^\alpha T_i' + \frac{1}{2\sqrt{n}} \sum_i \sum_j b_{ij}^\alpha T_i' T_j'$$
$$+ \frac{1}{6n} \sum_i \sum_j \sum_k c_{ijk}^\alpha T_i' T_j' T_k'$$
$$+ \frac{1}{\sqrt{n}} \Delta_\alpha + \frac{1}{n} \sum_i d_i^\alpha T_i' + R_\alpha' \quad .$$

It may be shown that R_α' is stochastically of smaller order than n^{-1} .

Let

$$Y_\alpha = \sqrt{n}(\hat{\theta}_\alpha - \theta_\alpha) - R_\alpha' \quad .$$

Then we have $\sqrt{n}(\hat{\theta}_\alpha - \theta_\alpha) - Y_\alpha = o_p(1/n)$, moreover, we have the following :

Theorem 5.1.1. The joint distributions of $\{\sqrt{n}(\hat{\theta}_\alpha - \theta_\alpha); \alpha=1,\cdots,p\}$ and $\{Y_\alpha; \alpha=1,\cdots,p\}$ coincide up to the order n^{-1}. More precisely, for any bounded continuous function φ , we have

$$\overline{\lim_{n\to\infty}}\, n \left| E_{\theta,n}[\varphi(Y_1,\ldots,Y_p)] - E_{\theta,n}[\varphi(\sqrt{n}(\hat{\theta}_1 - \theta_1),\ldots,\sqrt{n}(\hat{\theta}_p - \theta_p))] \right| = 0 \ .$$

Proof. Observing that $P_{\theta,n}\{\sqrt{n}\ |R_\alpha'| > \varepsilon\} = o(1/n)$ for every $\varepsilon > 0$ (which is not directly derived from the stochastic order of R' but can be easily proved from the fact that R_α' is of the same order as $|T_i - \mu_i|^4$), we can show that

$$\lim_{n\to\infty} n \left| E_{\theta,n}[\exp \sum_\alpha it_\alpha(\hat{\theta}_\alpha - \theta_\alpha)] - E_{\theta,n}[\exp \sum_\alpha it_\alpha Y_\alpha \ ] \right| = 0$$

uniformly in $(t_1, \ldots , t_p)$ in any compact subset of R^p, which establishes the theorem.

Corollary 5.1.1. For any measurable set $C \subset R^p$ whose boundary is of Lebesgue measure 0, we have

$$\overline{\lim_{n\to\infty}}\, n \left| P_{\theta,n}\{Y \in C\} - P_{\theta,n}\{\sqrt{n}(\hat{\theta} - \theta) \in C\} \right| = 0 \ .$$

From Theorem 5.1.1 we have the following :

Theorem 5.1.2. The joint distribution of $\{\sqrt{n}(\hat{\theta}_\alpha - \theta_\alpha)$; $\alpha = 1, \ldots, p\}$ is asymptotically equal to a continuous distribution up to the order n^{-1}.

Theorem 5.1.3. The joint distribution of $\{Y_\alpha ; \alpha = 1, \ldots, p\}$ admits Edgeworth expansion up to the order n^{-1}.

Proof. The theorem is established by algebraic evaluation of the joint characteristic function of Y_α's, the details of which are obmitted (See Bhattacharya and Ghosh [13]) .

Consequently the joint density of $\{\sqrt{n}(\hat{\theta}_\alpha - \theta_\alpha) ; \alpha = 1, \ldots, p\}$ can be expanded up to the order n^{-1} if we have joint cumulants of $\{Y_\alpha ; \alpha=1,\ldots,p\}$ up to the fourth.

In order to evaluate the joint cumulants, we introduce the following differential operator

$D_\alpha = \sum_\beta I^{\alpha\beta} \frac{\partial}{\partial\theta_\beta}$, where $I^{\alpha\beta}$ denote the elements of I^{-1} and the random variable

$$\begin{aligned} U_\alpha &= \sqrt{n} \sum_i \sum_\beta I^{\alpha\beta} \frac{\partial s_i}{\partial\theta_\beta}(T_i - \mu_i) \\ &= \sum_i (D_\alpha s_i) T_i' \\ &= D_\alpha\{n \log f(X, \theta)\} \quad . \end{aligned}$$

The following lemma will be frequently applied in the subsequent discussion.

Lemma 5.1.1. Suppose that Z_θ is a function of $T_1, \ldots, T_m$ and of θ , differentiable in θ and that

$$\frac{\partial}{\partial\theta} E_{\theta,n}(Z_\theta) = \int \frac{\partial}{\partial\theta} \left\{ Z_\theta \prod_i f(x_i, \theta) \right\} \prod_i d\mu(x_i)$$

holds. Then we have

$$E_{\theta,n}(U_\alpha Z_\theta) = \frac{1}{\sqrt{n}} D_\alpha E_{\theta,n}(Z_\theta) - \frac{1}{\sqrt{n}} E_{\theta,n}(D_\alpha Z_\theta) \quad .$$

Proof.

$$\begin{aligned} & D_\alpha E_{\theta,n}(Z_\theta) \\ &= \int (D_\alpha Z_\theta) \prod_i f(x_i,\theta) \prod_i d\mu(x_i) + \int Z_\theta D_\alpha \prod_i f(x_i,\theta) \prod_i d\mu(x_i) \\ &= E_{\theta,n}(D_\alpha Z_\theta) + \int Z_\theta D_\alpha (\sum_i \log f(x_i,\theta)) \prod_i f(x_i,\theta) \prod_i d\mu(x_i) \\ &= E_{\theta,n}(D_\alpha Z_\theta) + \sqrt{n}\, E_{\theta,n}(Z_\theta U_\alpha) \end{aligned}$$

from which the theorem is directly derived.

We shall use the following symbols.

$$\begin{aligned} J_{\alpha\beta\cdot\gamma} &= E_\theta\left[\frac{\partial^2}{\partial\theta_\alpha \partial\theta_\beta} \log f(X,\theta) \frac{\partial}{\partial\theta_\gamma} \log f(X,\theta) \right] \\ &= \sum_i \sum_j \sigma_{ij} \left(\frac{\partial^2 s_i}{\partial\theta_\alpha \partial\theta_\beta} \right) \left(\frac{\partial s_j}{\partial\theta_\gamma} \right) \quad ; \\ K_{\alpha\beta\gamma} &= E_\theta\left[\frac{\partial}{\partial\theta_\alpha} \log f(X,\theta) \frac{\partial}{\partial\theta_\beta} \log f(X,\theta) \frac{\partial}{\partial\theta_\gamma} \log f(X,\theta) \right] \\ &= \sum_i \sum_j \sum_k K_{ijk} \left(\frac{\partial s_i}{\partial\theta_\alpha} \right) \left(\frac{\partial s_j}{\partial\theta_\beta} \right) \left(\frac{\partial s_k}{\partial\theta_\gamma} \right) \quad ; \\ J^{\alpha\beta\cdot\gamma} &= \sum_{\alpha'} \sum_{\beta'} \sum_{\gamma'} I^{\alpha\alpha'} I^{\beta\beta'} I^{\gamma\gamma'} J_{\alpha'\beta'\cdot\gamma'} \quad ; \\ K^{\alpha\beta\gamma} &= \sum_{\alpha'} \sum_{\beta'} \sum_{\gamma'} I^{\alpha\alpha'} I^{\beta\beta'} I^{\gamma\gamma'} K_{\alpha'\beta'\gamma'} \quad ; \\ L_{\alpha\beta\gamma\cdot\delta} &= E_\theta\left[\frac{\partial^3}{\partial\theta_\alpha \partial\theta_\beta \partial\theta_\gamma} \log f(X,\theta) \frac{\partial}{\partial\theta_\delta} \log f(X,\theta) \right] \\ &= \sum_i \sum_j \sigma_{ij} \left(\frac{\partial^3 s_i}{\partial\theta_\alpha \partial\theta_\beta \partial\theta_\gamma} \right) \left(\frac{\partial s_j}{\partial\theta_\delta} \right) \quad ; \\ M_{\alpha\beta\cdot\gamma\delta} &= E_\theta\left[\frac{\partial^2}{\partial\theta_\alpha \partial\theta_\beta} \log f(X,\theta) \frac{\partial^2}{\partial\theta_\gamma \partial\theta_\delta} \log f(X,\theta) \right] - I_{\alpha\beta} I_{\gamma\delta} \\ &= \sum_i \sum_j \sigma_{ij} \left(\frac{\partial^2 s_i}{\partial\theta_\alpha \partial\theta_\beta} \right) \left(\frac{\partial^2 s_j}{\partial\theta_\gamma \partial\theta_\delta} \right) \quad ; \\ N_{\alpha\beta\cdot\gamma\delta} &= E_\theta\left[\frac{\partial^2}{\partial\theta_\alpha \partial\theta_\beta} \log f(X,\theta) \frac{\partial}{\partial\theta_\gamma} \log f(X,\theta) \frac{\partial}{\partial\theta_\delta} \log f(X,\theta) \right] \\ &\quad + I_{\alpha\beta} I_{\gamma\delta} \\ &= \sum_i \sum_j \sum_k K_{ijk} \left(\frac{\partial^2 s_i}{\partial\theta_\alpha \partial\theta_\beta} \right) \left(\frac{\partial s_j}{\partial\theta_\gamma} \right) \left(\frac{\partial s_k}{\partial\theta_\delta} \right) \quad ; \end{aligned}$$

$$H_{\alpha\beta\gamma\delta} = E_\theta[\frac{\partial}{\partial\theta_\alpha}\log f(X,\theta)\frac{\partial}{\partial\theta_\beta}\log f(X,\theta)\frac{\partial}{\partial\theta_\gamma}\log f(X,\theta)$$

$$\cdot \frac{\partial}{\partial\theta_\delta}\log f(X,\theta)] - (I_{\alpha\beta}I_{\gamma\delta}+I_{\alpha\gamma}I_{\beta\delta}+I_{\alpha\delta}I_{\beta\gamma})$$

$$= \sum_i \sum_j \sum_k \sum_l K_{ijkl}(\frac{\partial s_i}{\partial\theta_\alpha})(\frac{\partial s_j}{\partial\theta_\beta})(\frac{\partial s_k}{\partial\theta_\gamma})(\frac{\partial s_l}{\partial\theta_\delta}) \;;$$

$$L^{\alpha\beta\gamma\cdot\delta} = \sum_{\alpha'}\sum_{\beta'}\sum_{\gamma'}\sum_{\delta'} I^{\alpha\alpha'}I^{\beta\beta'}I^{\gamma\gamma'}I^{\delta\delta'}L_{\alpha'\beta'\gamma'\cdot\delta'} \;;$$

$$M^{\alpha\beta\cdot\gamma\delta} = \sum_{\alpha'}\sum_{\beta'}\sum_{\gamma'}\sum_{\delta'} I^{\alpha\alpha'}I^{\beta\beta'}I^{\gamma\gamma'}I^{\delta\delta'}M_{\alpha'\beta'\cdot\gamma'\delta'} \;;$$

$$N^{\alpha\beta\cdot\gamma\delta} = \sum_{\alpha'}\sum_{\beta'}\sum_{\gamma'}\sum_{\delta'} I^{\alpha\alpha'}I^{\beta\beta'}I^{\gamma\gamma'}I^{\delta\delta'}N_{\alpha'\beta'\cdot\gamma'\delta'} \;;$$

$$H^{\alpha\beta\gamma\delta} = \sum_{\alpha'}\sum_{\beta'}\sum_{\gamma'}\sum_{\delta'} I^{\alpha\alpha'}I^{\beta\beta'}I^{\gamma\gamma'}I^{\delta\delta'}H_{\alpha'\beta'\gamma'\delta'} \;.$$

Using these symbols we have

$$D_\alpha I_{\beta\gamma} = \sum_i\sum_j D_\alpha(\sigma_{ij}\frac{\partial s_i}{\partial\theta_\beta}\frac{\partial s_j}{\partial\theta_\gamma})$$

$$= \sum_i\sum_j (D_\alpha\sigma_{ij})(\frac{\partial s_i}{\partial\theta_\beta})(\frac{\partial s_j}{\partial\theta_\gamma}) + \sum_i\sum_j \sigma_{ij}(D_\alpha\frac{\partial s_i}{\partial\theta_\beta})(\frac{\partial s_j}{\partial\theta_\gamma})$$

$$+ \sum_i\sum_j \sigma_{ij}(\frac{\partial s_i}{\partial\theta_\beta})(D_\alpha\frac{\partial s_j}{\partial\theta_\gamma})$$

$$= \sum_{\alpha'} I^{\alpha\alpha'}\sum_i\sum_j\sum_k K_{ijk}(\frac{\partial s_i}{\partial\theta_\beta})(\frac{\partial s_j}{\partial\theta_\gamma})(\frac{\partial s_k}{\partial\theta_{\alpha'}})$$

$$+ \sum_{\alpha'} I^{\alpha\alpha'}\sum_i\sum_j \sigma_{ij}(\frac{\partial^2 s_i}{\partial\theta_{\alpha'}\partial\theta_\beta})(\frac{\partial s_j}{\partial\theta_\gamma})$$

$$+ \sum_{\alpha'} I^{\alpha\alpha'}\sum_i\sum_j \sigma_{ij}(\frac{\partial s_i}{\partial\theta_\beta})(\frac{\partial^2 s_j}{\partial\theta_{\alpha'}\partial\theta_\gamma})$$

$$= \sum_{\alpha'} I^{\alpha\alpha'}K_{\alpha'\beta\gamma} + \sum_{\alpha'} I^{\alpha\alpha'}J_{\alpha'\beta\cdot\gamma} + \sum_{\alpha'} I^{\alpha\alpha'}J_{\alpha'\gamma\cdot\beta} \;;$$

$$D_\alpha I^{\beta\gamma} = -\sum_{\beta'}\sum_{\gamma'} I^{\beta\beta'}I^{\gamma\gamma'}D_\alpha I_{\beta'\gamma'}$$

$$= -\sum_{\beta'}\sum_{\gamma'}\sum_{\alpha'} I^{\beta\beta'}I^{\gamma\gamma'}I^{\alpha\alpha'}(K_{\alpha'\beta'\gamma'} + J_{\alpha'\beta'\cdot\gamma'} + J_{\alpha'\gamma'\cdot\beta'})$$

$$= -K^{\alpha\beta\gamma} - J^{\alpha\beta\cdot\gamma} - J^{\alpha\gamma\cdot\beta} \;;$$

$$D_\alpha J_{\beta\gamma\cdot\delta} = \sum_i \sum_j (D_\alpha \sigma_{ij})\left(\frac{\partial^2 s_i}{\partial\theta_\beta \partial\theta_\gamma}\right)\left(\frac{\partial^2 s_j}{\partial\theta_\delta^2}\right)$$

$$+ \sum_i \sum_j \sigma_{ij}\left(D_\alpha \frac{\partial^2 s_i}{\partial\theta_\beta \partial\theta_\gamma}\right)\left(\frac{\partial^2 s_j}{\partial\theta_\gamma^2}\right)$$

$$+ \sum_i \sum_j \sigma_{ij}\left(\frac{\partial^2 s_i}{\partial\theta_\beta \partial\theta_\gamma}\right)\left(D_\alpha \frac{\partial^2 s_j}{\partial\theta_\delta^2}\right)$$

$$= \sum_{\alpha'} I^{\alpha\alpha'} \sum_i \sum_j \sum_k K_{ijk}\left(\frac{\partial^2 s_i}{\partial\theta_\beta \partial\theta_\gamma}\right)\left(\frac{\partial s_j}{\partial\theta_\delta}\right)\left(\frac{\partial s_k}{\partial\theta_{\alpha'}}\right)$$

$$+ \sum_{\alpha'} I^{\alpha\alpha'} \sum_i \sum_j \left(\frac{\partial^3 s_i}{\partial\theta_{\alpha'} \partial\theta_\beta \partial\theta_\gamma}\right)\left(\frac{\partial s_j}{\partial\theta_\delta}\right)$$

$$+ \sum_{\alpha'} I^{\alpha\alpha'} \sum_i \sum_j \left(\frac{\partial^2 s_i}{\partial\theta_\beta \partial\theta_\gamma}\right)\left(\frac{\partial^2 s_j}{\partial\theta_{\alpha'} \partial\theta_\delta}\right)$$

$$= \sum_{\alpha'} I^{\alpha\alpha'}(N_{\beta\gamma\cdot\alpha'\delta} + L_{\alpha'\beta\gamma\cdot\delta} + M_{\beta\gamma\cdot\alpha'\delta}) \ ;$$

$$D_\alpha K_{\beta\gamma\delta} = \sum_i \sum_j \sum_k (D_\alpha K_{ijk})\left(\frac{\partial s_i}{\partial\theta_\beta}\right)\left(\frac{\partial s_j}{\partial\theta_\gamma}\right)\left(\frac{\partial s_k}{\partial\theta_\delta}\right)$$

$$+ \sum_i \sum_j \sigma_{ij}\left(D_\alpha \frac{\partial s_i}{\partial\theta_\beta}\right)\left(\frac{\partial s_j}{\partial\theta_\gamma}\right)\left(\frac{\partial s_k}{\partial\theta_\delta}\right)$$

$$+ \sum_i \sum_j \sigma_{ij}\left(\frac{\partial s_i}{\partial\theta_\beta}\right)\left(D_\alpha \frac{\partial s_j}{\partial\theta_\gamma}\right)\left(\frac{\partial s_k}{\partial\theta_\delta}\right)$$

$$+ \sum_i \sum_j \sigma_{ij}\left(\frac{\partial s_i}{\partial\theta_\beta}\right)\left(\frac{\partial s_j}{\partial\theta_\gamma}\right)\left(D_\alpha \frac{\partial s_k}{\partial\theta_\delta}\right)$$

$$= \sum_{\alpha'} I^{\alpha\alpha'}(H_{\alpha'\beta\gamma\delta} + N_{\alpha'\beta\cdot\gamma\delta} + N_{\alpha'\gamma\cdot\beta\delta} + N_{\alpha'\delta\cdot\beta\gamma}) \ ;$$

$$D_\alpha J^{\beta\gamma\cdot\delta} = \sum_{\beta'} \sum_{\gamma'} \sum_{\delta'} (D_\alpha I^{\beta\beta'}) I^{\gamma\gamma'} I^{\delta\delta'} J_{\beta'\gamma'\cdot\delta'}$$

$$+ \sum_{\beta'} \sum_{\gamma'} \sum_{\delta'} I^{\beta\beta'} (D_\alpha I^{\gamma\gamma'}) I^{\delta\delta'} J_{\beta'\gamma'\cdot\delta'}$$

$$+ \sum_{\beta'} \sum_{\gamma'} \sum_{\delta'} I^{\beta\beta'} I^{\gamma\gamma'} (D_\alpha I^{\delta\delta'}) J_{\beta'\gamma'\cdot\delta'}$$

$$+ \sum_{\beta'} \sum_{\gamma'} \sum_{\delta'} I^{\beta\beta'} I^{\gamma\gamma'} I^{\delta\delta'} D_\alpha J_{\beta'\gamma'\cdot\delta'}$$

$$= - \sum_{\beta'} \sum_{\gamma'} \sum_{\delta'} (K^{\alpha\beta\beta'} + J^{\alpha\beta\cdot\beta'} + J^{\alpha\beta'\cdot\beta}) I^{\gamma\gamma'} I^{\delta\delta'} J_{\beta'\gamma'\cdot\delta'}$$

$$- \sum_{\beta'} \sum_{\gamma'} \sum_{\delta'} (K^{\alpha\gamma\gamma'} + J^{\alpha\gamma\cdot\gamma'} + J^{\alpha\gamma'\cdot\gamma}) I^{\beta\beta'} I^{\delta\delta'} J_{\beta'\gamma'\cdot\delta'}$$

$$- \sum_{\beta'} \sum_{\gamma'} \sum_{\delta'} (K^{\alpha\delta\delta'} + J^{\alpha\delta\cdot\delta'} + J^{\alpha\delta'\cdot\delta}) I^{\beta\beta'} I^{\gamma\gamma'} J_{\beta'\gamma'\cdot\delta'}$$

$$+\sum_{\alpha'}\sum_{\beta'}\sum_{\gamma'}\sum_{\delta'} I^{\alpha\alpha'} I^{\beta\beta'} I^{\gamma\gamma'} I^{\delta\delta'} (N_{\beta'\gamma'\cdot\alpha'\delta'} + L_{\alpha'\beta'\gamma'\cdot\delta'} + M_{\beta'\gamma'\cdot\alpha'\delta'}) \; .$$

Introducing still more notation as

$$J^{\alpha\beta}_{\cdot\gamma'} = \sum_{\alpha'}\sum_{\beta'} I^{\alpha\alpha'} I^{\beta\beta'} J_{\alpha'\beta'\cdot\gamma'} \; ;$$

$$J^{\alpha\cdot\gamma}_{\beta'} = \sum_{\alpha'}\sum_{\gamma'} I^{\alpha\alpha'} I^{\gamma\gamma'} J_{\alpha'\beta'\cdot\gamma'} \; ;$$

$$K^{\alpha\beta}_{\gamma'} = \sum_{\alpha'}\sum_{\beta'} I^{\alpha\alpha'} I^{\beta\beta'} K_{\alpha'\beta'\gamma'} \; ,$$

the above can be abbreviated as

(5.1.5)
$$\begin{aligned} D_\alpha J^{\beta\gamma\cdot\delta} = & -\sum_{\beta'} (K^{\alpha\beta\beta'} + J^{\alpha\beta\cdot\beta'} + J^{\alpha\beta'\cdot\beta}) J^{\gamma\cdot\delta}_{\beta'} \\ & -\sum_{\gamma'} (K^{\alpha\gamma\gamma'} + J^{\alpha\gamma\cdot\gamma'} + J^{\alpha\gamma'\cdot\gamma}) J^{\beta\cdot\delta}_{\gamma'} \\ & -\sum_{\delta'} (K^{\alpha\delta\delta'} + J^{\alpha\delta\cdot\delta'} + J^{\alpha\delta'\cdot\delta}) J^{\beta\gamma}_{\cdot\delta'} \\ & + N^{\beta\gamma\cdot\alpha\delta} + L^{\alpha\beta\gamma\cdot\delta} + M^{\alpha\delta\cdot\beta\gamma} \; ; \end{aligned}$$

(5.1.6)
$$\begin{aligned} D_\alpha K^{\beta\gamma\delta} = & \sum_{\beta'}\sum_{\gamma'}\sum_{\delta'} (D_\alpha I^{\beta\beta'}) I^{\gamma\gamma'} I^{\delta\delta'} K_{\beta'\gamma'\delta'} \\ & + \sum_{\beta'}\sum_{\gamma'}\sum_{\delta'} I^{\beta\beta'} (D_\alpha I^{\gamma\gamma'}) I^{\delta\delta'} K_{\beta'\gamma'\delta'} \\ & + \sum_{\beta'}\sum_{\gamma'}\sum_{\delta'} I^{\beta\beta'} I^{\gamma\gamma'} (D_\alpha I^{\delta\delta'}) K_{\beta'\gamma'\delta'} \\ & + \sum_{\beta'}\sum_{\gamma'}\sum_{\delta'} I^{\beta\beta'} I^{\gamma\gamma'} I^{\delta\delta'} (D_\alpha K_{\beta'\gamma'\delta'}) \\ = & -\sum_{\beta'} (K^{\alpha\beta\beta'} + J^{\alpha\beta\cdot\beta'} + J^{\alpha\beta'\cdot\beta}) K^{\gamma\delta}_{\beta'} \\ & -\sum_{\gamma'} (K^{\alpha\gamma\gamma'} + J^{\alpha\gamma\cdot\gamma'} + J^{\alpha\gamma'\cdot\gamma}) K^{\beta\delta}_{\gamma'} \\ & -\sum_{\delta'} (K^{\alpha\delta\delta'} + J^{\alpha\delta\cdot\delta'} + J^{\alpha\delta'\cdot\delta}) K^{\beta\gamma}_{\delta'} \\ & + H^{\alpha\beta\gamma\delta} + N^{\alpha\beta\cdot\gamma\delta} + N^{\alpha\gamma\cdot\beta\delta} + N^{\alpha\delta\cdot\beta\gamma} \; . \end{aligned}$$

We have

$$\begin{aligned} D_\alpha U_\beta & = D_\alpha D_\beta \sum_i s_i T_i' \\ & = \sum_i (D_\alpha D_\beta s_i) T_i' - \sqrt{n} \sum_i D_\beta s_i (D_\alpha \mu_i) \\ & = \sum_i (D_\alpha D_\beta s_i) T_i' - \sqrt{n} \sum_i \sum_j (D_\beta s_i) \sigma_{ij} (D_\alpha s_j) \\ & = \sum_i (D_\alpha D_\beta s_i) T_i' \\ & \quad -\sqrt{n} \sum_{\alpha'}\sum_{\beta'}\sum_i \sum_j I^{\alpha\alpha'} I^{\beta\beta'} \sigma_{ij} \left(\frac{\partial s_i}{\partial \theta_{\alpha'}}\right)\left(\frac{\partial s_j}{\partial \theta_{\beta'}}\right) \end{aligned}$$

$$= \sum_i (D_\alpha D_\beta s_i) T_i' - \sqrt{n} \sum_{\alpha'} \sum_{\beta'} I^{\alpha\alpha'} I^{\beta\beta'} I_{\alpha'\beta'}$$

$$= \sum_i (D_\alpha D_\beta s_i) T_i' - \sqrt{n}\, I^{\alpha\beta} \quad .$$

We shall denote

$$V_{\alpha\beta} = \sum_i (D_\alpha D_\beta s_i) T_i'$$

Then we have a second lemma :

Lemma 5.1.2. If $U_\beta Z_\theta$ and Z_θ both satisfy the conditions of the lemma 5.1.1 we have

$$E_{\theta,n}(U_\alpha U_\beta Z_\theta) = \frac{1}{\sqrt{n}} D_\alpha E_{\theta,n}(U_\beta Z_\theta) + I^{\alpha\beta} E_{\theta,n}(Z_\theta)$$

$$- \frac{1}{\sqrt{n}} E_{\theta,n}(V_{\alpha\beta} Z_\theta) - \frac{1}{\sqrt{n}} E_{\theta,n}(U_\alpha D_\alpha Z_\theta)$$

$$= I^{\alpha\beta} E_{\theta,n}(Z_\theta) - \frac{1}{\sqrt{n}} E_{\theta,n}(V_{\alpha\beta} Z_\theta) + \frac{1}{n} D_\alpha D_\beta E_{\theta,n}(Z_\theta)$$

$$- \frac{1}{n} D_\alpha E_{\theta,n}(D_\beta Z_\theta) - \frac{1}{n} D_\beta E_{\theta,n}(D_\alpha Z_\theta) + \frac{1}{n} E_{\theta,n}(D_\alpha D_\beta Z_\theta)$$

Proof. By lemma 5.1.1, we have

$$E_{\theta,n}(U_\alpha U_\beta Z_\theta) = \frac{1}{\sqrt{n}} D_\alpha E_{\theta,n}(U_\beta Z_\theta) - \frac{1}{\sqrt{n}} E_{\theta,n}(D_\alpha (U_\beta Z_\theta))$$

$$= \frac{1}{\sqrt{n}} D_\alpha E_{\theta,n}(U_\beta Z_\theta) - \frac{1}{\sqrt{n}} E_{\theta,n}[\ (V_{\alpha\beta} - \sqrt{n}\, I^{\alpha\beta}) Z_\theta\ ]$$

$$- \frac{1}{\sqrt{n}} E_{\theta,n}(U_\beta D_\alpha Z_\theta)$$

which establishes the first half of the lemma. The second half is derived by applying lemma 5.1.1 to the terms $E_{\theta,n}(U_\beta Z_\theta)$ and $E_{\theta,n}(U_\beta D_\alpha Z_\theta)$ in the above.

We shall derive the moments of U_α and $V_{\beta\gamma}$. Obviously it is shown that

$$E_\theta(U_\alpha)=0\ ;\quad E_\theta(V_{\alpha\beta})=0\ ;$$

$$E_\theta(U_\alpha U_\beta)=I^{\alpha\beta}.$$

We have

$$\begin{aligned}
V_{\alpha\beta} &= \sum_i (D_\alpha D_\beta S_i)T_i' \\
&= \sum_i\sum_{\beta'} D_\alpha\Big(I^{\beta\beta'}\frac{\partial s_i}{\partial\theta_{\beta'}}\Big)T_i' \\
&= \sum_i\sum_{\beta'}(D_\alpha I^{\beta\beta'})\Big(\frac{\partial s_i}{\partial\theta_{\beta'}}\Big)T_i' + \sum_i\sum_{\alpha'}\sum_{\beta'} I^{\alpha\alpha'}I^{\beta\beta'}\Big(\frac{\partial^2 s_i}{\partial\theta_{\alpha'}\partial\theta_{\beta'}}\Big)T_i' \\
&= -\sum_i\sum_{\beta'}(K^{\alpha\beta\beta'}+J^{\alpha\beta\cdot\beta'}+J^{\alpha\beta'\cdot\beta})\Big(\frac{\partial s_i}{\partial\theta_{\beta'}}\Big)T_i' \\
&\quad + \sum_i\sum_{\alpha'}\sum_{\beta'} I^{\alpha\alpha'}I^{\beta\beta'}\Big(\frac{\partial^2 s_i}{\partial\theta_{\alpha'}\partial\theta_{\beta'}}\Big)T_i'\ .
\end{aligned}$$

Hence it follows that

$$\begin{aligned}
E_\theta(V_{\alpha\beta}U_\gamma) &= -\sum_{\beta'}\sum_{\gamma'}(K^{\alpha\beta\beta'}+J^{\alpha\beta\cdot\beta'}\ J^{\alpha\beta'\cdot\beta})I^{\gamma\gamma'}I_{\beta'\gamma'} \\
&\quad + \sum_{\alpha'}\sum_{\beta'}\sum_{\gamma'} I^{\alpha\alpha'}I^{\beta\beta'}I^{\gamma\gamma'}J_{\alpha'\beta'\cdot\gamma'} \\
&= -K^{\alpha\beta\gamma}-J^{\alpha\beta\cdot\gamma}-J^{\alpha\gamma\cdot\beta}+J^{\alpha\beta\cdot\gamma} \\
&= -K^{\alpha\beta\gamma}-J^{\alpha\gamma\cdot\beta}\ ;
\end{aligned}$$

$$\begin{aligned}
(5.1.7)\quad E_\theta(V_{\alpha\beta}V_{\gamma\delta}) &= \sum_{\beta'}\sum_{\delta'}(K^{\alpha\beta\beta'}+J^{\alpha\beta\cdot\beta'}+J^{\alpha\beta'\cdot\beta})I_{\beta'\delta'} \\
&\quad\cdot(K^{\gamma\delta\delta'}+J^{\gamma\delta\cdot\delta'}+J^{\gamma\delta'\cdot\delta}) \\
&\quad -\sum_{\beta'}\sum_{\gamma'}\sum_{\delta'}(K^{\alpha\beta\beta'}+J^{\alpha\beta\cdot\beta'}+J^{\alpha\beta'\cdot\beta})I^{\gamma\gamma'}I^{\delta\delta'}J_{\gamma'\delta'\cdot\beta'} \\
&\quad -\sum_{\alpha'}\sum_{\beta'}\sum_{\delta'}(K^{\gamma\delta\delta'}+J^{\gamma\delta\cdot\delta'}+J^{\gamma\delta'\cdot\delta})I^{\alpha\alpha'}I^{\beta\beta'}J_{\alpha'\beta'\cdot\delta'} \\
&\quad +\sum_{\alpha'}\sum_{\beta'}\sum_{\gamma'}\sum_{\delta'} I^{\alpha\alpha'}I^{\beta\beta'}I^{\gamma\gamma'}I^{\delta\delta'}M_{\alpha'\beta'\cdot\gamma'\delta'} \\
&= \sum_{\delta'}(K^{\alpha\beta}_{\delta'}+J^{\alpha\beta}_{\cdot\delta'}+J^{\alpha\cdot\beta}_{\delta'})(K^{\gamma\delta\delta'}+J^{\gamma\delta\cdot\delta'}\ J^{\gamma\delta'\cdot\delta}) \\
&\quad -\sum_{\beta'}(K^{\alpha\beta\beta'}+J^{\alpha\beta\cdot\beta'}+J^{\alpha\beta'\cdot\beta})J^{\gamma\delta}_{\beta'}
\end{aligned}$$

$$- \sum_{\delta'} (K^{\gamma\delta\delta'} + J^{\gamma\delta\cdot\delta'} + J^{\gamma\delta'\cdot\delta}) J^{\alpha\beta}_{\cdot\delta'}$$
$$+ M^{\alpha\beta\cdot\gamma\delta}$$
$$= \sum_{\delta'} (K^{\alpha\beta}_{\delta'} + J^{\alpha\cdot\beta}_{\delta'})(K^{\gamma\delta\cdot\delta'} + J^{\gamma\delta'\cdot\delta})$$
$$- \sum_{\delta'} J^{\alpha\beta}_{\cdot\delta} J^{\gamma\delta\cdot\delta'} + M^{\alpha\beta\cdot\gamma\delta} ;$$

$$\sqrt{n}\, E_\theta(U_\alpha U_\beta U_\gamma) = K^{\alpha\beta\gamma} ;$$

$$(5.1.8) \quad \sqrt{n}\, E_\theta(V_{\alpha\beta} U_\gamma U_\delta) = - \sum_{\beta'} \sum_{\gamma'} \sum_{\delta'} (K^{\alpha\beta\beta'} + J^{\alpha\beta\cdot\beta'} + J^{\alpha\beta'\cdot\beta}) I^{\gamma\gamma'} I^{\delta\delta'} K_{\beta'\gamma'\cdot\delta'}$$
$$+ \sum_{\alpha'} \sum_{\beta'} \sum_{\gamma'} \sum_{\delta'} I^{\alpha\alpha'} I^{\beta\beta'} I^{\gamma\gamma'} I^{\delta\delta'} N_{\alpha'\beta'\cdot\gamma'\delta'}$$
$$= - \sum_{\beta'} (K^{\alpha\beta\beta'} + J^{\alpha\beta\cdot\beta'} + J^{\alpha\beta'\cdot\beta}) K^{\gamma\delta}_{\beta'} + N^{\alpha\beta\cdot\gamma\delta} ;$$

$$E_\theta(U_\alpha U_\beta U_\gamma U_\delta) = \frac{1}{n} H^{\alpha\beta\gamma\delta} + I^{\alpha\beta} I^{\gamma\delta} + I^{\alpha\gamma} I^{\beta\delta} + I^{\alpha\delta} I^{\beta\gamma} .$$

We shall obtain asymptotic joint cumulants of Z_α's making use of the above results. Note that the asymptotic cumulants of Z_α's are equal to those of $\sqrt{n}(\hat{\theta}_\alpha - \theta_\alpha)$ up to the order n^{-1} .

We can write

$$Y_\alpha = U_\alpha + \frac{1}{2\sqrt{n}} Q_\alpha + \frac{1}{n} W_\alpha .$$

And we have

$$E_\theta(Y_\alpha) = \frac{1}{2\sqrt{n}} E_\theta(Q_\alpha) + o(\frac{1}{n}) .$$

We shall denote $\frac{1}{2} E_\theta(Q_\alpha) = \mu_\alpha$, and

$$Z_\alpha = Y_\alpha - \frac{1}{\sqrt{n}} \mu_\alpha .$$

Then we have

$$D_\alpha Z_\beta = - \sqrt{n}\, D_\alpha \theta_\beta - \frac{1}{\sqrt{n}} D_\alpha \mu_\beta = - \sqrt{n}\, I^{\alpha\beta} - \frac{1}{\sqrt{n}} D_\alpha \mu_\beta .$$

First we shall obtain the variance of Z_α up to the order n^{-1}.

We have

$$E_\theta(Z_\alpha - U_\alpha)^2 = \frac{1}{4n} V_\theta(Q_\alpha) + o(\frac{1}{n})$$

and we have

$$\mathrm{Cov}_\theta(Z_\alpha, Z_\beta) = E_\theta(Z_\alpha Z_\beta) + o(\frac{1}{n})$$

$$= E_\theta(U_\alpha Z_\beta) + E_\theta(U_\beta Z_\alpha) - E_\theta(U_\alpha U_\beta) + E_\theta(Z_\alpha - U_\alpha)^2$$

$$+ o(\frac{1}{n}) .$$

Applying the lemma 5.1.1 we get

$$E_\theta(U_\alpha Z_\beta) = \frac{1}{\sqrt{n}} D_\alpha E_\theta(Z_\beta) - \frac{1}{\sqrt{n}} E_\theta(D_\alpha Z_\beta)$$

$$= I^{\alpha\beta} + \frac{1}{n} D_\alpha \mu_\beta .$$

$$E_\theta(U_\beta Z_\alpha) = I^{\alpha\beta} + \frac{1}{n} D_\beta \mu_\alpha .$$

Hence we have

$$\mathrm{Cov}_\theta(Z_\alpha , Z_\beta) = \mathrm{Cov}_\theta(Y_\alpha , Y_\beta)$$

$$= I^{\alpha\beta} + \frac{1}{n} D_\alpha \mu_\beta + \frac{1}{n} D_\beta \mu_\alpha + \frac{1}{4n} V_\theta(Q_\alpha) + o(\frac{1}{n}) .$$

For the third cumulants we shall refer to the following equation :

$$Z_\alpha Z_\beta Z_\gamma = \frac{1}{2}(U_\alpha Z_\beta Z_\gamma + U_\beta Z_\alpha Z_\gamma + U_\gamma Z_\alpha Z_\beta)$$

$$- \frac{1}{2} U_\alpha U_\beta U_\gamma + \frac{1}{2} U_\alpha (Z_\beta - U_\beta)(Z_\gamma - U_\gamma)$$

$$+ \frac{1}{2} U_\beta (Z_\alpha - U_\alpha)(Z_\gamma - U_\gamma) + \frac{1}{2} U_\gamma (Z_\alpha - U_\alpha)(Z_\beta - U_\beta)$$

$$+ (Z_\alpha - U_\alpha)(Z_\beta - U_\beta)(Z_\gamma - U_\gamma) .$$

Since the last four terms are of order of magnitude smaller than n^{-1}, we have

$$E_\theta(Z_\alpha Z_\beta Z_\gamma) = \frac{1}{2} E_\theta(U_\alpha Z_\beta Z_\gamma + U_\beta Z_\alpha Z_\gamma + U_\gamma Z_\alpha Z_\beta)$$

$$- \frac{1}{2} E_\theta(U_\alpha U_\beta U_\gamma) + o(\frac{1}{n}) .$$

Furthermore we derive that

$$E_\theta(U_\alpha Z_\beta Z_\gamma) = \frac{1}{\sqrt{n}} D_\alpha E_\theta(Z_\beta Z_\gamma) - \frac{1}{\sqrt{n}} E_\theta[(D_\alpha Z_\beta) Z_\gamma]$$

$$- \frac{1}{\sqrt{n}} E_\theta[Z_\beta (D_\alpha Z_\gamma)]$$

$$= \frac{1}{\sqrt{n}} D_\alpha I^{\beta\gamma} + o(\frac{1}{n})$$

$$= - \frac{1}{\sqrt{n}}(K^{\alpha\beta\gamma} + J^{\alpha\beta\cdot\gamma} + J^{\alpha\gamma\cdot\beta}) + o(\frac{1}{n}) .$$

Thus the third cumulant is given as

$$E_\theta(Z_\alpha Z_\beta Z_\gamma) = -\frac{1}{\sqrt{n}}(2K^{\alpha\beta\gamma} + J^{\alpha\beta\cdot\gamma} + J^{\alpha\gamma\cdot\beta} + J^{\beta\gamma\cdot\alpha}) + o(\frac{1}{n}) = \frac{1}{\sqrt{n}} \beta_{\alpha\beta\gamma} + o(\frac{1}{n}) .$$

For the fourth cumulants, we shall need the following equation,

$$\begin{aligned} Z_\alpha Z_\beta Z_\gamma Z_\delta &= \frac{2}{3}(U_\alpha Z_\beta Z_\gamma Z_\delta + U_\beta Z_\alpha Z_\gamma Z_\delta + U_\gamma Z_\alpha Z_\beta Z_\delta + U_\delta Z_\alpha Z_\beta Z_\gamma) \\ &- \frac{1}{3}(U_\alpha U_\beta Z_\gamma Z_\delta + U_\alpha U_\gamma Z_\beta Z_\delta + U_\alpha U_\delta Z_\beta Z_\gamma + U_\beta U_\gamma Z_\alpha Z_\delta + U_\beta U_\delta Z_\alpha Z_\gamma + U_\gamma U_\delta Z_\alpha Z_\beta) \\ &+ \frac{1}{3} U_\alpha U_\beta U_\gamma U_\delta + \frac{1}{3} U_\alpha (Z_\beta - U_\beta)(Z_\gamma - U_\gamma)(Z_\delta - U_\delta) \\ &+ \frac{1}{3} U_\beta (Z_\alpha - U_\alpha)(Z_\gamma - U_\gamma)(Z_\delta - U_\delta) + \frac{1}{3} U_\gamma (Z_\alpha - U_\alpha)(Z_\beta - U_\beta)(Z_\delta - U_\delta) \\ &+ \frac{1}{3} U_\delta (Z_\alpha - U_\alpha)(Z_\beta - U_\beta)(Z_\gamma - U_\gamma) + (Z_\alpha - U_\alpha)(Z_\beta - U_\beta)(Z_\gamma - U_\gamma)(Z_\delta - U_\delta) . \end{aligned}$$

The last four terms in the above are of order of magnitude less than n^{-1}, hence we have

$$\begin{aligned} (5.1.9)\quad E_\theta(Z_\alpha Z_\beta Z_\gamma Z_\delta) &= \frac{2}{3} E_\theta(U_\alpha Z_\beta Z_\gamma Z_\delta + U_\beta Z_\alpha Z_\gamma Z_\delta + U_\gamma Z_\alpha Z_\beta Z_\delta + U_\delta Z_\alpha Z_\beta Z_\gamma) \\ &- \frac{1}{3} E_\theta(U_\alpha U_\beta Z_\gamma Z_\delta + U_\alpha U_\gamma Z_\beta Z_\delta + U_\alpha U_\delta Z_\beta Z_\gamma + U_\beta U_\gamma Z_\alpha Z_\delta \\ &+ U_\beta U_\delta Z_\alpha Z_\gamma + U_\gamma U_\delta Z_\alpha Z_\beta) + \frac{1}{3} E_\theta(U_\alpha U_\beta U_\gamma U_\delta) + o(\frac{1}{n}) \end{aligned}$$

Applying the lemma 5.1.1, it follows that

$$\begin{aligned} E_\theta(U_\alpha Z_\beta Z_\gamma Z_\delta) &= \frac{1}{\sqrt{n}} D_\alpha E_\theta(Z_\beta Z_\gamma Z_\delta) - \frac{1}{\sqrt{n}} E_\theta((D_\alpha Z_\beta) Z_\gamma Z_\delta) \\ &- \frac{1}{\sqrt{n}} E_\theta((D_\alpha Z_\gamma) Z_\beta Z_\delta) - \frac{1}{\sqrt{n}} E_\theta((D_\alpha Z_\delta) Z_\beta Z_\gamma) \\ &= - \frac{1}{n} D_\alpha (2K^{\beta\gamma\delta} + J^{\beta\gamma\cdot\delta} + J^{\beta\delta\cdot\gamma} + J^{\gamma\delta\cdot\beta}) \\ &+ (I^{\alpha\beta} + \frac{1}{n} D_\alpha \mu_\beta) E_\theta(Z_\gamma Z_\delta) + (I^{\alpha\gamma} + \frac{1}{n} D_\alpha \mu_\gamma) E_\theta(Z_\beta Z_\delta) \\ &+ (I^{\alpha\delta} + \frac{1}{n} D_\alpha \mu_\delta) E_\theta(Z_\beta Z_\gamma) + o(\frac{1}{n}) \\ &= - \frac{1}{n} D_\alpha (2K^{\beta\gamma\delta} + J^{\beta\gamma\cdot\delta} + J^{\beta\delta\cdot\gamma} + J^{\gamma\delta\beta}) \end{aligned}$$

$$+ I^{\alpha\beta} E_\theta(Z_\gamma Z_\delta) + I^{\alpha\gamma} E_\theta(Z_\beta Z_\delta) + I^{\alpha\delta} E_\theta(Z_\beta Z_\gamma)$$
$$+ \frac{1}{n}(I^{\gamma\delta} D_\alpha \mu_\beta + I^{\beta\delta} D_\alpha \mu_\gamma + I^{\beta\gamma} D_\alpha \mu_\delta) + o(\frac{1}{n}) .$$

It also follows that

$$E_\theta(U_\alpha U_\beta Z_\gamma Z_\delta) = I^{\alpha\beta} E_\theta(Z_\gamma Z_\delta) - \frac{1}{\sqrt{n}} E_\theta(V_{\alpha\beta} Z_\gamma Z_\delta)$$
$$+ \frac{1}{n} D_\alpha D_\beta E_\theta(Z_\gamma Z_\delta) - \frac{1}{n} D_\alpha E_\theta((D_\beta Z_\gamma) Z_\delta) - \frac{1}{n} D_\alpha E_\theta(Z_\gamma D_\beta Z_\delta)$$
$$- \frac{1}{n} D_\beta E_\theta((D_\alpha Z_\gamma) Z_\delta) - \frac{1}{n} D_\beta E_\theta(Z_\gamma D_\alpha Z_\delta)$$
$$+ \frac{1}{n} E_\theta((D_\alpha D_\beta Z_\gamma) Z_\delta) + \frac{1}{n} E_\theta((D_\alpha Z_\gamma)(D_\beta Z_\delta))$$
$$+ \frac{1}{n} E_\theta((D_\beta Z_\gamma)(D_\alpha Z_\delta)) + \frac{1}{n} E_\theta((D_\alpha D_\beta Z_\delta) Z_\gamma)$$
$$= I^{\alpha\beta} E_\theta(Z_\gamma Z_\delta) - \frac{1}{\sqrt{n}} E_\theta(V_{\alpha\beta} Z_\gamma Z_\delta) + \frac{1}{n} D_\alpha D_\beta I^{\gamma\delta}$$
$$+ (I^{\alpha\gamma} + \frac{1}{n} D_\alpha \mu_\gamma)(I^{\beta\delta} + \frac{1}{n} D_\beta \mu_\delta)$$
$$+ (I^{\beta\gamma} + \frac{1}{n} D_\beta \mu_\gamma)(I^{\alpha\delta} + \frac{1}{n} D_\alpha \mu_\delta) + o(\frac{1}{n}) .$$

We shall show that

$$E_\theta(V_{\alpha\beta} Z_\gamma Z_\delta) = E_\theta(V_{\alpha\beta} U_\gamma U_\delta) + \frac{1}{\sqrt{n}} E_\theta(V_{\alpha\beta} V_{\gamma\delta}) + \frac{1}{\sqrt{n}} E_\theta(V_{\alpha\beta} V_{\delta\gamma}) + o(\frac{1}{\sqrt{n}}) .$$

In order to prove this, first note that

$$E_\theta(V_{\alpha\beta} Z_\gamma Z_\delta) = E_\theta(V_{\alpha\beta} U_\gamma Z_\delta) + E_\theta(V_{\alpha\beta} U_\delta Z_\gamma)$$
$$- E_\theta(V_{\alpha\beta} U_\gamma U_\delta) + o(\frac{1}{\sqrt{n}}) .$$

And we have

$$E_\theta(U_\gamma V_{\alpha\beta} Z_\delta) = \frac{1}{\sqrt{n}} D_\gamma E_\theta(V_{\alpha\beta} Z_\delta) - \frac{1}{\sqrt{n}} E_\theta((D_\gamma V_{\alpha\beta}) Z_\delta)$$
$$- \frac{1}{\sqrt{n}} E_\theta(V_{\alpha\beta} D_\gamma Z_\delta)$$
$$= \frac{1}{\sqrt{n}} D_\gamma E_\theta(V_{\alpha\beta} Z_\delta) - \frac{1}{\sqrt{n}} E_\theta((D_\gamma Z_{\alpha\beta}) Z_\delta)) + o(\frac{1}{\sqrt{n}})$$
$$= \frac{1}{\sqrt{n}} D_\gamma E_\theta(V_{\alpha\beta} U_\delta) - \frac{1}{\sqrt{n}} E_\theta((D_\gamma V_{\alpha\beta}) U_\delta)) + o(\frac{1}{\sqrt{n}}) .$$

On the other hand,

$$E_\theta(U_\gamma V_{\alpha\beta} U_\delta) = \frac{1}{\sqrt{n}} D_\gamma E_\theta(V_{\alpha\beta} U_\delta) - \frac{1}{\sqrt{n}} E_\theta((D_\gamma V_{\alpha\beta}) U_\delta)$$
$$- \frac{1}{\sqrt{n}} E_\theta(V_{\alpha\beta} V_{\gamma\delta}) + o(\frac{1}{\sqrt{n}}) \quad \text{holds.}$$

It follows that

$$E_\theta(U_\gamma V_{\alpha\beta} Z_\delta) = E_\theta(U_\gamma V_{\alpha\beta} U_\delta) + E_\theta(V_{\alpha\beta} V_{\gamma\delta}) + o(\frac{1}{\sqrt{n}})$$

From this the desired relation is obtained.

Thus we get

$$\begin{aligned} E_\theta(U_\alpha U_\beta Z_\gamma Z_\delta) = {} & I^{\alpha\beta} E_\theta(Z_\gamma Z_\delta) - \frac{1}{\sqrt{n}} E_\theta(V_{\alpha\beta} U_\gamma U_\delta) \\ & - \frac{1}{n} E_\theta(V_{\alpha\beta} V_{\gamma\delta}) - \frac{1}{n} E_\theta(V_{\alpha\beta} V_{\delta\gamma}) \\ & + I^{\alpha\gamma} I^{\beta\delta} + I^{\alpha\delta} I^{\beta\gamma} + \frac{1}{n} D_\alpha D_\beta I^{\gamma\delta} \\ & + \frac{1}{n}(I^{\alpha\gamma} D_\beta \mu_\delta + I^{\beta\gamma} D_\alpha \mu_\delta + I^{\alpha\delta} D_\beta \mu_\gamma + I^{\beta\delta} D_\alpha \mu_\gamma) + o(\frac{1}{n}) . \end{aligned}$$

Substituting (5.1.5) $\sim$ (5.1.8) into the terms $E_\theta(V_{\alpha\beta} U_\gamma U_\delta)$, $E_\theta(V_{\alpha\beta} V_{\gamma\delta})$, $E_\theta(V_{\alpha\beta} V_{\delta\gamma})$ and $D_\alpha D_\beta I^{\gamma\delta} = -D_\alpha(K^{\beta\gamma\delta} + J^{\beta\gamma\cdot\delta} + J^{\beta\delta\cdot\gamma})$, we obtain after some algebraic manipulations,

$$\begin{aligned} E_\theta(U_\alpha U_\beta Z_\gamma Z_\delta) = {} & I^{\alpha\beta} E_\theta(Z_\gamma Z_\delta) + I^{\alpha\gamma} I^{\beta\delta} + I^{\alpha\delta} I^{\beta\gamma} \\ & + \frac{1}{n}(I^{\alpha\gamma} D_\beta \mu_\delta + I^{\alpha\delta} D_\beta \mu_\gamma + I^{\beta\gamma} D_\alpha \mu_\delta + I^{\beta\delta} D_\alpha \mu_\delta) \\ & - \frac{1}{n}(H^{\alpha\beta\gamma\delta} + 2N^{\alpha\beta\cdot\gamma\delta} + N^{\alpha\gamma\cdot\beta\delta} + N^{\alpha\delta\cdot\beta\gamma} \\ & + N^{\beta\gamma\cdot\alpha\delta} + N^{\beta\delta\cdot\alpha\gamma} + L^{\alpha\beta\gamma\cdot\delta} + L^{\alpha\beta\delta\cdot\gamma} + 2M^{\alpha\beta\cdot\gamma\delta} + M^{\alpha\gamma\cdot\beta\delta} \\ & + M^{\alpha\delta\cdot\beta\gamma}) \\ & + \frac{1}{n} \sum_{\gamma'} (K^{\alpha\gamma\gamma'} + J^{\alpha\gamma\cdot\gamma'} + J^{\alpha\gamma'\cdot\gamma})(K^{\beta\delta}_{\gamma'} + J^{\beta\delta}_{\gamma'} + J^{\beta\cdot\delta}_{\gamma'}) \\ & + \frac{1}{n} \sum_{\delta'} (K^{\alpha\delta\delta'} + J^{\alpha\delta\cdot\delta'} + J^{\alpha\delta'\cdot\delta})(K^{\beta\gamma}_{\delta'} + J^{\beta\gamma}_{\delta'} + J^{\beta\cdot\gamma}_{\delta'}) \\ & + \frac{1}{n} \sum_{\gamma'} J^{\alpha\beta\cdot\gamma'}(2K^{\beta\delta}_{\gamma'} + 2J^{\gamma\delta\cdot\gamma'} + J^{\gamma\gamma'\cdot\delta} + J^{\gamma'\delta\cdot\gamma}) + o(\frac{1}{n}) . \end{aligned}$$

By a similar calculation we have

$$\begin{aligned} E_\theta(U_\alpha Z_\beta Z_\gamma Z_\delta) = {} & I^{\alpha\beta} E_\theta(Z_\gamma Z_\delta) + I^{\alpha\gamma} E_\theta(Z_\beta Z_\delta) + I^{\alpha\delta} E_\theta(Z_\beta Z_\gamma) \\ & + \frac{1}{n}(I^{\gamma\delta} D_\alpha \mu_\beta + I^{\beta\delta} D_\alpha \mu_\gamma + I^{\beta\gamma} D_\alpha \mu_\beta) \\ & - \frac{1}{n}(2H^{\alpha\beta\gamma\delta} + 2N^{\alpha\beta\cdot\gamma\delta} + 2N^{\alpha\gamma\cdot\beta\delta} + 2N^{\alpha\delta\cdot\beta\gamma} + N^{\beta\gamma\cdot\alpha\delta} \end{aligned}$$

$$+ N^{\beta\delta\cdot\alpha\gamma} + N^{\gamma\delta\cdot\alpha\beta} + L^{\alpha\beta\gamma\cdot\delta} + L^{\alpha\beta\delta\cdot\gamma} + L^{\alpha\gamma\delta\cdot\beta}$$

$$+ M^{\alpha\delta\cdot\beta\gamma} + M^{\alpha\gamma\cdot\beta\delta} + M^{\alpha\beta\cdot\gamma\delta})$$

$$+ \frac{1}{n} \sum_{\beta'} (K^{\alpha\beta\beta'} + J^{\alpha\beta\cdot\beta'} + J^{\alpha\beta'\cdot\beta})(2K^{\gamma\delta}_{\beta'} + J^{\gamma\delta}_{\beta'} + J^{\gamma\cdot\delta}_{\beta'} + J^{\delta\cdot\gamma}_{\beta'})$$

$$+ \frac{1}{n} \sum_{\gamma'} (K^{\alpha\gamma\gamma'} + J^{\alpha\gamma\cdot\gamma'} + J^{\alpha\gamma'\cdot\gamma})(2K^{\beta\delta}_{\gamma'} + J^{\beta\delta}_{\gamma'} + J^{\beta\cdot\delta}_{\gamma'} + J^{\delta\cdot\beta}_{\gamma'})$$

$$+ \frac{1}{n} \sum_{\delta'} (K^{\alpha\delta\delta'} + J^{\alpha\delta\cdot\delta'} + J^{\alpha\delta'\cdot\delta})(2K^{\beta\gamma}_{\delta'} + J^{\beta\gamma}_{\delta'} + J^{\beta\cdot\gamma}_{\delta'} + J^{\gamma\cdot\beta}_{\delta'})$$

$$+ o(\frac{1}{n}) .$$

Substituting all these forms together with

$$E_\theta(U_\alpha U_\beta U_\gamma U_\delta) = \frac{1}{n} H^{\alpha\beta\gamma\delta} + I^{\alpha\beta} I^{\gamma\delta} + I^{\alpha\gamma} I^{\beta\delta} + I^{\alpha\delta\cdot\beta\gamma}$$

into (5.1.9) we have the fourth cumulant as

$$E_\theta(Z_\alpha Z_\beta Z_\gamma Z_\delta) - E_\theta(Z_\alpha Z_\beta)E_\theta(Z_\gamma Z_\delta) - E_\theta(Z_\alpha Z_\gamma)E_\theta(Z_\beta Z_\delta) - E_\theta(Z_\alpha Z_\delta)E_\theta(Z_\beta Z_\gamma)$$

$$= (I^{\alpha\beta} - E_\theta(Z_\alpha Z_\beta))(I^{\gamma\delta} - E_\theta(Z_\gamma Z_\delta)) + (I^{\alpha\gamma} - E_\theta(Z_\alpha Z_\gamma))(I^{\beta\delta} - E_\theta(Z_\beta Z_\delta))$$

$$+ (I^{\alpha\delta} - E_\theta(Z_\alpha Z_\delta))(I^{\beta\gamma} - E_\theta(Z_\beta Z_\gamma))$$

$$- \frac{1}{n}(3H^{\alpha\beta\gamma\delta} + 2N^{\alpha\beta\cdot\gamma\delta} + 2N^{\alpha\gamma\cdot\beta\delta} + 2N^{\alpha\delta\cdot\beta\gamma} + 2N^{\beta\gamma\cdot\alpha\delta} + 2N^{\beta\delta\cdot\alpha\gamma}$$

$$+ 2N^{\gamma\delta\cdot\alpha\beta} + L^{\alpha\beta\gamma\cdot\delta} + L^{\alpha\beta\delta\cdot\gamma} + L^{\alpha\gamma\delta\cdot\beta} + L^{\beta\gamma\delta\cdot\alpha})$$

$$+ \frac{1}{2n}(\Delta^{\alpha\beta\cdot\gamma\delta} + \Delta^{\beta\alpha\cdot\gamma\delta} + \Delta^{\alpha\gamma\cdot\beta\delta} + \Delta^{\gamma\alpha\cdot\beta\delta} + \Delta^{\beta\delta\cdot\alpha\gamma} + \Delta^{\delta\beta\cdot\alpha\gamma} + \Delta^{\gamma\delta\cdot\alpha\beta}$$

$$+ \Delta^{\delta\gamma\cdot\alpha\beta} + \Delta^{\beta\gamma\cdot\alpha\delta} + \Delta^{\gamma\beta\cdot\alpha\delta} + \Delta^{\alpha\delta\cdot\beta\gamma} + \Delta^{\delta\alpha\cdot\beta\gamma})$$

$$+ o(\frac{1}{n}) ,$$

where

$$\Delta^{\alpha\beta\cdot\gamma\delta} = \sum_{\beta'} (K^{\alpha\beta\cdot\beta'} + J^{\alpha\beta\cdot\beta'} + J^{\alpha\beta'\cdot\beta})(2K^{\gamma\delta}_{\beta'} + J^{\gamma\cdot\delta}_{\beta'} + J^{\delta\cdot\gamma}_{\beta'})$$

$$= \sum_{\xi}\sum_{\alpha'}\sum_{\beta'}\sum_{\gamma'}\sum_{\delta'}\sum_{\xi'} I^{\alpha\alpha'} I^{\beta\beta'} I^{\gamma\gamma'} I^{\delta\delta'} I^{\xi\xi'} (K_{\alpha'\beta'\xi'} + J_{\alpha'\beta'\cdot\xi'} + J_{\alpha'\xi'\cdot\gamma'})$$

$$\cdot (2K_{\gamma'\delta'\xi'} + J_{\gamma'\xi'\cdot\delta'} + J_{\alpha'\xi'\cdot\gamma'}) .$$

The first three terms in the above are of order smaller than n^{-1}. Then the 4th cumulant is given by

(5.1.10) $$- \frac{3}{n} H^{\alpha\beta\gamma\delta} - \frac{2}{n}(N^{\alpha\beta\cdot\gamma\delta} + N^{\alpha\gamma\cdot\beta\delta} + N^{\alpha\delta\cdot\beta\gamma} + N^{\beta\gamma\cdot\alpha\delta} + N^{\beta\delta\cdot\alpha\gamma} + N^{\gamma\delta\cdot\alpha\beta})$$

$$- \frac{1}{n}(L^{\alpha\beta\gamma\cdot\delta} + L^{\alpha\beta\delta\cdot\gamma} + L^{\alpha\gamma\delta\cdot\beta} + L^{\beta\gamma\delta\cdot\alpha})$$

$$+ \frac{1}{2n}(\Delta^{\alpha\beta\cdot\gamma\delta} + \Delta^{\beta\alpha\cdot\gamma\delta} + \Delta^{\alpha\gamma\cdot\beta\delta} + \Delta^{\gamma\alpha\cdot\beta\delta} + \Delta^{\beta\delta\cdot\alpha\gamma} + \Delta^{\delta\beta\cdot\alpha\gamma}$$

$$+ \Delta^{\gamma\delta\cdot\alpha\beta} + \Delta^{\delta\gamma\cdot\alpha\beta} + \Delta^{\beta\gamma\cdot\alpha\delta} + \Delta^{\gamma\beta\cdot\alpha\delta} + \Delta^{\alpha\delta\cdot\beta\gamma} + \Delta^{\delta\alpha\cdot\beta\gamma})$$

$$+ o(\frac{1}{n}) = \frac{1}{n}\beta_{\alpha\beta\gamma\delta} + o(\frac{1}{n}) \quad \text{(say)}$$

Summarizing the computations we have the following theorem.

Theorem 5.1.4. The asymptotic cumulants of $\{\sqrt{n}(\hat{\theta}_\alpha - \theta_\alpha); \alpha=1,\cdots,p\}$ up to the order n^{-1} are given by

$$E(\sqrt{n}(\hat{\theta}_\alpha - \theta_\alpha)) = \frac{1}{\sqrt{n}}\mu_\alpha + o(\frac{1}{n}) ;$$

$$\mathrm{Cov}_\theta(\sqrt{n}(\hat{\theta}_\alpha - \theta_\alpha), \sqrt{n}(\hat{\theta}_\beta - \theta_\beta)) = I^{\alpha\beta} + \frac{1}{n}D_\alpha\mu_\beta + \frac{1}{n}D_\beta\mu_\alpha + \frac{1}{4n}\mathrm{Cov}_\theta(Q_\alpha, Q_\beta) + o(\frac{1}{n}) ;$$

$$K_3(\sqrt{n}(\hat{\theta}_\alpha - \theta_\alpha), \sqrt{n}(\hat{\theta}_\beta - \theta_\beta), \sqrt{n}(\hat{\theta}_\gamma - \theta_\gamma)) = \frac{1}{\sqrt{n}}\beta_{\alpha\beta\gamma} + o(\frac{1}{n}) ;$$

$$K_4(\sqrt{n}(\hat{\theta}_\alpha - \theta_\alpha), \sqrt{n}(\hat{\theta}_\beta - \theta_\beta), \sqrt{n}(\hat{\theta}_\gamma - \theta_\gamma), \sqrt{n}(\hat{\theta}_\delta - \theta_\delta)) = \frac{1}{n}\beta_{\alpha\beta\gamma\delta} + o(\frac{1}{n}),$$

where $\beta_{\alpha\beta\gamma}$ and $\beta_{\alpha\beta\gamma\delta}$ are given above.

We also have the following :

Theorem 5.1.5. The joint distribution of $\{\sqrt{n}(\hat{\theta}_\alpha - \theta_\alpha); \alpha=1,\ldots,p\}$ is equivalent up to the order n^{-1} to a continuous distribution, whose density is given by :

$$f(y_1, \ldots, y_p) = \frac{|I|^{1/2}}{(\sqrt{2\pi})^p} \left[\exp - \frac{1}{2}\left\{\sum_\alpha \sum_\beta I_{\alpha\beta}\, y_\alpha y_\beta\right\}\right]$$

$$\cdot\left\{ 1 + \frac{1}{6\sqrt{n}} \sum_\alpha \sum_\beta \sum_\gamma \beta_{\alpha\beta\gamma} H_{\alpha\beta\gamma} + \frac{1}{\sqrt{n}} \sum_\alpha \mu_\alpha H_\alpha \right.$$

$$+ \frac{1}{72n} \sum_\alpha \sum_\beta \sum_\gamma \sum_{\alpha'} \sum_{\beta'} \sum_{\gamma'} \beta_{\alpha\beta\gamma} \beta_{\alpha'\beta'\gamma'} H_{\alpha\beta\gamma\alpha'\beta'\gamma'}$$

$$+ \frac{1}{24n} \sum_\alpha \sum_\beta \sum_\gamma \sum_\delta (\beta_{\alpha\beta\gamma\delta} + 4\mu_\alpha \beta_{\beta\gamma\delta}) H_{\alpha\beta\gamma\delta}$$

$$+ \frac{1}{2n} \sum_\alpha \sum_\beta (\mu_\alpha \mu_\beta + D_\alpha \mu_\beta + D_\beta \mu_\alpha) H_{\alpha\beta}$$

$$\left. + \frac{1}{4n} \sum_\alpha \sum_\beta \operatorname{Cov}(Q_\alpha, Q_\beta) H_{\alpha\beta} \right\} + o\left(\frac{1}{n}\right),$$

where $H_{\alpha\beta\cdots}$ are (multivariate) Hermite polynomials defined by

$$H_\alpha = \sum_\beta I_{\alpha\beta} y_\beta \quad ;$$

$$H_{\alpha\beta} = H_\alpha H_\beta - I_{\alpha\beta} \quad ;$$

$$H_{\alpha\beta\gamma} = H_\alpha H_\beta H_\gamma - I_{\alpha\beta} H_\gamma - I_{\alpha\gamma} H_\beta - I_{\beta\gamma} H_\alpha \quad ,$$

etc.

Remark : The third and the fourth cumulants are equal for all ER BAN estimators up to the order n^{-1}.

Now we impose some higher order conditions on ER estimators. Either of the following three conditions may be taken into consideration.

a) higher order asymptotic unbiasedness : $\mu_\alpha = 0$, $\alpha = 1, \ldots, p$;

b) higher order asymptotic coordinate-wise median unbiasedness :

$$\lim_{n\to\infty} \sqrt{n} \left| P_{\theta,n}\{\hat{\theta}_\alpha \leq \theta_\alpha\} - \frac{1}{2} \right| = 0 \ , \quad \alpha = 1, \ldots, p.$$

Then μ_α should be expressed in terms of 3rd cumulants, i.e. $\mu_\alpha = \frac{1}{6} \beta_{\alpha\alpha\alpha}$ ($\alpha = 1, \ldots, p$).

c) higher order asymptotic mode unbiasedness :

the asymptotic density has its mode at the origin ;
again μ_α should be expressed in terms of 3rd cumulants.
Now we can consider the class S of estimators which are ER BAN estimators satisfying one of the above conditions.

It should be remarked that for a given g which satisfies the condition

$$\frac{\partial}{\partial T_i} g \Big|_{T_i=\mu_i} = D_\alpha s_i$$

we can determine h so that

$$\mu_\alpha = \frac{1}{2} \sum_i \sum_j b_{ij} \sigma_{ij} + h(\mu_1, \ldots, \mu_m) ,$$

where $b_{ij} = \frac{\partial^2}{\partial T_i \partial T_j} g \Big|_{T_i=\mu_i}$.

satisfies the conditions stated above. Therefore for any g there is an ER estimator which belongs to the class S.

Now for the estimators in S, the asymptotic distribution up to the order n^{-1} is determined except for the term

$$\frac{1}{4n} Cov_\theta(Q_\alpha , Q_\beta) .$$

Therefore we shall minimize the matrix

$$\Lambda = \{\tau_{\alpha\beta}\} ;$$

$$\tau_{\alpha\beta} = Cov_\theta(Q_\alpha , Q_\beta) = nE_\theta(Z_\alpha - U_\alpha)(Z_\beta - U_\beta) + o(1) .$$

$\sqrt{n}(Z_\alpha - U_\alpha)$ must satisfy the following conditions

$$\sqrt{n}\, E_\theta(U_\beta U_\gamma (Z_\alpha - U_\alpha)) = D_\beta E_\theta (U_\gamma (Z_\alpha - U_\alpha)) - E_\theta((D_\beta U_\gamma)(Z_\alpha - U_\alpha))$$
$$- E_\theta(U_\gamma D_\beta (Z_\alpha - U_\alpha))$$
$$= E_\theta(U_\gamma V_{\beta\alpha}) + o(1) = -K^{\alpha\beta\gamma} - J^{\beta\gamma\cdot\alpha} + o(1) ;$$

$$\sqrt{n}\, E_\theta(U_\beta V_{\gamma\delta}(Z_\alpha - U_\alpha)) = D_\beta E_\theta(V_{\gamma\delta}(Z_\alpha - U_\alpha))$$

$$- E_\theta((D_\beta V_{\gamma\delta})(Z_\alpha - U_\alpha)) - E_\theta(V_{\gamma\delta} D_\beta (Z_\alpha - U_\alpha))$$

$$= E_\theta(V_{\gamma\delta} V_{\beta\alpha}) + o(1) .$$

Under these conditions $n\mathrm{Cov}_\theta(Z_\alpha - U_\alpha , Z_\beta - U_\beta)$ is minimized (asymptotically) when it is expressed as

$$\sqrt{n}\,(Z_\alpha - U_\alpha) = \sum_\beta \sum_\gamma \lambda_{\alpha\beta\gamma} U_\beta U_\gamma + \sum_\alpha \sum_\beta \sum_\gamma \sum_\delta \lambda_{\alpha\beta\gamma\delta} U_\beta V_{\gamma\delta} + o(1) ,$$

where $\lambda_{\alpha\beta\gamma}$ and $\lambda_{\alpha\beta\gamma\delta}$ are lagrangian multipliers determined so that the conditions are satisfied.

Now instead of explicitly determining $\lambda_{\alpha\beta\gamma}$ and $\lambda_{\alpha\beta\gamma\delta}$, we shall show that the above equation is actually satisfied when g is equal to the MLE.

The MLE is determined by the equations

$$\text{(5.1.11)} \qquad \sum_i \frac{\partial s_i}{\partial \theta_\alpha}(T_i - \mu_i) = 0 , \quad \alpha = 1, \ldots , p .$$

Since the mapping from $(s_1, \ldots , s_m)$ to $(\mu_1, \ldots , \mu_m)$ is 1 : 1 and continuous, $\hat{\theta}_0$ is consistent provided that $(s_1(\theta_1) \ldots s_m(\theta_1)) \neq (s_1(\theta_2) \ldots s_m(\theta_2))$ when $\theta_1 \neq \theta_2$. Moreover if for all θ , and for any t_α , $\left(\sum_\alpha t_\alpha \frac{\partial s_1}{\partial \theta_\alpha} , \ldots , \sum_\alpha t_\alpha \frac{\partial s_m}{\partial \theta_\alpha}\right) \neq (0,\ldots,0)$, it is seen that

$$P_{\theta,n}\left\{ \| \hat{\theta} - \theta \| > \varepsilon \right\} = o(n^{-k}) \text{ for any } k .$$

Consequently we can expand the estimator around the true value of θ .

Differentiating (5.1.11) with respect to T_i, we have

$$\text{(5.1.12)} \qquad \sum_\alpha \frac{\partial \hat{\theta}_\alpha}{\partial T_i} \left\{ \sum_j \frac{\partial^2 s_j}{\partial \theta_\alpha \partial \theta_\beta} (T_j - \mu_j) - \sum_j \frac{\partial s_j}{\partial \theta_\beta} \frac{\partial \mu_j}{\partial \theta_\alpha} \right\} + \frac{\partial s_i}{\partial \theta_\beta} = 0 ,$$

$$\beta = 1 , \ldots , p .$$

Putting $T_i = \mu_i$ we have

$$- \sum_{\alpha} I_{\alpha\beta} \frac{\partial \hat{\theta}_\alpha}{\partial T_i} + \frac{\partial s_i}{\partial \theta_\beta} = 0 \ , \quad \beta = 1, \ldots , p \ ;$$

$$\frac{\partial \hat{\theta}_\alpha}{\partial T_i} = \sum_{\beta} I^{\alpha\beta} \frac{\partial s_i}{\partial \theta_\beta}$$

which is to be expected since MLE is a BAN estimator.

Differentiating (5.1.12) again with respect to T_j, we have

$$\sum_{\alpha} \frac{\partial^2 \hat{\theta}_\alpha}{\partial T_i \partial T_j} \left\{ \sum_{k} \frac{\partial^2 s_k}{\partial \theta_\alpha \partial \theta_\beta} (T_k - \mu_k) - \sum_{k} \frac{\partial s_k}{\partial \theta_\alpha} \frac{\partial \mu_k}{\partial \theta_\beta} \right\}$$

$$+ \sum_{\alpha} \sum_{\gamma} \left(\frac{\partial \hat{\theta}_\alpha}{\partial T_i}\right) \left(\frac{\partial \hat{\theta}_\gamma}{\partial T_j}\right) \left\{ \sum_{k} \frac{\partial^3 s_k}{\partial \theta_\alpha \partial \theta_\beta \partial \theta_\gamma} (T_k - \mu_k) - \sum_{k} \frac{\partial^2 s_k}{\partial \theta_\alpha \partial \theta_\beta} \frac{\partial \mu_k}{\partial \theta_\gamma} \right.$$

$$\left. - \sum_{k} \frac{\partial^2 s_k}{\partial \theta_\beta \partial \theta_\gamma} \frac{\partial \mu_k}{\partial \theta_\alpha} - \sum_{k} \frac{\partial s_k}{\partial \theta_\beta} \frac{\partial^2 \mu_k}{\partial \theta_\alpha \partial \theta_\gamma} \right\}$$

$$+ \sum_{\alpha} \frac{\partial \hat{\theta}_\alpha}{\partial T_i} \frac{\partial^2 s_i}{\partial \theta_\alpha \partial \theta_\beta} + \sum_{\gamma} \frac{\partial \hat{\theta}_\gamma}{\partial T_j} \frac{\partial^2 s_j}{\partial \theta_\gamma \partial \theta_\beta} = 0 \ ,$$

$$\beta = 1, \ldots , p \ .$$

Substituting $T_i = \mu_i$, $\dfrac{\partial \hat{\theta}_\alpha}{\partial T_i} = D_\alpha s_i$, $\displaystyle\sum_{k} \frac{\partial^2 s_k}{\partial \theta_\alpha \partial \theta_\beta} \frac{\partial \mu_k}{\partial \theta_\gamma} = J_{\alpha\beta \cdot \gamma}$,

$$\sum_{k} \frac{\partial s_k}{\partial \theta_\beta} \frac{\partial^2 \mu_k}{\partial \theta_\alpha \partial \theta_\gamma} = J_{\alpha\gamma \cdot \beta} + K_{\alpha\beta\gamma} \ ,$$

it follows that

$$- \sum_{\alpha} \left(\frac{\partial^2 \hat{\theta}_\alpha}{\partial T_i \partial T_j}\right) I^{\alpha\beta} - \sum_{\alpha} \sum_{\gamma} (D_\alpha s_i)(D_\gamma s_j)(J_{\alpha\beta \cdot \gamma} + J_{\alpha\gamma \cdot \beta} + J_{\beta\gamma \cdot \alpha} + K_{\alpha\beta\gamma})$$

$$+ \sum_{\alpha} (D_\alpha s_i) \frac{\partial^2 s_j}{\partial \theta_\alpha \partial \theta_\beta} + \sum_{\alpha} (D_\alpha s_j) \frac{\partial^2 s_i}{\partial \theta_\alpha \partial \theta_\beta} = 0 \ .$$

Hence we have

$$\frac{\partial^2 \hat{\theta}_\alpha}{\partial T_i \partial T_j} = - \sum_{\beta} \sum_{\gamma} \sum_{\delta} I^{\alpha\beta} (J_{\beta\gamma \cdot \delta} + J_{\beta\delta \cdot \gamma} + J_{\gamma\delta \cdot \beta} + K_{\beta\gamma\delta})(D_\gamma s_i)(D_\delta s_j)$$

$$+ \sum_{\beta} \sum_{\gamma} I^{\alpha\beta} \left(\frac{\partial^2 s_j}{\partial \theta_\beta \partial \theta_\gamma}\right)(D_\gamma s_i) + \sum_{\beta} \sum_{\gamma} I^{\alpha\beta} \left(\frac{\partial^2 s_i}{\partial \theta_\beta \partial \theta_\gamma}\right)(D_\gamma s_j) \ .$$

We have

$$\sum_j \sum_\beta I^{\alpha\beta}\left(\frac{\partial^2 s_j}{\partial\theta_\beta\partial\theta_\gamma}\right)T_j' = \sum_{\gamma'} I_{\gamma\gamma'}V_{\alpha\gamma'} + \sum_{\beta'}\sum_{\gamma'} I_{\gamma\gamma'}(K^{\alpha\beta'\gamma'} + J^{\alpha\beta'\cdot\gamma'} + J^{\alpha\gamma'\cdot\beta'})\sum_i\left(\frac{\partial s_i}{\partial\theta_{\beta'}}\right)T_i'$$

$$= \sum_\delta I_{\gamma\delta}V_{\alpha\delta} + \sum_{\alpha'}\sum_\beta I^{\alpha\alpha'}(K_{\alpha'\beta\gamma} + J_{\alpha'\beta\cdot\gamma} + J_{\alpha'\gamma\cdot\beta})U_\beta \quad .$$

Consequently,

$$\sum_i\sum_j\left(\frac{\partial^2\hat\theta_\alpha}{\partial T_i\partial T_j}\right)T_i'T_j' = Q_\alpha = \sum_{\alpha'}\sum_\beta\sum_\gamma I^{\alpha\alpha'}(K_{\alpha'\beta\gamma} + 2J_{\alpha'\beta\gamma} - J_{\beta\gamma\cdot\alpha'})U_\beta U_\gamma$$

$$+ 2\sum_\delta\sum_\gamma I_{\gamma\delta}V_{\alpha\delta}U_\gamma$$

which has the form required for the minimizing solution of Λ .

Thus if Λ^* denotes the variance-covariance matrix of Q_α corresponding to the (extended) MLE and Λ be that of any other estimator in the set S, we have

$$\Lambda - \Lambda^* \geqq 0 \quad \text{(non-negative definite)} .$$

From this the following theorem is established :

Theorem 5.1.6. Let $\hat\theta^*$ be the extended MLE in S, and $\hat\theta$ be any other estimator in S and C any convex set in R^p containing the origin. Then it holds that

$$\varliminf_{n\to\infty} n[\, P_{\theta,n}\{\sqrt{n}(\hat\theta^* - \theta)\in C\} - P_{\theta,n}\{\sqrt{n}(\hat\theta - \theta)\in C\}\,] \geqq 0 \; .$$

We also have established :

Theorem 5.1.7. Suppose that L is a bounded, negatively unimodal function around the origin, i.e. for any c the set $\{u : L(u) \leqq c\} \subset R^p$ is convex and contains the origin. Let $\hat\theta^*$ be the extended MLE and $\hat\theta$ be any other estimator in S. Then

$$\varlimsup_{n\to\infty} n[\, E_{\theta,n}[L(\sqrt{n}(\hat\theta^* - \theta))] - E_{\theta,n}[L(\sqrt{n}(\hat\theta - \theta))]\,] \leqq 0 \; .$$

5.2. Third order asymptotic efficiency of maximum likelihood estimators in general cases

Summary :

In order to generalize the results of the previous section, it is necessary to define appropriate classes of estimators among which asymptotic distributions can be compared. As was shown in Chapter 4 it is possible to establish the second order asymptotic efficiency of the modified MLE within the class of second order AMU estimators without any further restriction on the class of estimators to be considered. Contrary to this, in the previous section we restricted the class of estimators to be special types of functions of the sufficient statistic which we called the extended regular estimators. As will be shown later we can not usually have third order asymptotically efficient estimators without such restriction. In more general cases where sufficient statistics of finite dimension do not exist, the argument of the previous section can not be applied directly.

We shall establish the third order asymptotic efficiency of the modified MLE and also of the modified generalized Bayes estimator within the classes $\mathbb{C}$ and $\mathbb{D}$ of estimators which are defined below. For simplicity's sake we shall first consider the real parameter (one dimensional) case. We assume that the sample is independently and identically distributed with all necessary regularity conditions to be specified later.

First we shall restrict our attention to the class of (first order) asymptotic efficient estimators which we shall call the class $\mathbb{A}$. It has been well established that any estimator $\hat{\theta}_n$ in the class $\mathbb{A}$ can be expressed as

$$\sqrt{n}(\hat{\theta}_n - \theta) = \frac{1}{I(\theta)\sqrt{n}} \frac{\partial L_n}{\partial \theta} + o_p(1) ,$$

where $L_n = \sum_i \log f(X_i, \theta)$.

Secondly, we call the class of estimators which are second order AMU and second order asymptotically efficient as <u>class $\mathbb{B}$</u>. Then any estimator $\hat{\theta}_n$ which is second order AMU and for which the distribution of $\sqrt{n}(\hat{\theta}_n - \theta)$ admits the Edgeworth expansion up to the order $n^{-1/2}$ belongs to the class $\mathbb{B}$.

Now we generalize the concept one step further. We call the <u>class $\mathbb{C}$</u> the class of estimators $\hat{\theta}_n$ which are third order AMU and for which the distribution of $\sqrt{n}(\hat{\theta}_n - \theta)$ admits the Edgeworth expansion up to the order n^{-1} and

$$\sqrt{n}(\hat{\theta}_n - \theta) = \frac{1}{I(\theta)\sqrt{n}} \frac{\partial L_n}{\partial \theta} + \frac{1}{\sqrt{n}} Q(\theta) + o_p(\frac{1}{n}) ,$$

where $Q(\theta)$ is a quantity of stochastic order 1.

We say the estimator $\hat{\theta}_n$ belongs to the <u>class $\mathbb{D}$</u> if in the above we have

$$E[\frac{\partial L_n}{\partial \theta} Q(\theta)^2] = 0 ,$$

where E stands for the asymptotic mean.

Note that Lemma 5.1.1 can be generalized straight-forwardly as the following lemma.

<u>Lemma 5.2.1.</u> Suppose that Z_θ is a function of $X_1, \dots, X_n$ and θ and is differentiable in θ. Then

$$E_\theta(UZ_\theta) = \frac{1}{\sqrt{n}} \frac{\partial}{\partial \theta} E_\theta(Z_\theta) - \frac{1}{\sqrt{n}} E_\theta(\frac{\partial Z_\theta}{\partial \theta}) ,$$

where $U = \frac{\sqrt{n}}{I(\theta)} \frac{\partial}{\partial \theta} \log f(X, \theta)$, provided that the differentiation under the integral sign of $E_\theta(Z_\theta)$ is allowed.

In the following discussion we assume that expectation of the relevant quantities always exist and the asymptotic value of the expectation and the asymptotic mean are the same at least up to the required order.

Remark : The problem of non-existence of the relevant moments vis-à-vis asymptotic moments can be evaded in the following way. Suppose that we have a smooth and monotone increasing transformation $\eta = \Psi(\theta)$ so that η is bounded. In stead of θ we may consider η as the parameter to be estimated. Since η is bounded we may assume that all the estimators are also bounded hence have moments of any order, and we can easily formulate the condition for which the asymptotic moments and the moments of the asymptotic distribution are equal. If we consider the estimation of η instead of θ the asymptotic theory based on moments is applicable in a straightforward way. Transforming back to the estimator of θ by $\hat{\theta} = \Psi^{-1}(\hat{\eta})$ and considering the monotoniticy of the transformation of the distribution, the results are also applicable to the estimation of θ .

Let us first consider the class $\mathbb{C}$. Suppose that $\sqrt{n}(\hat{\theta}_n - \theta)$ is expanded as

$$\sqrt{n}(\hat{\theta}_n - \theta) = \frac{Z_1(\theta)}{I(\theta)} + \frac{1}{\sqrt{n}}Q(\theta) + o_p(\frac{1}{n}) ,$$

where $Q(\theta) = O_p(1)$ and $Z_1(\theta) = \frac{1}{\sqrt{n}} \sum_{i=1}^{n} \frac{\partial}{\partial\theta} \log f(X_i, \theta)$.

Then we have the following lemma.

Lemma 5.2.2.

$$E_\theta[Z_1(\theta)Q(\theta)] = o(1) .$$

Proof.

Define

$$T_\theta = \sqrt{n}(\hat{\theta}_n - \theta) - \frac{Z_1(\theta)}{I(\theta)} = \frac{1}{\sqrt{n}}Q(\theta) + o_p(\frac{1}{\sqrt{n}}) .$$

Then we have

$$\frac{1}{I(\theta)}E\,[Z_1(\theta)T_\theta] = \frac{1}{\sqrt{n}}\,\frac{\partial}{\partial\theta}E_\theta(T_\theta) - \frac{1}{\sqrt{n}}E_\theta(\frac{\partial T_\theta}{\partial\theta})$$

$$= \frac{1}{n}\,\frac{\partial}{\partial\theta}E_\theta[Q\;(\theta)]+1-\frac{1}{\sqrt{n}}E_\theta[\frac{\partial}{\partial\theta}\left\{\frac{Z_1(\theta)}{I(\theta)}\right\}] + o_p(\frac{1}{\sqrt{n}})$$

$$= o_p(\frac{1}{\sqrt{n}}) \;.$$

From which we have

$$E_\theta[Z_1(\theta)Q(\theta)] = \sqrt{n}\;E\;[Z_1(\theta)T_\theta] + o_p(1)$$

$$= o_p(1)\;.$$

Thus we complete the proof.

Thus we have the asymptotic cumulants of $\sqrt{n}(\hat{\theta}_n-\theta)$ up to the order n^{-1} as

$$E_\theta[\sqrt{n}\;(\hat{\theta}_n-\theta)] = \frac{1}{\sqrt{n}}\,\mu_1(\theta) + \frac{1}{n}\,\mu_2(\theta) + o(\frac{1}{n}) \quad ;$$

$$V_\theta[\sqrt{n}\;(\hat{\theta}_n-\theta)] = \frac{1}{I(\theta)} + \frac{1}{n}\,v_2(\theta) + o(\frac{1}{n}) \quad ;$$

$$K_3[\sqrt{n}\;(\hat{\theta}_n-\theta)] = \frac{1}{\sqrt{n}}\,\beta_3(\theta) + \frac{1}{n}\,\gamma_3(\theta) + o(\frac{1}{n}) \;;$$

$$K_4[\sqrt{n}\;(\hat{\theta}_n-\theta)] = \frac{1}{n}\,\beta_4(\theta) + o(\frac{1}{n}) \;.$$

We can show in a way exactly similar to the previous section that

$$\beta_3(\theta) = -\,\frac{3J(\theta)+2K(\theta)}{I(\theta)^3} \;;$$

$$\beta_4(\theta) = -\,\frac{1}{I(\theta)^4}\left\{3H^*(\theta)+12N^*(\theta)+4L(\theta)\right\} + \frac{12}{I(\theta)^5}\left\{2J(\theta)+K(\theta)\right\}\left\{J(\theta)+K(\theta)\right\},$$

where

$$L(\theta)=E_\theta[\left\{\frac{\partial^3}{\partial\theta^3}\log\,f(x,\theta)\right\}\left\{\frac{\partial}{\partial\theta}\log\,f(x,\theta)\right\}] \quad ;$$

$$N^*(\theta)=E_\theta[\;\left\{\frac{\partial^2}{\partial\theta^2}\log\,f(x,\theta)+I(\theta)\right\}\left\{\frac{\partial}{\partial\theta}\log\,f(x,\theta)\right\}^2] \quad ;$$

$$H^*(\theta)=E_\theta[\;\left\{\frac{\partial}{\partial\theta}\log\,f(x,\theta)\right\}^4] - 3I(\theta)^2 \;.$$

Since the odd order cumulants affect only the asymmetric aspects of the asymptotic distribution, the asymptotic expansion of $P_{\theta,n}\{\sqrt{n}|\hat{\theta}_n-\theta|<a\}$ is independent of them. Hence we have the following theorem.

Theorem 5.2.1. Let $\hat{\theta}^*_{ML}$ be the modified MLE in the class $\mathbb{C}$ and $\hat{\theta}_n$ be any other estimator in the class $\mathbb{C}$. Then it holds that

$$\lim_{n\to\infty} n[P_{\theta,n}\{\sqrt{n}|\hat{\theta}^*_{ML}-\theta|<a\} - P_{\theta,n}\{\sqrt{n}|\hat{\theta}_n-\theta|<a\}] \geqq 0$$

for all $a>0$.

For the class $\mathbb{D}$ we further have that $\gamma_3(\theta)=0$ and $\mu_1(\theta)=\beta_3(\theta)/6$, hence all the asymptotic moments are determined up to the order n^{-1} except v_2. Since v_2 is minimized for the modified MLE we have the following theorem.

Theorem 5.2.2. Let $\hat{\theta}^{**}_{ML}$ be the modified MLE in the class $\mathbb{D}$ and $\hat{\theta}_n$ be any other estimator in the class $\mathbb{D}$. Then it holds that

$$\lim_{n\to\infty} n[P_{\theta,n}\{-a<\sqrt{n}(\hat{\theta}^{**}_{ML}-\theta)<b\} - P_{\theta,n}\{-a<\sqrt{n}(\hat{\theta}_n-\theta)<b\}] \geqq 0$$

for all $a>0$ and all $b>0$.

5.3 Second and third order asymptotic efficiency of the generalized Bayes estimator

In this section the following is shown : For all symmetric loss functions, the generalized Bayes estimator is second order asymptotically efficient in the class $\mathbb{A}_2$ of all second order asymptotically median unbiased (AMU) estimators. It is third order asymptotically efficient in the resticted class $\mathbb{D}$ of estimators. When the loss function is not symmetric, the generalized Bayes estimator is second order asymptotically efficient in the class $\mathbb{A}_2$ but not third order asymptotically efficient in a restricted wider class than the class $\mathbb{D}$. The results are extended to multiparameter cases ([49]).

The expansion of a generalized Bayes estimator with respect to a loss function of the type $L(\theta)=|\theta|^a$ $(a \geqq 1)$ was obtained by Gusev [23]. His result can be extended to all symmetric loss functions. It was shown generally by Pfanzagl and Wefelmeyer [34] and Akahira and Takeuchi [8], [46], [47], [50] that there exist second order asymptotically efficient estimators but not third order asymptotically efficient estimators in the class of all AMU estimators. But it was also shown in sections 5.1 and 5.2 that if we have asymptotically efficient estimators among an appropriately chosen restricted class of estimators, we have higher order asymptotically efficient estimators among the restricted class of estimators and that the maximum likelihood estimator belongs to the class of higher order asymptotically efficient estimators.

We assume that for each $\theta \in \Theta$ P_θ is absolutely continuous with respect to σ-finite measure μ. We denote a density $dP_\theta/d\mu$ by $f(x,\theta)$.

Let $X_1, X_2, \ldots, X_n, \ldots$ be a sequence of i.i.d. random variables with the density $f(x,\theta)$. Suppose that for almost all $x[\mu]$, $f(x,\theta)$ is continuously differentiable. The joint density is given by $\prod_{i=1}^n f(x_i,\theta)$. In the subsequent discussion we shall deal with the case when $\Theta=R^1$ of the previous section. Let $L_n(u)$ be a bounded non-negative and monotone increasing function of $|u|$ and $\pi(\theta)$ be a non-negative function. Define a posterior density $p_n(\theta \mid \tilde{x}_n)$ and a posterior risk $r_n(d \mid \tilde{x}_n)$ by

$$p_n(\theta \mid \tilde{x}_n)=\left\{\prod_{i=1}^n f(x_i,\theta)\right\}\pi(\theta)\left[\int_\Theta \left(\prod_{i=1}^n f(x_i,\theta)\right)\pi(\theta)d\theta\right]^{-1}.$$

and

$$r_n(d \mid \tilde{x}_n)=\int_\Theta L_n(d-\theta)p_n(\theta \mid \tilde{x}_n)d\theta ,$$

respectively, where $\tilde{x}_n=(x_1, x_2, \ldots , x_n)$.

Now suppose that $\lim_{n\to\infty} L_n(u/\sqrt{n})=L^*(u)$ for all real numbers u.

We define

$$r_n^*(d \mid \tilde{x}_n) = \int_{\Theta} L^*(\sqrt{n}(d-\theta))p_n(\theta \mid \tilde{x}_n)d\theta \quad .$$

An estimator $\hat{\theta}_n$ is called a generalized Bayes estimator with respect to a loss function L^* and a prior density π if

$$r_n^*(\hat{\theta}_n \mid \tilde{x}_n) = \inf_{d\in\Theta} r_n^*(d \mid \tilde{x}_n) \ .$$

Then $\hat{t}_n=\sqrt{n}(\hat{\theta}_n-\theta)$ may also be called a generalized Bayes estimator w.r.t. L^* and π .

Since

$$\lim_{n\to\infty}\left| \inf_{d\in\Theta}\int_{\Theta} L_n(d-\theta)\tilde{p}(\theta)d\theta - \inf_{d\in\Theta}\int_{\Theta} L^*(\sqrt{n}(d-\theta))\tilde{p}(\theta)d\theta \right| = 0$$

uniformly in every posterior density $\tilde{p}(\theta)$, it follows that for a generalized Bayes estimator $\hat{\theta}_n$

$$\lim_{n\to\infty}\left| \inf_{d\in\Theta} r_n(d \mid \tilde{x}_n) - r_n^*(\hat{\theta}_n \mid \tilde{x}_n)\right| = 0 \ .$$

Suppose that $X_1, X_2, \ldots , X_n, \ldots$ is a sequence of i.i.d. random variables with a density $f(x,\theta)$ satisfying (A.3.1.1) $\sim$ (A.3.1.3) and (A.4.1.1) $\sim$ (A.4.1.3).

It will be shown that generalized Bayes estimator w.r.t. a loss function and a prior density is second order asymptotically efficient. Let θ_0 be the true value of $\theta\ (\in \Theta)$.

Further we assume the following :

(A.5.3.1) $\pi(\theta)$ is twice differentiable in θ .

Then we have

$$p_n(\theta \mid \tilde{X}_n)/p_n(\theta_0 \mid \tilde{X}_n)$$

$$= \exp\left[\log\left\{ p_n(\theta\mid \tilde{X}_n)/p_n(\theta_0 \mid \tilde{X}_n)\right\}\right]$$

$$= \exp\left[\sum_{i=1}^{n}\log\left\{ f(X_i,\theta)/f(X_i,\theta_0)\right\} + \log\left\{\pi(\theta)/\pi(\theta_0)\right\}\right]$$

$$= \exp\left[\sum_{i=1}^{n}\log f(X_i,\theta) - \sum_{i=1}^{n}\log f(X_i,\theta_0)+\log\pi(\theta)-\log\pi(\theta_0)\right]$$

$$= \exp\left[\left\{\sum_{i=1}^{n}\frac{\partial}{\partial\theta}\log f(X_i,\theta_0)\right\}(\theta-\theta_0)+\frac{1}{2}\left\{\sum_{i=1}^{n}\frac{\partial^2}{\partial\theta^2}\log f(X_i,\theta_0)\right\}\right.$$

$$(\theta-\theta_0)^2+\frac{1}{6}\left\{\sum_{i=1}^{n}\frac{\partial^3}{\partial\theta^3}\log f(X_i,\theta^*)\right\}(\theta-\theta_0)^3+\frac{\pi'(\theta_0)}{\pi(\theta_0)}(\theta-\theta_0)$$

$$\left.+\,o_p\left(\frac{1}{\sqrt{n}}\right)\right]$$

$$= \exp\left[\sqrt{n}\,Z_1(\theta_0)\,(\theta-\theta_0)+\frac{1}{2}\left\{\sqrt{n}\,Z_2(\theta_0)-nI(\theta_0)\right\}(\theta-\theta_0)^2\right.$$

$$\left.+\,\frac{n}{6}W(\theta^*)(\theta-\theta_0)^3+\frac{\pi'(\theta_0)}{\pi(\theta_0)}(\theta-\theta_0)+o_p\left(\frac{1}{\sqrt{n}}\right)\right],$$

where $\tilde{X}_n=(X_1, X_2, \ldots, X_n)$ and $|\theta^*-\theta_0|\leq|\theta-\theta_0|$.

Since $W(\theta)$ converges in probability to $-3J(\theta)-K(\theta)$, it follows that

$$p_n(\theta|\tilde{X}_n)/p_n(\theta_0|\tilde{X}_n)$$

$$= \exp\left[\left\{Z_1(\theta_0)\right\}\sqrt{n}(\theta-\theta_0)+\frac{1}{2}\left\{\frac{Z_2(\theta_0)}{\sqrt{n}}-I(\theta_0)\right\}\left\{\sqrt{n}(\theta-\theta_0)\right\}^2\right.$$

$$\left.-\frac{1}{6\sqrt{n}}\left\{3J(\theta_0)+K(\theta_0)\right\}\left\{\sqrt{n}(\theta-\theta_0)\right\}^3+\frac{\pi'(\theta_0)}{\sqrt{n}\,\pi(\theta_0)}\left\{\sqrt{n}(\theta-\theta_0)\right\}+o_p\left(\frac{1}{\sqrt{n}}\right)\right]$$

Putting $t=\sqrt{n}(\theta-\theta_0)$ we obtain

(5.3.1) $p_n(\theta|\tilde{X}_n)/p(\theta_0|\tilde{X}_n)$

$$= \exp\left[Z_1(\theta_0)t+\frac{1}{2}\left\{\frac{Z_2(\theta_0)}{\sqrt{n}}-I(\theta_0)\right\}t^2-\frac{3J(\theta_0)+K(\theta_0)}{6\sqrt{n}}t^3+\frac{\pi'(\theta_0)}{\sqrt{n}\,\pi(\theta_0)}t\right.$$

$$\left.+\,o_p\left(\frac{1}{\sqrt{n}}\right)\right]$$

$$=\exp\left[-\frac{I(\theta_0)}{2}\left\{t-\frac{Z_1(\theta_0)}{I(\theta_0)}\right\}^2+\frac{Z_1(\theta_0)^2}{2I(\theta_0)}+\frac{\pi'(\theta_0)}{\sqrt{n}\,\pi(\theta_0)}t+\frac{Z_2(\theta_0)}{2\sqrt{n}}t^2\right.$$

$$\left.-\,\frac{3J(\theta_0)+K(\theta_0)}{6\sqrt{n}}t^3+o_p\left(\frac{1}{\sqrt{n}}\right)\right]$$

$$= \left\{\exp\frac{Z_1(\theta_0)^2}{2I(\theta_0)}\right\}\left[\exp-\frac{I(\theta_0)}{2}\left\{t-\frac{Z_1(\theta_0)}{I(\theta_0)}\right\}^2\right]$$

$$\cdot \exp\left\{ \frac{\pi'(\theta_0)}{\sqrt{n}\,\pi(\theta_0)} t + \frac{Z_2(\theta_0)}{2\sqrt{n}} t^2 - \frac{3J(\theta_0)+K(\theta_0)}{6\sqrt{n}} t^3 + o_p\left(\frac{1}{\sqrt{n}}\right) \right\} .$$

$$= \left\{ \exp \frac{Z_1(\theta_0)^2}{2I(\theta_0)} \right\} \left[\exp - \frac{I(\theta_0)}{2} \left\{ t - \frac{Z_1(\theta_0)}{I(\theta_0)} \right\}^2 \right]$$

$$\cdot \left\{ 1 + \frac{\pi'(\theta_0)}{\sqrt{n}\,\pi(\theta_0)} + \frac{Z_2(\theta_0)}{2\sqrt{n}} t^2 - \frac{3J(\theta_0)+K(\theta_0)}{6\sqrt{n}} t^3 + o_p\left(\frac{1}{\sqrt{n}}\right) \right\}$$

$$= q_n(t, \theta_0 \mid \tilde{x}_n) \text{ (say)} .$$

Let $\hat{t}_n = \sqrt{n}(\hat{\theta}_n - \theta_0)$. Then the posterior risk is given by

$$r_n^*(\hat{\theta}_n \mid \tilde{x}_n) = \frac{1}{\sqrt{n}} p_n(\theta_0 \mid \tilde{x}_n) \int_{-\infty}^{\infty} L^*(\hat{t}_n - t) q_n(t, \theta_0 \mid \tilde{x}_n) dt . \tag{5.3.2}$$

Further we assume the following :

(A.5.3.2) $L^*(u)$ is a convex function ;

(A.5.3.3) $\int_{-\infty}^{\infty} L^*(-u) q_n(u+t, \theta_0 \mid \tilde{x}_n) du$ is partially differentiable with respect to t under the integral sign.

By (5.3.2) and the assumption (A.5.3.2) it is seen that the generalized Bayes estimator $\hat{t}_n$ w.r.t. $L^*(\cdot)$ and $\pi(\cdot)$ is given as a solution u of the equation $(d/du)\int_{-\infty}^{\infty} L^*(u-t) q_n(t, \theta_0 \mid \tilde{x}_n) dt = 0$. Since by (A.5.3.3)

$$\frac{d}{du} \int_{-\infty}^{\infty} L^*(u-t) q_n(t, \theta_0 \mid \tilde{x}_n) dt$$

$$= \frac{d}{du} \int_{-\infty}^{\infty} L^*(-t) q_n(u+t, \theta_0 \mid \tilde{x}_n) dt$$

$$= \frac{d}{dt} \int_{-\infty}^{\infty} L^*(-u) q_n(t+u, \theta_0 \mid \tilde{x}_n) du$$

$$= \int_{-\infty}^{\infty} L^*(-u) \left\{ \frac{d}{dt} q_n(t+u, \theta_0 \mid \tilde{x}_n) \right\} du ,$$

the generalized Bayes estimator $\hat{t}_n$ is obtained as the solution of the equation

$$\int_{-\infty}^{\infty} L^*(-u) \left\{ \frac{d}{dt} q_n(\hat{t}_n + u, \theta_0 \mid \tilde{x}_n) \right\} du = 0 . \tag{5.3.3}$$

Since $t=\sqrt{n}(\theta-\theta_0)$ and $\hat{t}_n=\sqrt{n}(\hat{\theta}_n-\theta_0)$, $\hat{\theta}_n$ may be called to be the generalized Bayes estimator.

From (5.3.1) and (5.3.3) we have

$$0=\int_{-\infty}^{\infty} L^*(-u)e^{-\frac{I(\theta_0)}{2}\left\{\hat{t}_n+u-\frac{Z_1(\theta_0)}{I(\theta_0)}\right\}^2}\Big[-I(\theta_0)\left\{\hat{t}_n+u-\frac{Z_1(\theta_0)}{I(\theta_0)}\right\}$$

$$\cdot\left\{1+\frac{\pi'(\theta_0)}{\sqrt{n}\,\pi(\theta_0)}(\hat{t}_n+u)+\frac{Z_2(\theta_0)}{2\sqrt{n}}(\hat{t}_n+u)^2-\frac{3J(\theta_0)+K(\theta_0)}{6\sqrt{n}}\right.$$

$$\left.\cdot(t_n+u)^3+o_p\left(\frac{1}{\sqrt{n}}\right)\right\}+\frac{\pi'(\theta_0)}{\sqrt{n}\,\pi(\theta_0)}+\frac{Z_2(\theta_0)}{\sqrt{n}}(\hat{t}_n+u)$$

$$-\frac{3J(\theta_0)+K(\theta_0)}{2\sqrt{n}}(\hat{t}_n+u)^2+o_p\left(\frac{1}{\sqrt{n}}\right)\Big]du$$

Putting $\hat{u}=\hat{t}_n-\frac{Z_1}{I}$, we obtain

$$(5.3.4)\quad 0=\int_{-\infty}^{\infty} L^*(-u)e^{-\frac{I(\theta_0)}{2}(u+\hat{u})^2}\Big[-I(\theta_0)(u+\hat{u})\left\{1+\frac{\pi'(\theta_0)}{\sqrt{n}\,\pi(\theta_0)}\left(u+\hat{u}+\frac{Z_1(\theta_0)}{I(\theta_0)}\right)\right.$$

$$+\frac{Z_2(\theta_0)}{2\sqrt{n}}\left(u+\hat{u}+\frac{Z_1(\theta_0)}{I(\theta_0)}\right)^2-\frac{3J(\theta_0)+K(\theta_0)}{6\sqrt{n}}\left(u+\hat{u}+\frac{Z_1(\theta_0)}{I(\theta_0)}\right)^3$$

$$\left.+o_p\left(\frac{1}{\sqrt{n}}\right)\right\}+\frac{\pi'(\theta_0)}{\sqrt{n}\,(\theta_0)}+\frac{Z_2(\theta_0)}{\sqrt{n}}\left(u+\hat{u}+\frac{Z_1(\theta_0)}{I(\theta_0)}\right)-\frac{3J(\theta_0)+K(\theta_0)}{2\sqrt{n}}$$

$$\cdot\left(u+\hat{u}+\frac{Z_1(\theta_0)}{I(\theta_0)}\right)^2+o_p\left(\frac{1}{\sqrt{n}}\right)\Big]du$$

$$=\int_{-\infty}^{\infty} L^*(-u)e^{-\frac{I(\theta_0)}{2}u^2}\left\{1-I(\theta_0)u\hat{u}+o_p\left(\frac{1}{\sqrt{n}}\right)\right\}$$

$$\Big[-I(\theta_0)u-I(\theta_0)\hat{u}-\frac{I(\theta_0)\pi'(\theta_0)}{\sqrt{n}\,\pi(\theta_0)}(u+\hat{u})\left\{u+\hat{u}+\frac{Z_1(\theta_0)}{I(\theta_0)}\right\}$$

$$-\frac{I(\theta_0)Z_2(\theta_0)}{2\sqrt{n}}(u+\hat{u})\left\{u^2+\hat{u}^2+\frac{Z_1(\theta_0)^2}{I(\theta_0)^2}+2u\hat{u}+\frac{2Z_1(\theta_0)}{I(\theta_0)}\hat{u}+\frac{2Z_1(\theta_0)}{I(\theta_0)}u\right\}$$

$$+\frac{I(\theta_0)\{3J(\theta_0)+K(\theta_0)\}}{6\sqrt{n}}(u+\hat{u})\left\{u^3+\hat{u}^3+\frac{Z_1(\theta_0)^3}{I(\theta_0)^3}+3u^2\hat{u}\right.$$

$$+3u^2\frac{Z_1(\theta_0)}{I(\theta_0)}+3\hat{u}^2\frac{Z_1(\theta_0)}{I(\theta_0)}+3\hat{u}^2u+3\frac{Z_1(\theta_0)^2}{I(\theta_0)^2}u+3\frac{Z_1(\theta_0)^2}{I(\theta_0)^2}\hat{u}$$

$$+6u\hat{u}\frac{Z_1(\theta_0)}{I(\theta_0)}\Big\}+\frac{\pi'(\theta_0)}{\sqrt{n}\,\pi(\theta_0)}+\frac{Z_2(\theta_0)}{\sqrt{n}}u+\frac{Z_2(\theta_0)}{\sqrt{n}}\hat{u}+\frac{Z_1(\theta_0)Z_2(\theta_0)}{I(\theta_0)\sqrt{n}}$$

$$-\frac{3J(\theta_0)+K(\theta_0)}{2\sqrt{n}}\Big\{u^2+\hat{u}^2+\frac{Z_1(\theta_0)^2}{I(\theta_0)^2}+2u\hat{u}+\frac{2Z_1(\theta_0)}{I(\theta_0)}\hat{u}+\frac{2Z_1(\theta_0)}{I(\theta_0)}u\Big\}$$

$$+o_p\Big(\frac{1}{\sqrt{n}}\Big)\Big]\,du\ .$$

(I) Symmetric loss functions

We assume

(A.5.3.4) $L^*(u)$ is a symmetric (about zero) loss function.

We define

$$(5.3.5)\qquad \varphi_k=\int_{-\infty}^{\infty}L^*(u)u^k e^{-\frac{I(\theta_0)}{2}u^2}du\ . \qquad (k=2j\ ;\ j=0,1,2,\ldots)\ .$$

Note that

$$\int_{-\infty}^{\infty}L^*(u)u^{2j+1}e^{-\frac{I(\theta_0)}{2}u^2}du=0\qquad (j=0,1,2,\ldots)\ .$$

It follows from (5.3.4), (5.3.5) and the assumption (A.5.3.4) that

$$0=-I\varphi_0\hat{u}-\frac{I(\theta_0)\pi'(\theta_0)}{\sqrt{n}\,\pi(\theta_0)}\varphi_2-\frac{I(\theta_0)Z_2(\theta_0)}{2\sqrt{n}}\Big\{2\varphi_2\hat{u}+\frac{2Z_1(\theta_0)}{I(\theta_0)}\varphi_2+\varphi_2\hat{u}+\frac{Z_1(\theta_0)^2}{I(\theta_0)^2}\varphi_0\hat{u}\Big\}$$

$$+\frac{I(\theta_0)\{3J(\theta_0)+K(\theta_0)\}}{6\sqrt{n}}\Big\{\varphi_4+3\varphi_2\frac{Z_1(\theta_0)^2}{I(\theta_0)^2}+6\varphi_2\hat{u}\frac{Z_1(\theta_0)}{I(\theta_0)}+o_p(1)\Big\}$$

$$+\frac{\pi'(\theta_0)}{\sqrt{n}\pi(\theta_0)}\varphi_0+\frac{Z_2(\theta_0)}{\sqrt{n}}\varphi_0\hat{u}+\frac{Z_1(\theta_0)Z_2(\theta_0)}{I(\theta_0)\sqrt{n}}\varphi_0$$

$$-\frac{3J(\theta_0)+K(\theta_0)}{2\sqrt{n}}\Big\{\varphi_2+\frac{Z_1(\theta_0)^2}{I(\theta_0)^2}\varphi_0+\frac{2Z_1(\theta_0)}{I(\theta_0)}\varphi_0\hat{u}\Big\}+I(\theta_0)^2\varphi_2\hat{u}+o_p\Big(\frac{1}{\sqrt{n}}\Big)\ .$$

Since

$$I(\theta_0)\{\varphi_0 - I(\theta_0)\varphi_2\}\hat{u} = \frac{\pi'(\theta_0)}{\pi(\theta_0)\sqrt{n}}\{\varphi_0 - I(\theta_0)\varphi_2\} + \frac{3J(\theta_0)+K(\theta_0)}{6\sqrt{n}}\{I(\theta_0)\varphi_4 - 3\varphi_2\}$$
$$+ \frac{Z_1(\theta_0)Z_2(\theta_0)}{I(\theta_0)\sqrt{n}}\{\varphi_0 - I(\theta_0)\varphi_2\}$$
$$- \frac{3J(\theta_0)+K(\theta_0)}{2I(\theta_0)^2\sqrt{n}} Z_1(\theta_0)^2\{\varphi_0 - I(\theta_0)\varphi_2\} + o_p\left(\frac{1}{\sqrt{n}}\right)$$

and $\varphi_0 - I(\theta_0)\varphi_2 \neq 0$,
it follows that

$$\hat{u} = \frac{\pi'(\theta_0)}{I(\theta_0)\pi(\theta_0)\sqrt{n}} + \frac{I(\theta_0)\varphi_4 - 3\varphi_2}{\varphi_0 - I(\theta_0)\varphi_2} \cdot \frac{3J(\theta_0)+K(\theta_0)}{6I(\theta_0)\sqrt{n}} + \frac{Z_1(\theta_0)Z_2(\theta_0)}{I(\theta_0)^2\sqrt{n}}$$
$$- \frac{3J(\theta_0)+K(\theta_0)}{2I(\theta_0)^3\sqrt{n}} Z_1(\theta_0)^2 + o_p\left(\frac{1}{\sqrt{n}}\right) . \tag{5.3.6}$$

Hence we have

$$\hat{t}_n = \frac{Z_1(\theta_0)}{I(\theta_0)} + \hat{u} \quad , \tag{5.3.7}$$

where $\hat{u}$ is given by (5.3.6). Since $\hat{t}_n = \sqrt{n}(\hat{\theta}_n - \theta_0)$, we modify $\hat{\theta}_n$ to be second order AMU and denote it by $\hat{\theta}^*_n$.
From Theorem 4.1.2, (5.3.6) and (5.3.7) it follows that the MLE $\hat{\theta}^*_{ML}$ is asymptotically equivalent to the generalized Bayes estimator $\hat{\theta}^*_n$ up to order $n^{-1/2}$. By Theorem 4.1.3 it is seen that $\hat{\theta}^*_n$ is second order asymptotically efficient. It follows from (5.3.7) that the generalized Bayes estimator $\hat{\theta}^*_n$ belongs to the class $\mathbb{D}$ which is defined in section 5.2. Then the asymptotic distribution of $\hat{\theta}^*_n$ is equivalent to that of the MLE $\hat{\theta}^*_{ML}$ up to order n^{-1}. Since $\hat{\theta}^*_{ML}$ is third order asymptotically efficient in the class $\mathbb{D}$, $\hat{\theta}^*_n$ is also. Let $\mathbb{A}_k$ be the class of the all k-th order AMU estimators.

We have established :

Theorem 5.3.1. Under the assumptions (A.3.1.1) $\sim$ (A.3.1.3), (A.4.1.1) $\sim$ (A.4.1.3) and (A.5.3.1) $\sim$ (A.5.3.4), the generalized Bayes estimator $\hat{\theta}_n^*$ is second order asymptotically efficient in the class $\mathbb{A}_2$ and also third order asymptotically efficient in the class $\mathbb{D}$.

Remark : In the location parameter case, $L^*(u)=u^2$, $\pi(\theta)\equiv 1$, the generalized Bayes estimator is reduced to the Pitman estimator. It follows by Theorem 5.3.1 that Pitman estimator is second order asymptotically efficient if propertly adjusted to be asymptotically median unbiased.

(II) Asymmetric loss functions.

In the case when the loss function $L^*(\cdot)$ is asymmetric we shall obtain a similar result as the symmetric case. Further we assume the following :

(A.5.3.5) $L^*(u)$ is an asymmetric loss function ;

(A.5.3.6) There exists a c satisfying

$$\int_{-\infty}^{\infty} L^*(c-u)ue^{-\frac{I(\theta_0)}{2}u^2}du=0 .$$

Then we define

$$\Psi_k = \int_{-\infty}^{\infty} L^*(c-u)u^k e^{-\frac{I(\theta_0)}{2}u^2} du \quad (k=0,1,2,\dots) .$$

Put $\hat{u}=\hat{t}_n - \frac{Z_1}{I}$ and $\hat{v}=\hat{u}+c$.

In a similar way as the symmetric loss function case we have from (5.3.4) and the assumptions (A.5.3.5) and (A.5.3.6)

$$(5.3.8) \quad \hat{v}= \frac{\pi'(\theta_0)}{I(\theta_0)\pi(\theta_0)\sqrt{n}}+\frac{I(\theta_0)\Psi_4-3\Psi_2}{\Psi_0-I(\theta_0)\Psi_2} \cdot \frac{3J(\theta_0)+K(\theta_0)}{6I(\theta_0)\sqrt{n}}+\frac{Z_1(\theta_0)Z_2(\theta_0)}{I(\theta_0)^2\sqrt{n}}$$

$$- \frac{3J(\theta_0)+K(\theta_0)}{2I(\theta_0)^3\sqrt{n}} Z_1(\theta_0)^2 - \frac{\Psi_3}{2\{\Psi_0 - I(\theta_0)\Psi_2\}\sqrt{n}}\left\{Z_2(\theta_0) - \frac{3J(\theta_0)+K(\theta_0)}{I(\theta_0)} Z_1(\theta_0)\right\}$$

$$+ o_p\left(\frac{1}{\sqrt{n}}\right) .$$

Hence we have

$$(5.3.9) \qquad \hat{t}_n = \frac{Z_1(\theta_0)}{I(\theta_0)} - c + \hat{v} \quad ,$$

where $\hat{v}$ is given by (5.3.8). Since $\hat{t}_n = \sqrt{n}(\hat{\theta}_n - \theta)$, we modify $\hat{\theta}_n$ to be second order AMU and denote it by $\hat{\theta}_n^*$.

Since there exists the fifth term on the right-hand side of (5.3.8), the generalized Bayes estimator $\hat{\theta}_n^*$ does not belong to $\mathbb{D}$ but to $\mathbb{C}$ which is defined in section 5.2. However it follows from Theorem 4.1.2, (5.3.8) and (5.3.9) that the asymptotic distribution of $\hat{\theta}_n^*$ is equivalent to that of the MLE $\hat{\theta}_{ML}$ up to order $n^{-1/2}$. Since the fifth term of the right-hand side of (5.3.8) depends on the specific choice of the loss function, the modified generalized Bayes estimator $\hat{\theta}_n^{**}$ ($\in \mathbb{A}_3$) is not third order asymptotically efficient in the class $\mathbb{C}$.

Hence we have established the following theorem.

Theorem 5.3.2. Under the assumptions (A.3.1.1) $\sim$ (A.3.1.3), (A.4.1.1) $\sim$ (A.4.1.3), (A.5.3.1) $\sim$ (A.5.3.3), (A.5.3.5) and (A.5.3.6), the generalized Bayes estimator $\hat{\theta}_n^*$ is second order asymptotically efficient in the class $\mathbb{A}_2$ but $\hat{\theta}_n^{**}$($\in \mathbb{A}_3$) is not third order asymptotically efficient in the class $\mathbb{C}$.

5.4. Third order asymptotic efficiency of maximum likelihood estimators in multivariate linear regression models

Summary :

In section 5.1 it has been shown that the maximum likelihood estimator is asymptotically best among all best asymptotically normal estimators considering the asymptotic distribution up to the order n^{-1}, if it is properly adjusted for the bias.

More or less similar results are obtained for all i.i.d. cases, but can be easily extended to the case of multivariate normal linear regression models, and then to simultaneous linear equation models. The purpose of this section is to outline general results and some detailed results for the single equation case.

Suppose that we are give a multivariate normal linear regression model in which the data X is given in a pxn matrix form which is expressed as

$$\underset{(p\times n)}{X} = \underset{(p\times 1)}{\Pi}\ \underset{(1\times n)}{Z} + \underset{(p\times n)}{U} .$$

Here each column vector of X represents a p-dimensional vector-valued observation, that of Z a p-dimensional vector of the independent or explanatory variables, that of U the vector of the errors or disturbances. We assume that the columns of U are mutually independently distributed with mean vector $\underset{\sim}{0}$ and variance-covariance matrix Σ.

We consider the following situation : Π is a function of an unknown real parameter θ , that is, $\Pi = \Pi(\theta)$. It is assumed that $\theta_1 \neq \theta_2$ implies $\Pi(\theta_1) \neq \Pi(\theta_2)$ and that $\Pi(\theta)$ is continuous and continuously differentiable up to a required order with respect to θ . Under these assumptions it is well known that the least square estimators $\hat{\Pi}$, $\hat{\Sigma}$ of Π and Σ ,

$$\hat{\Pi} = XZ'(ZZ')^{-1} \quad ;$$

$$\hat{\Sigma} = (X - \hat{\Pi}Z)(X - \hat{\Pi}Z)'/(n-q)$$

together form a sufficient statistic. Therefore according to the well-known theory of sufficient statistics, we may restrict estimators to the class of the functions of $\hat{\Pi}$ and $\hat{\Sigma}$.

We shall consider the case when n increases independently of p and q which remain fixed. In order to make the algebra simpler, we shall let $n'=n-q$ go to infinity, and consider the situation when

$$Z_n Z_n'/n/ = M + o(1/n') \quad ,$$

where $Z_n=(z_1, \ldots , z_n)$ represents the matrix formed by the first $n=n'+q$ vectors of the independent variables. Similarly we denote by X_n the pxn matrix formed by the first n' vectors of the observations. We shall denote

$$\hat{\Pi}_{n'} = X_n Z_n'(Z_n Z_n')^{-1} \quad ;$$

$$\hat{\Sigma}_{n'} = (X_n - \hat{\Pi}_{n'} Z_n)(X_n - \hat{\Pi}_{n'} Z_n)'/n' \quad .$$

Hereafter we shall simply use $\hat{\Pi}_n$ and $\hat{\Sigma}_n$ instead of $\hat{\Pi}_{n'}$ and $\Sigma_{n'}$ without any fear of confusion.

A sequence of estimators $\{\hat{\theta}_n\}$ or an estimator for short is called a <u>regular estimator</u> if it can be expressed as

$$\hat{\theta}_n = g(\hat{\Pi}_n , \hat{\Sigma}_n) \quad ,$$

where g is a continuous and continuously differentaible function independent of n. Since $\hat{\Pi}_n \to \Pi$ and $\hat{\Sigma}_n \to \Sigma$ in probability as n goes to infinity, we have

$$\hat{\theta}_n \to g(\Pi(\theta), \Sigma)$$

in probability.

Hence the necessary and sufficient condition for $\hat{\theta}_n$ be consistent is

(5.4.1) $\quad g(\Pi(\theta) , \Sigma) \equiv \theta$.

When (5.4.1) holds, we can expand $\hat{\theta}_n$ as

(5.4.2) $\quad \hat{\theta}_{n\alpha} = \theta_\alpha + \sum_i \sum_j \frac{\partial g}{\partial \pi_{ij}}(\hat{\pi}_{ij} - \pi_{ij}) + \sum_i \sum_k \frac{\partial g}{\partial \sigma_{ik}}(\hat{\sigma}_{ik} - \sigma_{ik}) + R_\alpha$,

$\alpha=1, \ldots, p$, where $\hat{\theta}_{n\alpha}$ and θ_α denote the α-th component of $\hat{\theta}_n$ and θ, respectively.

We observe that $\sqrt{n}(\hat{\pi}_{ij}-\pi_{ij})$ are normally distributed with mean 0 and variance-convariance

$$E[\{\sqrt{n}(\hat{\pi}_{ij}-\pi_{ij})\}\{\sqrt{n}(\hat{\pi}_{hk}-\pi_{hk})\}] = \sigma_{ih}m^{jk} + o(1/n),$$

where m^{jk} is the element of the matrix M^{-1}, and that $\sqrt{n}(\hat{\sigma}_{ik}-\sigma_{ik})$ are asymptotically normally distributed with variance-covariance

$$E[\{\sqrt{n}(\hat{\sigma}_{ih}-\sigma_{ih})\}\{\sqrt{n}(\hat{\sigma}_{i'h'}-\sigma_{i'h'})\}] = \sigma_{ii'}\sigma_{hh'}+\sigma_{ih'}\sigma_{i'h}$$

and that $\hat{\pi}_{ij}$ and $\hat{\sigma}_{hk}$ are independent.

Hence $\sqrt{n}(\hat{\theta}_{n\alpha}-\theta_\alpha)$ is asymptotically normally distributed with the variance-covariance

$$(5.4.3)\quad \tilde{E}[\{\sqrt{n}(\hat{\theta}_{n\alpha}-\theta_\alpha)\}\{\sqrt{n}(\hat{\theta}_{n\beta}-\theta_\beta)\}]$$

$$= \sum_i\sum_j\sum_h\sum_k \frac{\partial g_\alpha}{\partial \pi_{ij}}\frac{\partial g_\beta}{\partial \pi_{hk}}\sigma_{ih}m^{jk}$$

$$+\sum_i\sum_h\sum_{i'}\sum_{h'}\frac{\partial g_\alpha}{\partial \sigma_{ih}}\frac{\partial g_\beta}{\partial \sigma_{i'h'}}(\sigma_{ii'}\sigma_{hh'}+\sigma_{ih'}\sigma_{i'h})+o(1),$$

$\alpha,\beta=1, \ldots, p$, where $\tilde{E}$ designates the asymptotic mean or expectation in terms of the asymptotic distribution.

Differentiating (5.4.2) with respect to θ we have

$$(5.4.4)\quad \sum_i\sum_j \frac{\partial \pi_{ij}}{\partial \theta_\beta}\frac{\partial g_\alpha}{\partial \pi_{ij}} = \delta_{\alpha\beta},$$

where $\delta_{\alpha\beta}$ denotes Kronecker's delta.

Under the condition (5.4.4) the asymptotic variance-covariance matrix whose components are given by (5.4.3) is minimized when it is expressed as

$$(5.4.5)\quad \frac{\partial g_\alpha}{\partial \pi_{ij}} = \sum_\beta \lambda_{\alpha\beta}\sum_h\sum_k \sigma^{ih} m_{jk}\frac{\partial \pi_{hk}}{\partial \theta_\beta};$$

(5.4.6) $\dfrac{\partial g}{\partial \sigma_{ij}} = 0$,

where $\lambda_{\alpha\beta}$ is a lagragian multiplier determined so that the condition is satisfied.

We have the log-likelihood function as

$$\log L = \text{const.} - \frac{1}{2}\,\text{trace}\,\Sigma^{-1}(X-\Pi Z)(X-\Pi Z)' - \frac{n}{2}\log|\Sigma| .$$

Hence by differentiation we get

(5.4.7)
$$\frac{\partial \log L}{\partial \theta_\beta} = \text{trace}\,\Sigma^{-1}(X-\Pi Z)Z'\,\frac{\partial \Pi}{\partial \theta_\beta}$$
$$= \text{trace}\,\Sigma^{-1}(\hat{\Pi}-\Pi)(ZZ')\,\frac{\partial \Pi}{\partial \theta_\beta}$$
$$= \sum_i \sum_h \sum_j \sum_k \sigma^{ih}\, m_{jk}\, \frac{\partial \Pi_{hk}}{\partial \theta_\beta}(\hat{\pi}_{ij} - \pi_{ij}) .$$

Substituting the above in (5.4.2) we have

$$\hat{\theta}_{n\alpha} = \theta_\alpha + \sum_\beta \lambda_{\alpha\beta} \frac{\partial \log L}{\partial \theta_\beta} + R_\alpha$$

for $\alpha = 1, \ldots, p$.

On the other hand, after some straightforward manipulation from (5.4.7) we have

$$E_\theta\left(\frac{\partial \log L}{\partial \theta_\alpha} \frac{\partial \log L}{\partial \theta_\beta} \right) = n \sum_i \sum_j \sum_h \sum_k \sigma^{ih}\, m_{jk} \frac{\partial \pi_{ij}}{\partial \theta_\alpha} \frac{\partial \pi_{hk}}{\partial \theta_\beta}$$

$$= n\, I_{\alpha\beta} ,$$

where $I_{\alpha\beta}$ denotes the (α,β)-element of the Fisher information matrix. From (5.4.4) and (5.4.5) we have

$$\sum_\beta \lambda_{\alpha\beta} \sum_i \sum_j \sum_h \sum_k \sigma^{ih}\, m_{jk} \frac{\partial \pi_{hk}}{\partial \theta_\beta} \frac{\partial \pi_{ij}}{\partial \theta_\gamma} \quad \delta_{\alpha\gamma}$$

$$= \sum_\beta \lambda_{\alpha\beta} I_{\beta\gamma}$$

$$= \delta_{\alpha\gamma}$$

which implies that $(\lambda_{\alpha\beta})=(I_{\alpha\beta})^{-1}$, and that

$$E_\theta[(\hat{\theta}_{n\alpha}-\theta_\alpha)(\hat{\theta}_{n\beta}-\theta_\beta)] = \lambda_{\alpha\beta} + o(1) \quad .$$

We shall call an estimator $\hat{\theta}_n$ satisfying the condition (5.4.5) with $(\lambda_{\alpha\beta})=(I_{\alpha\beta})^{-1}$ as a regular best asymptotically normal (RBAN) estimator. (The asymptotic normality of $\hat{\theta}_n$ should be almost obvious.)

Without that condition we get a class of RBAN estimators in the following manner. Let $\lambda_1(\theta), \ldots, \lambda_p(\theta)$ be the solutions of the determinant equation

$$\left| \frac{1}{n}(X-\Pi(\theta)Z)(X-\Pi(\theta)Z)' - \lambda\hat{\Sigma} \right| = 0$$

and let $\hat{\theta}_n$ be the value of θ which minimizes $G(\lambda_1(\theta),\ldots,\lambda_p(\theta))$, where G is a continuously differentiable symmetric function of p real variables, monotone increasing in each term, and satisfying the condition

$$\left.\frac{\partial G}{\partial \lambda_i}\right|_{\lambda_1=1, \ldots, \lambda_p=1} > 0 \quad .$$

By putting $G(\lambda_1, \ldots, \lambda_p)= \prod_i \lambda_i$ we get the ML estimator which in this case is reduced to minimizing

$$\left| (X - \Pi(\theta)Z)(X - \Pi(\theta)Z)' \right|$$

If we put $G(\lambda_1,\ldots,\lambda_p)= \sum\lambda_i$ we have an estimator which minimizes

$$\operatorname{trace} \hat{\Sigma}^{-1}(X- \Pi(\theta)Z)(X- \Pi(\theta)Z)'$$

which is sometimes called the minimum chi-square estimator.

In order to compare RBAN estimators we need the higher order asymptotic expansion of their distributions. But before direct comparison we have to generalize our class of estimators so that the estimators be higher order asymptotically median unbiased. An estimator $\hat{\theta}_n$ which can be expressed as

$$\hat{\theta}_n = g(\hat{\Pi}, \hat{\Sigma}) + \frac{1}{n}h(\hat{\Pi}, \hat{\Sigma}),$$

where g and h are functions independent of n, g is three times, h is once continuously differentiable in their argument is called an <u>extended regular estimator</u> (<u>ER estimator</u> for short).
For an ER estimator, if the first term g represents an RBAN estimator, then $\hat{\theta}_n$ is also asymptotically efficient and it is called an <u>ERBAN estimator</u>.

For any given RBAN estimator $\hat{\theta}_n = g(\hat{\Pi}, \hat{\Sigma})$ which is three times continuously differentiable, we can construct an ERBAN estimator

$$\hat{\theta}_n^* = \hat{\theta}_n + \frac{1}{n}h(\hat{\Pi}, \hat{\Sigma})$$

so that each component of $\hat{\theta}_n^*$ is third order asymptotically median unbiased. Thus from the ML estimator $\hat{\theta}_{ML}$ we have the modified ML estimator

$$\hat{\theta}_{ML}^* = \hat{\theta}_{ML} + \frac{1}{n}h_0(\hat{\Pi}, \hat{\Sigma})$$

which is third order asymptotically median unbiased.
Now we shall prove the following theorem.

<u>Theorem 5.4.1.</u> For any ERBAN estimator $\hat{\theta}_n$ we have

$$\underline{\lim}_{n\to\infty} n[P_{\theta,n}\{\sqrt{n}(\hat{\theta}_{ML}^* - \theta) \in C\} - P_{\theta,n}\{\sqrt{n}(\hat{\theta}_n - \theta) \in C\}] \geqq 0$$

for all θ and for any convex set C in R^p containing the origin.

Since the proof of this theorem is basically parallel to that of Theorem 5.1.6 we shall only give its outline.

First we expand an extended regular estimator $\hat{\theta}_n$ in a Taylor series as

$$\hat{\theta}_{n\alpha} = \theta_\alpha + \frac{1}{\sqrt{n}}L_\alpha + \frac{1}{n}(Q_\alpha + h_\alpha) + \frac{1}{n\sqrt{n}}W_\alpha + R_\alpha,$$

$\alpha = 1, \ldots, p$, where $L_\alpha/\sqrt{n}$ is a linear function in terms of $\sqrt{n}(\hat{\pi}_{ij} - \pi_{ij})$, Q_α/n is a quadratic function in $\sqrt{n}(\hat{\pi}_{ij} - \pi_{ij})$ and $\sqrt{n}(\hat{\sigma}_{hk} - \sigma_{hk})$, h_α is the value of the α-th component of h when $\hat{\Pi} = \Pi$ and $\hat{\Sigma} = \Sigma$, and $W_\alpha/n\sqrt{n}$ is the sum of a cubic function and a linear function of $\sqrt{n}(\hat{\pi}_{ij}-\pi_{ij})$ and $\sqrt{n}(\hat{\sigma}_{ij}-\sigma_{ij})$. The remainder term R_α is stochastically of smaller order of magnitude than $n^{-3/2}$, and also the probability that $|R_\alpha|$ is of magnitude $n^{-1/2}$ is smaller than n^{-1}. Hence the asymptotic distribution of $\sqrt{n}(\hat{\theta}_n-\theta)$ is not affected up to the order of n^{-1} if we ignore the remainder term of R_α.

Then it is easily shown (See also Sargan [39]) that the joint distribution of

$$(5.4.8) \qquad L_\alpha + \frac{1}{\sqrt{n}}(Q_\alpha + h_\alpha) + \frac{1}{n}W_\alpha$$

(α=1, ... , p) has a density which can be expanded in a multivariate Edgeworth series determined by its joint cumulants, and the asymptotic distribution of $\sqrt{n}(\hat{\theta}_{n\alpha}-\theta_\alpha)$ coincides with that of (5.4.8) up to the order n^{-1}.

First we remark that

$$E[L_\alpha + \frac{1}{\sqrt{n}}(Q_\alpha + h_\alpha) + \frac{1}{n}W_\alpha]$$

$$= \frac{1}{\sqrt{n}}\left\{E(Q_\alpha) + h_\alpha\right\} + o(\frac{1}{n}) \quad .$$

Now we define

$$\mu_\alpha = E(Q_\alpha) + h_\alpha \ ;$$

$$(5.4.9) \qquad T_\alpha = L_\alpha + \frac{1}{\sqrt{n}}(Q_\alpha + h_\alpha) + \frac{1}{n}W_\alpha - \frac{1}{\sqrt{n}}\mu_\alpha$$

$$= L_\alpha + \frac{1}{\sqrt{n}}\left\{Q_\alpha - E(Q_\alpha)\right\} + \frac{1}{n}W_\alpha$$

$$= L_\alpha + \frac{1}{\sqrt{n}}Q^*_\alpha + \frac{1}{n}W_\alpha \ .$$

When $\hat{\theta}_n$ is an ERBAN estimator, L_α is expressed as

$$L_\alpha = \sum_\beta \lambda_{\alpha\beta}U_\beta$$

$$U_\beta = \frac{1}{\sqrt{n}}\frac{\partial}{\partial\theta_\beta}\log L \ ,$$

where $(\lambda_{\alpha\beta}) = (I_{\alpha\beta})^{-1}$.

For the calculation of moments of T_α , the following lemma is extensively used.

Lemma 5.4.1. Let T_θ be a quantity depending on $X_1, \ldots , X_n$ and θ , differentiable with respect to θ and with finite second order moment. Then we have

$$E_\theta(U_\alpha T_\theta) = \frac{1}{\sqrt{n}} \frac{\partial}{\partial\theta_\alpha} E_\theta(T_\theta) - \frac{1}{\sqrt{n}} E_\theta(\frac{\partial}{\partial\theta_\alpha} T_\theta) .$$

Proof. Differentiation of

$$E_\theta(T_\theta) = \int T_\theta \prod_{i=1}^{n} f(x_i, \theta) \prod_{i=1}^{n} d\mu(x_i)$$

yields that

$$\frac{\partial}{\partial\theta_\alpha} E_\theta(T_\theta) = E_\theta(\frac{\partial}{\partial\theta_\alpha} T_\theta) + E_\theta(T_\theta \frac{\partial}{\partial\theta_\alpha} \log L)$$

$$= E_\theta(\frac{\partial}{\partial\theta_\alpha} T_\theta) + \sqrt{n}\, E_\theta(T_\theta U_\theta)$$

for which the lemma follows immediately.

Differentiation under the integral sign is guaranteed by the existence of the second order moment of T_θ .

From the lemma we also have

$$E_\theta(L_\alpha T_\theta) = \frac{1}{\sqrt{n}} \sum_\beta \lambda_{\alpha\beta} \left\{ \frac{\partial}{\partial\theta_\beta} E_\theta(T_\theta) - E_\theta(\frac{\partial}{\partial\theta_\beta} T_\theta) \right\} .$$

Applying (5.4.9) we get the asymptotic covariance T_α and T_β as follows.

(5.4.10) $$E_\theta(T_\alpha T_\beta) = E_\theta[(T_\alpha - L_\alpha)(T_\beta - L_\beta)] + E_\theta(L_\alpha T_\beta) + E_\theta(L_\beta T_\alpha) - E_\theta(L_\alpha L_\beta)$$

$$= \frac{1}{n} E_\theta(Q^*_\alpha Q^*_\beta) + E_\theta(L_\alpha T_\beta) + E_\theta(L_\beta T_\alpha) - I^{\alpha\beta} + o(\frac{1}{n}) \quad ;$$

$$E_\theta(L_\alpha T_\beta) = \frac{1}{\sqrt{n}} \sum_\gamma \lambda_{\alpha\gamma} \left\{ \frac{\partial}{\partial\theta_\gamma} E_\theta(T_\beta) - E_\theta(\frac{\partial}{\partial\theta_\gamma} T_\beta) \right\} .$$

Since we have $E_\theta(T_\beta)=0$, $\frac{\partial}{\partial\theta_\gamma} E_\theta(T_\beta)=0$ and

$$T_\beta = \sqrt{n}(\hat{\theta}_{n\beta} - \theta_\beta) - \frac{1}{\sqrt{n}} \mu_\beta + o_p(\frac{1}{n\sqrt{n}})$$

we get

$$(5.4.11)\qquad \frac{\partial}{\partial\theta_\gamma}T_\beta = -\sqrt{n}\,\delta_{\gamma\beta} - \frac{1}{\sqrt{n}}\frac{\partial\mu_\beta}{\partial\theta_\gamma} + o_p\left(\frac{1}{n\sqrt{n}}\right)$$

hence

$$(5.4.12)\qquad E_\theta(L_\alpha T_\beta) = \lambda_{\alpha\beta} + \frac{1}{n}\sum_\gamma \lambda_{\alpha\gamma}\frac{\partial\mu_\beta}{\partial\theta_\gamma} + o\left(\frac{1}{n}\right) .$$

Similarly we have

$$E_\theta(L_\beta T_\alpha) = \lambda_{\beta\alpha} + \frac{1}{n}\sum_\gamma \lambda_{\beta\gamma}\frac{\partial\mu_\alpha}{\partial\theta_\gamma} + o\left(\frac{1}{n}\right) .$$

Substituting $\lambda_{\alpha\beta} = \lambda_{\beta\alpha} = I^{\alpha\beta}$, we have from (5.4.10)

$$E_\theta(T_\alpha T_\beta) = I^{\alpha\beta} + \frac{1}{n}\sum_\gamma\left(I^{\alpha\gamma}\frac{\partial\mu_\beta}{\partial\theta_\gamma} + I^{\beta\gamma}\frac{\partial\mu_\alpha}{\partial\theta_\gamma}\right)$$

$$+ \frac{1}{n}E_\theta(Q^*_\alpha Q^*_\beta) + o\left(\frac{1}{n}\right) ,$$

$I^{\alpha\beta}$ denotes (α,β)-element of the inverse of the Fisher information matrix $I=(I_{\alpha\beta})$.

For the third order moment, first we note that

$$(5.4.13)\qquad E_\theta(T_\alpha T_\beta T_\gamma) = \frac{1}{2}E_\theta(L_\alpha T_\beta T_\gamma + L_\beta T_\gamma T_\alpha + L_\gamma T_\alpha T_\beta)$$

$$+ \frac{1}{2}E_\theta[(T_\alpha - L_\alpha)(T_\beta - L_\beta)(T_\gamma - L_\gamma)]$$

$$+ \frac{1}{2}E_\theta[L_\alpha(T_\beta - L_\beta)(T_\gamma - L_\gamma)]$$

$$+ \frac{1}{2}E_\theta[L_\beta(T_\alpha - L_\alpha)(T_\gamma - L_\gamma)]$$

$$+ \frac{1}{2}E_\theta[L_\gamma(T_\alpha - L_\alpha)(T_\beta - L_\beta)]$$

$$- \frac{1}{2}E_\theta(L_\alpha L_\beta L_\gamma) .$$

We have

$$E_\theta(L_\alpha T_\beta T_\gamma) = \frac{1}{\sqrt{n}}\sum_\delta \lambda_{\alpha\delta}\left\{\frac{\partial}{\partial\theta_\delta}E_\theta(T_\beta T_\gamma) - E_\theta\left(\frac{\partial}{\partial\theta_\delta}T_\beta T_\gamma\right)\right\}$$

$$= \frac{1}{\sqrt{n}}\sum_\delta \lambda_{\alpha\delta}\frac{\partial}{\partial\theta_\delta}I^{\beta\gamma} + o\left(\frac{1}{n}\right) ;$$

$$E_\theta[(T_\alpha - L_\alpha)(T_\beta - L_\beta)(T_\gamma - I_\gamma)] = o\left(\frac{1}{n}\right) ;$$

$$E_\theta[L_\alpha(T_\beta - L_\beta)(T_\gamma - L_\gamma)] = \frac{1}{n}E(L_\alpha Q^*_\beta Q^*_\gamma) + o\left(\frac{1}{n}\right)$$

$$= o\left(\frac{1}{n}\right) ;$$

$$E_\theta(L_\alpha L_\beta L_\gamma) = 0$$

etc., and substituting these in (5.4.13) we obtain

(5.4.14) $$E_\theta(T_\alpha T_\beta T_\gamma) = \frac{1}{2\sqrt{n}} \sum_\delta \left(I^{\alpha\delta} \frac{\partial}{\partial\theta_\delta} I^{\beta\gamma} + I^{\beta\delta} \frac{\partial}{\partial\theta_\delta} I^{\alpha\gamma} + I^{\gamma\delta} \frac{\partial}{\partial\theta_\delta} I^{\alpha\beta}\right) + o\left(\frac{1}{n}\right) .$$

We further have

$$\frac{\partial}{\partial\theta_\delta} I^{\beta\gamma} = -\sum_{\beta'}\sum_{\gamma'} I^{\beta\beta'} I^{\gamma\gamma'} \frac{\partial}{\partial\theta_\delta} I_{\beta'\gamma'}$$
$$= -2\sum_{\beta'}\sum_{\gamma'} I^{\beta\beta'} I^{\gamma\gamma'} \left(\sum_i \sum_j \sum_h \sum_k \sigma^{ih} m_{jk} \frac{\partial^2 \pi_{ij}}{\partial\theta_\delta \partial\theta_{\beta'}} \frac{\partial \pi_{hk}}{\partial\theta_{\gamma'}}\right) .$$

Thus it has been proved that the third order cumulant is equal to $K_{\alpha\beta\gamma}/\sqrt{n}$ up to the order n^{-1} for all the ERBAN estimators, where $K_{\alpha\beta\gamma}$ is given by (5.4.14).

For the fourth order cumulant we make use of the following relation.

(5.4.15) $$\begin{aligned} T_\alpha T_\beta T_\gamma T_\delta = {} & \frac{2}{3}(L_\alpha T_\beta T_\gamma T_\delta + L_\beta T_\alpha T_\gamma T_\delta + L_\gamma T_\alpha T_\beta T_\delta + L_\delta T_\alpha T_\beta T_\gamma) \\ & - \frac{1}{3}(L_\alpha L_\beta T_\gamma T_\delta + L_\alpha L_\delta T_\beta T_\gamma + L_\beta L_\gamma T_\alpha T_\delta + L_\beta L_\delta T_\alpha T_\gamma + L_\gamma L_\delta T_\alpha T_\beta \\ & + L_\alpha L_\gamma T_\beta T_\delta) + \frac{1}{3}\{L_\alpha (T_\alpha - L_\alpha)(T_\gamma - L_\gamma)\ (T_\delta - L_\delta) + L_\beta (T_\alpha - L_\alpha) \\ & (T_\gamma - L_\gamma)(T_\delta - L_\delta) + L_\gamma (T_\alpha - L_\alpha)(T_\beta - L_\beta)(T_\delta - L_\delta) + L_\delta (T_\alpha - L_\alpha) \\ & (T_\beta - L_\beta)(T_\gamma - L_\gamma)\} + (T_\alpha - L_\alpha)(T_\beta - L_\beta)(T_\gamma - L_\gamma)(T_\delta - L_\delta) \\ & + \frac{1}{3} L_\alpha L_\beta L_\gamma L_\delta \\ = {} & \frac{2}{3}(L_\alpha T_\beta T_\gamma T_\delta + L_\beta T_\alpha T_\gamma T_\delta + L_\gamma T_\alpha T_\beta T_\delta + L_\delta T_\alpha T_\beta T_\gamma) \\ & - \frac{1}{3}(L_\alpha L_\beta T_\gamma T_\delta + L_\alpha L_\gamma T_\beta T_\delta + L_\alpha L_\delta T_\beta T_\gamma + L_\beta L_\gamma T_\alpha T_\delta + L_\beta L_\delta T_\alpha T_\gamma \\ & + L_\gamma L_\delta T_\alpha T_\beta) + \frac{1}{3} L_\alpha L_\beta L_\gamma L_\delta + o\left(\frac{1}{n}\right) . \end{aligned}$$

For the terms in the above expression we have

(5.4.16) $$E_\theta(L_\alpha T_\beta T_\gamma T_\delta) = \frac{1}{\sqrt{n}} \sum_{\alpha'} \lambda_{\alpha\alpha'} \left\{ \frac{\partial}{\partial\theta_{\alpha'}} E_\theta(T_\beta T_\gamma T_\delta)\right.$$

$$- E_\theta\left(\frac{\partial}{\partial\theta_{\alpha'}} T_\beta T_\gamma T_\delta \right) \Big\}$$

$$= \frac{1}{n} \sum_{\alpha'} I^{\alpha\alpha'} \frac{\partial}{\partial\theta_{\alpha'}} K_{\beta\gamma\delta} + I^{\alpha\beta} E_\theta(T_\gamma T_\delta) + I^{\alpha\gamma} E_\theta(T_\beta T_\delta)$$

$$+ I^{\alpha\delta} E_\theta(T_\beta T_\gamma) + \frac{1}{n} \sum_{\alpha'} I^{\alpha\alpha'} \left(\frac{\partial \mu_\beta}{\partial\theta_{\alpha'}} I^{\gamma\delta} + \frac{\partial \mu_\gamma}{\partial\theta_{\alpha'}} I^{\beta\delta} \right.$$

$$\left. + \frac{\partial \mu_\delta}{\partial\theta_{\alpha'}} I^{\beta\gamma} \right) + o\left(\frac{1}{n}\right) .$$

And we have

$$(5.4.17) \quad E(L_\alpha L_\beta T_\gamma T_\delta) = \frac{1}{\sqrt{n}} \sum_{\alpha'} I^{\alpha\alpha'} \Big\{ \frac{\partial}{\partial\theta_{\alpha'}} E_\theta(L_\beta T_\gamma T_\delta) - E_\theta\left(\frac{\partial}{\partial\theta_{\alpha'}} (L_\beta T_\gamma T_\delta) \right) \Big\} ;$$

$$\frac{\partial}{\partial\theta_{\alpha'}} E_\theta(L_\beta T_\gamma T_\delta) = \frac{1}{\sqrt{n}} \frac{\partial}{\partial\theta_{\alpha'}} \left(\sum_{\beta'} I^{\beta\beta'} \frac{\partial}{\partial\theta_{\beta'}} I^{\gamma\delta} \right) + o\left(\frac{1}{n}\right) ;$$

$$(5.4.18) \quad E_\theta\left[\frac{\partial}{\partial\theta_{\alpha'}} (L_\beta T_\gamma T_\delta) \right] = E_\theta\left(\frac{\partial L_\beta}{\partial\theta_{\alpha'}} T_\gamma T_\delta \right) + E_\theta\left(\frac{\partial T_\gamma}{\partial\theta_{\alpha'}} L_\beta T_\delta \right)$$

$$+ E_\theta\left(\frac{\partial T_\delta}{\partial\theta_{\alpha'}} L_\beta T_\gamma \right) .$$

Now we have

$$\frac{\partial L_\beta}{\partial\theta_{\alpha'}} = \frac{1}{\sqrt{n}} \sum_\alpha I^{\beta\alpha} \frac{\partial^2}{\partial\theta_{\alpha'}\partial\theta_\alpha} \log L + \frac{1}{\sqrt{n}} \sum_\alpha \left(\frac{\partial}{\partial\theta_{\alpha'}} I^{\beta\alpha} \right) \left(\frac{\partial}{\partial\theta_\alpha} \log L \right) ;$$

$$\frac{\partial^2}{\partial\theta_{\alpha'}\partial\theta_\alpha} \log L = -nI_{\alpha\alpha'} + n \sum_i \sum_j \sum_h \sum_k \sigma^{ih} m_{jk} \frac{\partial^2 \pi_{hk}}{\partial\theta_\alpha \partial\theta_{\alpha'}} (\hat{\pi}_{ij} - \pi_{ij}) .$$

Therefore we can write

$$\frac{\partial L_\beta}{\partial\theta_{\alpha'}} = -\sqrt{n}\, \delta_{\beta\alpha'} + \sum_\alpha I^{\beta\alpha} U_{\alpha\alpha'} - \sum_\delta \sum_{\delta'} I^{\beta\delta} (J_{\delta'\alpha'.\delta} + J_{\delta\alpha'.\delta'}) L_{\delta'} ,$$

where $U_{\alpha\alpha'}$ is a linear function in terms of $\sqrt{n}(\hat{\pi}_{ij} - \pi_{ij})$.

Hence it follows that

$$(5.4.19) \quad E_\theta\left(\frac{\partial L_\beta}{\partial\theta_{\alpha'}} T_\gamma T_\delta \right) = -\sqrt{n}\, \delta_{\alpha'\beta} E_\theta(T_\gamma T_\delta) + \sum_\alpha I^{\beta\alpha} E_\theta(U_{\alpha\alpha'} T_\gamma T_\delta)$$

$$= -\sqrt{n}\, \delta_{\alpha'\beta} E_\theta(T_\gamma T_\delta) + \sum_\alpha I^{\beta\alpha} E_\theta(U_{\alpha\alpha'} L_\gamma T_\delta + U_{\alpha\alpha'} L_\delta T_\gamma)$$

$$- \sum_{\beta'} \sum_{\xi} I^{\beta\beta'} (J_{\xi\alpha'.\beta'} + J_{\beta'\alpha'.\xi}) E_\theta(L_\xi T_\gamma T_\delta)$$

$$+ o\left(\frac{1}{n}\right) .$$

We have

$$(5.4.20) \quad E_\theta(U_{\alpha\alpha'} L_\gamma T_\delta) = \frac{1}{\sqrt{n}} \sum_{\gamma'} I^{\gamma\gamma'} \Big\{ \frac{\partial}{\partial\theta_{\gamma'}} E_\theta(U_{\alpha\alpha'} T_\delta) - E_\theta\left(\frac{\partial}{\partial\theta_{\gamma'}} U_{\alpha\alpha'} T_\delta \right) \Big\}$$

$$= \frac{1}{\sqrt{n}} \sum_{\gamma'} I^{\gamma\gamma'} \Big\{ \frac{\partial}{\partial\theta_{\gamma'}} E_\theta(U_{\alpha\alpha'} L_\delta) - E_\theta(U_{\alpha\alpha'\gamma'} L_\delta) \Big\} + o\left(\frac{1}{n}\right) ,$$

where $U_{\alpha\alpha'\gamma'}$ is defined as

$$U_{\alpha\alpha'\gamma'} = \sqrt{n} \sum_i \sum_j \sum_k \sum_h \sigma^{ih} m_{jk} \frac{\partial^3 \pi_{hk}}{\partial\theta_\alpha \partial\theta_{\alpha'} \partial\theta_{\gamma'}} (\hat{\pi}_{ij} - \pi_{ij}) .$$

It is easily seen that

$$E_\theta(U_{\alpha\alpha'} L_\delta) = \sum_{\delta'} \left(\sum_i \sum_j \sum_k \sum_h \sigma^{ij} m_{jk} \frac{\partial^2 \pi_{hk}}{\partial\theta_\alpha \partial\theta_{\alpha'}} \frac{\partial \pi_{ij}}{\partial\theta_{\delta'}} \right) I^{\delta\delta'} ;$$

$$\frac{\partial}{\partial\theta_{\gamma'}} E_\theta(U_{\alpha\alpha'} L_\delta) = \sum_{\delta'} \left(\sum_i \sum_j \sum_k \sum_h \sigma^{ij} m_{jk} \frac{\partial^3 \pi_{hk}}{\partial\theta_\alpha \partial\theta_{\alpha'} \partial\theta_{\gamma'}} \frac{\partial^2 \pi_{ij}}{\partial\theta_\delta \partial\theta_{\gamma'}} \right.$$

$$\left. + \sum_i \sum_j \sum_k \sum_h \sigma^{ij} m_{jk} \frac{\partial^2 \pi_{hk}}{\partial\theta_\alpha \partial\theta_{\alpha'}} \frac{\partial^2 \pi_{ij}}{\partial\theta_\delta \partial\theta_{\gamma'}} \right) I^{\delta\delta'}$$

$$- \sum_\xi (J_{\gamma'}^{\delta\cdot\xi} + J_{\gamma'}^{\xi\cdot\delta}) J_{\alpha\alpha'\cdot\xi} ;$$

$$E_\theta(U_{\alpha\alpha'\gamma'} L_\delta) = \sum_{\delta'} \left(\sum_i \sum_j \sum_k \sum_h \sigma^{ij} m_{jk} \frac{\partial^3 \pi_{hk}}{\partial\theta_\alpha \partial\theta_{\alpha'} \partial\theta_{\gamma'}} \frac{\partial \pi_{ij}}{\partial\theta_\delta} \right) I^{\delta\delta'} .$$

Hence it follows that

$$E_\theta(U_{\alpha\alpha'} L_\gamma T_\delta) = \frac{1}{\sqrt{n}} \sum_{\gamma'} \sum_{\delta'} I^{\gamma\gamma'} I^{\delta\delta'} \sum_i \sum_j \sum_h \sum_k \sigma^{ij} m_{ij} \frac{\partial^2 \pi_{ij}}{\partial\theta_\alpha \partial\theta_{\alpha'}} \frac{\partial^2 \pi_{hk}}{\partial\theta_{\delta'} \partial\theta_{\gamma'}}$$

$$- \frac{1}{\sqrt{n}} \sum_{\gamma'} I^{\gamma\gamma'} \sum_\xi (J_{\gamma'}^{\delta\cdot\xi} + J_{\gamma'}^{\xi\cdot\delta}) J_{\alpha'\alpha\cdot\xi}$$

which we shall denote as $\frac{1}{\sqrt{n}} \sum_{\gamma'} \sum_{\delta'} I^{\gamma\gamma'} I^{\delta\delta'} M_{\alpha\alpha'\cdot\gamma'\delta'}$.

From (5.4.11) and (5.4.12) we have

$$E_\theta\left(\frac{\partial T_\delta}{\partial\theta_{\alpha'}} L_\beta T_\gamma \right) = - \left(\sqrt{n}\, \delta_{\alpha\delta} + \frac{1}{\sqrt{n}} \frac{\partial \mu_\delta}{\partial\theta_{\alpha'}} \right) \left(I^{\beta\gamma} + \frac{1}{n} \sum_{\gamma'} I^{\beta\gamma'} \frac{\partial \mu_\gamma}{\partial\theta_{\gamma'}} \right) .$$

Substituting (5.4.19) and (5.4.20) in (5.4.18) we have

$$(5.4.21) \quad E_\theta\left[\frac{\partial}{\partial\theta_{\alpha'}} (L_\beta T_\gamma T_\delta) \right] = - \sqrt{n}\, \delta_{\alpha'\beta} E_\theta(T_\gamma T_\delta)$$

$$+ \frac{2}{\sqrt{n}} \sum_{\beta'} \sum_{\gamma'} \sum_{\delta'} I^{\beta\beta'} I^{\gamma\gamma'} I^{\delta\delta'} \Big\{ 2M_{\alpha'\beta'\cdot\gamma'\delta'}$$

$$- \sum_\xi \sum_{\xi'} I^{\xi'\xi} (2J_{\gamma'\delta'\cdot\xi'} + J_{\gamma'\xi'\cdot\delta'} + J_{\delta'\xi'\cdot\gamma'}) J_{\alpha'\beta'\cdot\xi}$$

$$+ \sum_\xi \sum_{\xi'} I^{\xi'\xi} (J_{\alpha'\beta'\cdot\xi'} + J_{\alpha'\xi'\cdot\beta'}) (J_{\gamma'\xi\cdot\delta'} + J_{\delta'\xi\cdot\gamma'})$$

$$- \left(\sqrt{n}\, \delta_{\alpha'\delta} + \frac{1}{\sqrt{n}} \frac{\partial \mu_\delta}{\partial\theta_{\alpha'}} \right) \left(I^{\beta\gamma} + \frac{1}{n} \sum_{\gamma'} I^{\beta\gamma'} \frac{\partial \mu_\gamma}{\partial\theta_{\gamma'}} \right)$$

$$-(\sqrt{n}\,\delta_{\alpha'\gamma}+\frac{1}{\sqrt{n}}\frac{\partial\mu_\gamma}{\partial\theta_{\alpha'}})(I^{\beta\delta}+\frac{1}{n}\sum_{\delta'}I^{\beta\delta'}\frac{\partial\mu_\delta}{\partial\theta_{\delta'}})$$

$$+\ o(\frac{1}{n})\ .$$

Substituting (5.4.21) back in (5.4.17) we obtain

$$\text{(5.4.22)}\quad E_\theta(L_\alpha L_\beta T_\gamma T_\delta)=\frac{1}{n}\sum_{\alpha'}I^{\alpha\alpha'}\frac{\partial}{\partial\theta_{\alpha'}}(\sum_{\beta'}I^{\beta\beta'}\frac{\partial}{\partial\theta_{\beta'}}I^{\gamma\delta})$$
$$+\ I^{\alpha\beta}E(T_\gamma T_\delta)+I^{\alpha\delta}I^{\beta\gamma}+I^{\alpha\gamma}I^{\beta\delta}+\frac{1}{n}(\sum_{\alpha'}I^{\alpha\alpha'}I^{\beta\gamma}\frac{\partial\mu_\delta}{\partial\theta_{\alpha'}}$$
$$+\sum_{\alpha'}I^{\alpha\alpha'}I^{\beta\delta}\frac{\partial\mu_\gamma}{\partial\theta_{\alpha'}}+\sum_{\beta'}I^{\alpha\gamma}I^{\beta\beta'}\frac{\partial\mu_\delta}{\partial\theta_{\beta'}}$$
$$+\sum_{\beta'}I^{\alpha\delta}I^{\beta\beta'}\frac{\partial\mu_\gamma}{\partial\theta_{\beta'}})-\frac{1}{n}\Big\{2M^{\alpha\beta\cdot\gamma\delta}\sum_{\xi}(2J^{\gamma\delta\cdot\xi}+J^{\gamma\xi\cdot\delta}+J^{\delta\xi\cdot\gamma})J_\xi^{\alpha\cdot\beta}$$
$$+\sum_{\xi}(J^{\alpha\beta}_{\cdot\xi}+J^{\alpha\cdot\beta}_{\xi})(J^{\gamma\xi\cdot\delta}+J^{\delta\xi\cdot\gamma})\Big\}\ .$$

Noting that

$$E_\theta(L_\alpha L_\beta L_\gamma L_\delta)\ =\ I^{\alpha\beta}I^{\gamma\delta}+I^{\alpha\delta}I^{\beta\gamma}+I^{\alpha\gamma}I^{\beta\delta}\ ;$$

$$K_{\alpha\beta\gamma}=\frac{1}{2}\sum_{\delta}(I^{\alpha\delta}\frac{\partial}{\partial\theta_\delta}I^{\beta\gamma}+I^{\beta\delta}\frac{\partial}{\partial\theta_\delta}I^{\alpha\gamma}+I^{\gamma\delta}\frac{\partial}{\partial\theta_\delta}I^{\alpha\beta})$$

and substituting (5.4.16) and (5.4.22) in (5.4.15) we finally establish that the fourth order moment is given as

$$E_\theta(T_\alpha T_\beta T_\gamma T_\delta)-E_\theta(T_\alpha T_\beta)E_\theta(T_\gamma T_\delta)-E_\theta(T_\alpha T_\gamma)E_\theta(T_\beta T_\delta)-E_\theta(T_\alpha T_\delta)E_\theta(T_\beta T_\gamma)$$
$$=\frac{1}{4n}\ (\sum_{\alpha'}I^{\alpha\alpha'}\frac{\partial}{\partial\theta_{\alpha'}}K_{\beta\gamma\delta}+\sum_{\beta'}I^{\beta\beta'}\frac{\partial}{\partial\theta_{\beta'}}K_{\alpha\gamma\delta}+\sum_{\gamma'}I^{\gamma\gamma'}\frac{\partial}{\partial\theta_{\gamma'}}K_{\alpha\beta\delta}$$
$$+\sum_{\delta'}I^{\delta\delta'}\frac{\partial}{\partial\theta_{\delta'}}K_{\alpha\beta\gamma})+\frac{8}{3n}\ (M^{\alpha\beta\cdot\gamma\delta}+M^{\alpha\gamma\cdot\beta\delta}+M^{\alpha\delta\cdot\beta\gamma})$$
$$-\ \frac{1}{3n}\eta^{\alpha\beta\gamma\delta}+o(\frac{1}{n})$$
$$=\frac{1}{n}K_{\alpha\beta\gamma\delta}+o(\frac{1}{n})\ ,$$

where $\eta^{\alpha\beta\gamma\delta}$ is a complicated sum of terms of the form $\sum_\xi J^{**\xi}J^{**\cdot}_\xi$. The most important point here is that $K_{\alpha\beta\gamma\delta}$ defined in the above is identical for all ERBAN estimators. We have thus established the following theorem.

Theorem 5.4.2. The joint density of $\{\sqrt{n}(\hat{\theta}_{n\alpha}-\theta_\alpha)\ ;\alpha=1,\ldots,p\}$ where $\hat{\theta}_n$ is an ERBAN estimator can be expanded up to the order n^{-1} as

$$
\begin{aligned}
f(y_1,\ldots,y_p) = {} & \frac{|I|^{1/2}}{(\sqrt{2\pi})^p}\left[\exp - \frac{1}{2}\left\{\sum_\alpha\sum_\beta I_{\alpha\beta}y_\alpha y_\beta\right\}\right] \\
& \cdot\left\{1 + \frac{1}{6\sqrt{n}}\sum_\alpha\sum_\beta\sum_\gamma K_{\alpha\beta\gamma}H_{\alpha\beta\gamma} + \frac{1}{\sqrt{n}}\sum_\alpha \mu_\alpha H_\alpha\right. \\
& + \frac{1}{72n}\sum_\alpha\sum_\beta\sum_\gamma\sum_{\alpha'}\sum_{\beta'}\sum_{\gamma'} K_{\alpha\beta\gamma}K_{\alpha'\beta'\gamma'}H_{\alpha\beta\gamma\alpha'\beta'\gamma'} \\
& + \frac{1}{24n}\sum_\alpha\sum_\beta\sum_\gamma\sum_\delta(K_{\alpha\beta\gamma\delta} + 4\mu_\alpha K_{\beta\gamma\delta})H_{\alpha\beta\gamma\delta} \\
& + \frac{1}{2n}\sum_\alpha\sum_\beta\left(\mu_\alpha\mu_\beta + \sum_\gamma I^{\alpha\gamma}\frac{\partial\mu_\beta}{\partial\theta_\gamma} + \sum_\gamma I^{\beta\gamma}\frac{\partial\mu_\alpha}{\partial\theta_\gamma}\right)H_{\alpha\beta} \\
& \left. + \frac{1}{4n}\sum_\alpha\sum_\beta \mathrm{Cov}\,(Q_\alpha\,,\,Q_\beta)H_{\alpha\beta}\right\} + o\left(\frac{1}{n}\right),
\end{aligned}
$$

where $H_{\alpha\beta\cdots}$ are (multivariate) Hermite polynomials defined by

$$
\begin{aligned}
H_\alpha &= \sum_\beta I_{\alpha\beta}y_\beta\ ; \\
H_{\alpha\beta} &= H_\alpha H_\beta - I_{\alpha\beta}\ ; \\
H_{\alpha\beta\gamma} &= H_\alpha H_\beta H_\gamma - I_{\alpha\beta}H_\gamma - I_{\alpha\gamma}H_\beta - I_{\beta\gamma}H_\alpha\ ,
\end{aligned}
$$

etc.

Now we shall prove that if we denote the term Q^*_α of the ML estimator by $\tilde{Q}^*_\alpha$, and of any other ERBAN estimator by Q^*_α, we have

$$
(5.4.23)\quad \sum_\alpha\sum_\beta\left\{E_\theta(Q^*_\alpha Q^*_\beta) - E_\theta(\tilde{Q}^*_\alpha\tilde{Q}^*_\beta)\right\}t_\alpha t_\beta \geqq 0
$$

for all real numbers $t_1,\ \ldots\ ,\ t_p$.

In order to prove (5.4.23) we first note that

$$
\begin{aligned}
E_\theta(U_\beta U_\gamma Q^*_\alpha) &= \sqrt{n}\,E_\theta\,[U_\beta U_\gamma(T_\alpha - L_\alpha)] + o(1) \\
&= \sqrt{n}\,E_\theta(U_\beta U_\gamma T_\alpha) + o(1) \\
&= \frac{\partial}{\partial\theta_\beta}E_\theta(U_\gamma T_\alpha) - E_\theta\left(\frac{\partial U_\gamma}{\partial\theta_\beta}T_\alpha\right) - E_\theta\left(U_\gamma\frac{\partial}{\partial\theta_\beta}T_\alpha\right) + o(1) \\
&= -\,E_\theta(U_{\beta\gamma}L_\alpha) + o(1)\ ;
\end{aligned}
$$

$$E_\theta(U_\beta U_{\gamma\delta} Q^*_\alpha) = \sqrt{n}\, E_\theta(U_\beta U_{\gamma\delta} T_\alpha) + o(1)$$

$$= \frac{\partial}{\partial\theta_\beta} E_\theta(U_{\gamma\delta} T_\alpha) - E_\theta\left(\frac{\partial U_{\gamma\delta}}{\partial\theta_\beta} T_\alpha\right) + o(1)$$

$$= \frac{\partial}{\partial\theta_\beta} E_\theta(U_{\gamma\delta} L_\alpha) - E_\theta(U_{\beta\gamma\delta} L_\alpha) + o(1)$$

which are asymptotically fixed for all ERBAN estimators. Therefore the quadratic form $\sum_\alpha \sum_\beta E_\theta(Q^*_\alpha Q^*_\beta t_\alpha t_\beta) = E_\theta(\sum_\alpha Q^*_\alpha t_\alpha)^2$ is asymptotically minimized when

$$Q^*_\alpha = \sum_\beta \sum_\gamma \lambda^\alpha_{\beta\gamma} U_\beta U_\gamma + \sum_\beta \sum_\gamma \sum_\delta \lambda^\alpha_{\beta\gamma\delta} U_\beta U_{\gamma\delta} + \lambda^\alpha ,$$

where λ^α, $\lambda^\alpha_{\beta\gamma}$ and $\lambda^\alpha_{\beta\gamma\delta}$ are constants.

Chapter 6

Discretized Likelihood Methods

Summary :

Suppose that $X_1, X_2, \dots, X_n, \dots$ is a sequence of i.i.d. random variables with a density $f(x,\theta)$. Let c_n be a maximum order of consistency. We consider a solution $\hat{\theta}_n$ of the discretized likelihood equation

$$\sum_{i=1}^{n} \log f(X_i, \hat{\theta}_n + rc_n^{-1}) - \sum_{i=1}^{n} \log f(X_i, \hat{\theta}_n) = a_n(\hat{\theta}_n, r)$$

where $a_n(\theta, r)$ is chosen so that $\hat{\theta}_n$ is asymptotically median unbiased (AMU). Then the solution $\hat{\theta}_n$ is called a discretized likelihood estimator (DLE). In this chapter it is shown in comparison with DLE that a maximum likelihood estimator (MLE) is second order asymptotically efficient but not always third order asymptotically efficient in the regular case. Further, it shall be seen that the asymptotic efficiency (including higher order cases) may be systematically discussed by discretized likelihood methods.

Suppose that $X_1, X_2, \dots, X_n, \dots$ is a sequence of i.i.d. random variables with a density $f(x, \theta)$. Let c_n be a maximum order of consistent estimator of θ. Consider the solution $\hat{\theta}_n$ of the discretized likelihood equation

$$(6.0.1) \qquad \sum_{i=1}^{n} \log f(X_i, \hat{\theta}_n + rc_n^{-1}) - \sum_{i=1}^{n} \log f(X_i, \hat{\theta}_n) = a_n(\hat{\theta}_n, r) ,$$

where $a_n(\theta, r)$ is chosen so that $\hat{\theta}_n$ is asymptotically median unbiased (AMU) ([44]) (the possibility of which will be shown in the context). Then the solution $\hat{\theta}_n$ is called a discretized likelihood estimator (DLE). If for each real number r,

$$\sum_{i=1}^{n} \log f(X_i, \theta + rc_n^{-1}) - \sum_{i=1}^{n} \log f(X_i, \theta)$$

is locally monotone in θ , then the asymptotic distribution of the DLE $\hat{\theta}_n$ attains the bound of the asymptotic distributions (discussed below) of AMU estimators of θ at r. It is easily seen that there is at least one estimator which attains the bound. In regular cases with $c_n = \sqrt{n}$ the left-hand side of (6.0.1) may be expanded as

(6.0.2) $$\frac{\partial}{\partial\theta} \sum_{i=1}^{n} \log f(X_i, \hat{\theta}_n) + \frac{r}{2\sqrt{n}} \frac{\partial^2}{\partial\theta^2} \sum_{i=1}^{n} \log f(X_i, \hat{\theta}_n) + \ldots = a_n(\hat{\theta}_n, r).$$

We derive from (6.0.2) the order to which the maximum likelihood estimator (MLE) is asymptotically efficient. In this chapter it is shown that an MLE is second order asymptotically efficient but not third order asymptotically efficient.

The motivation for the definition of the DLE is the following ; when we test the hypothesis $\theta = \theta_0 + tc_n^{-1}$ against $\theta = \theta_0$, the most powerful test is given by rejecting the hypothesis if

$$\sum_{i=1}^{n} \log f(X_i, \theta_0 + rc_n^{-1}) - \sum_{i=1}^{n} \log f(X_i, \theta_0) < k_n .$$

Hence if an estimator $\hat{\theta}_r$ is defined so that the event $\hat{\theta}_r > \theta_0$ is equivalent to the above inequality (at least asymptotically up to some order), then $\hat{\theta}_r$ is efficient (asymptotically up to some order) for specified choice of r. Therefore if $\hat{\theta}_r$ can be defined independently of r, then we shall show it is efficient (asymptotically up to the above mentioned order), and if not, we can establish that there does not exist any efficient (in the same sense) estimator. It shall also be seen that asymptotic efficiency (including higher order cases) may be systematically discussed by the discretized likelihood method.

6.1. Discretized likelihood estimator (DLE)

Let Ⓗ be an open set in a Euclidean space R^1.

We assume that for each $\theta \in$ Ⓗ P_θ is absolutely continuous with respect to σ-finite measure μ .

We denote a density $dP_\theta/d\mu$ by $f(x,\theta)$. Let $L(\theta\,;\,\tilde{x}_n)$ be a likelihood function, that is, $L(\theta\,;\,\tilde{x}_n)=\prod_{i=1}^{n} f(x_i,\theta)$, where $\tilde{x}_n=(x_1, x_2, \ldots, x_n)$. For each k=1,2,..., a $\{c_n\}$-consistent estimator $\hat{\theta}_n$ is called discretized likelihood estimator (DLE) if for each real number r, $\hat{\theta}_n$ satisfies the discretized likelihood equation

(6.1.1) $\log L(\hat{\theta}_n+rc_n^{-1}\,;\,\tilde{x}_n) - \log L(\hat{\theta}_n\,;\,\tilde{x}_n) = a_n(\hat{\theta}_n, r)$,

where $a_n(\theta, r)$ is a function in θ and r and it also depends on n. The function $a_n(\theta, r)$ is not now defined but will be determined in the sequel so that the solution obtained from the above equation will be asymptotically median unbiased up to k-th order.

It should be noted that DLE $\hat{\theta}_n$ is required to be $\{c_n\}$-consistent and we do not claim that the solution of equation (6.1.1) be $\{c_n\}$-consistent. We implicitely claim that there exists a solution of the equation in the $O(c_n^{-1})$-neighborhood of the true value.

In practice situation we have to obtain the DLE first by finding a $\{c_n\}$-consistent estimator $\tilde{\theta}_n$ in some way or another and then finding a solution of the equation in the neighborhood of $\tilde{\theta}_n$.

Suppose that for given function $a_n(\theta, r)$.

(6.1.2) $\log L(\theta+rc_n^{-1}, \tilde{x}_n) - \log L(\theta, \tilde{x}_n) - a_n(\theta, r)$

is locally monotone in θ with probability larger than $1-o(c_n^{-(k-1)})$. For regular cases the particular form of $a_n(\theta, r)$ will be given later (e.g. page 193 etc.). For the present it is only necessary to remark that $a_n(\theta, r)$ has smaller order of magnitude than the previous terms of (6.1.2).

Then the k-th order asymptotic distribution of the DLE $\hat{\theta}_{DL}$ attains the bound of the k-th order asymptotic distributions of k-th order AMU estimators of θ at r. Indeed, it follows by the monotonicity of (6.1.2) that for any $\vartheta\in\Theta$, there exists a positive number δ such that

$$(6.1.3)\qquad \limsup_{n\to\infty}\ \sup_{\theta:|\theta-\vartheta|<\delta} c_n^{k-1}\Big|P_{\theta,n}\{\hat\theta_n > \theta - rc_n^{-1}\} - P_{\theta,n}\{\log L(\theta;\tilde x_n) - \log L(\theta - rc_n^{-1};\tilde x_n) > a_n(\theta - rc_n^{-1}, r)\}\Big| = 0 .$$

Letting $\theta_0 (\in \Theta)$ be arbitrary but fixed we consider the problem of testing hypothesis $H:\theta=\theta_0 - rc_n^{-1}$ $(r>0)$ against alternative $K:\theta=\theta_0$. Putting $A_{\hat\theta_n,\theta}=\{c_n(\hat\theta_n-\theta)>-r\}$ we have $P_{\theta_0-rc_n^{-1},n}(A_{\hat\theta_n,\theta_0}) = 1/2 + o(c_n^{-(k-1)})$. Let $\mathcal{U}_k$ be the class of the all k-th order AMU estimators. Set $\Phi_{1/2}=\{\{\phi_n\} : E_{\theta_0-rc_n^{-1},n}(\phi_n) = 1/2 + o(c_n^{-(k-1)}),\ 0\leqq\phi_n(\tilde x_n)\leqq 1$ for all $\tilde x_n\in\mathcal{X}^{(n)}$ $(n=1,2,\dots)\}$.
Note that every sequence $\chi_{A_{\hat\theta_n,\theta_0}}(\tilde x_n)$ of indicators of the set $A_{\hat\theta_n,\theta_0}$ with the estimators $\hat\theta_n$ in $\mathcal{U}_k$ is contained in $\Phi_{1/2}$.
In order to obtain the upper bound of $\overline{\lim}_{n\to\infty} P_{\theta_0,n}(A_{\hat\theta_n,\theta_0})$ in $\mathcal{U}_k$, it is sufficient to find the sequence $\{\phi_n^*\}$ of tests which maximize $\overline{\lim}_{n\to\infty} E_{\theta_0,n}(\phi_n)$ in $\Phi_{1/2}$. It is seen by the Neyman-Pearson fundamental lemma that ϕ_n^* has the rejection S_n satisfying

$$\sum_{i=1}^{n}\log\frac{f(X_i,\theta_0-rc_n^{-1})}{f(X_i,\theta_0)} < k_n ,$$ where k_n is some constant.

Then it follows from (6.1.3) that the upper bound of $\overline{\lim}_{n\to\infty} P_{\theta_0,n}(A_{\hat\theta_n,\theta_0})$ in $\mathcal{U}_k$ is given by $\overline{\lim}_{n\to\infty} E_{\theta_0,n}(\phi_n^*)$. Hence the k-th order asymptotic distribution of the DLE $\hat\theta_{DL}$ attains the bound of the k-th order asymptotic distributions of k-th order AMU estimators at $-r$.
In a similar way as the case when $r>0$, we also obtain for $r<0$ the upper bound of $\overline{\lim}_{n\to\infty} P_{\theta_0,n}(A^c_{\hat\theta_n,\theta_0})$ in $\mathcal{U}_k$ of the same form.
Hence the desired result also holds for the case $r<0$.
In later sections it will be seen that $\hat\theta_{DL}$ is asymptotically efficient up to second order. Note that the DLE usually depends on r in cases of more than third order.

In the subsequent discussion we shall deal with the case when $c_n=\sqrt{n}$.

6.2. Second order asymptotic efficiency of DLE

Suppose that $X_1, X_2, \dots, X_n$ is a sequence of i.i.d. random variables with a density $f(x, \theta)$ satisfying (A.3.1.1), (A.3.1.3) and (A.4.1.1) $\sim$ (A.4.1.3).

We have shown in section 4.1 that an MLE is second order asymptotically efficient.

We shall obtain the same results as Theorems 4.1.2 and 4.1.3. for the DLE. We further assume the following :

(A.6.2.1) For given function $a_n(\theta, r)$

$$\log L(\theta + rc_n^{-1}, \tilde{x}_n) - \log L(\theta, \tilde{x}_n) - a_n(\theta, r)$$

is locally monotone in θ with probability larger than $1-o(n^{-1})$.

Remark : This is true in most situations since $(1/n)(\partial^2/\partial\theta^2)$ $\log L(\theta, \tilde{x}_n)$ is asymptotically equal to $-I(\theta)(<0)$ and $a_n(\theta, r)$ has smaller order than n^{-1}, and is usually of constant order $(O(1))$ as is shown below.

Let $\hat{\theta}_n$ be a DLE.

Since

$$\sum_{i=1}^{n} \log f(X_i, \hat{\theta}_n + rc_n^{-1/2}) - \sum_{i=1}^{n} \log f(X_i, \hat{\theta}_n) = a_n ,$$

it follows by Taylor expansion that

$$\frac{1}{n} \sum_{i=1}^{n} \frac{\partial}{\partial\theta} \log f(X_i, \hat{\theta}_n) + \frac{r}{2n} \sum_{i=1}^{n} \frac{\partial^2}{\partial\theta^2} \log f(X_i, \hat{\theta}_n)$$

$$+ \frac{r^2}{6n\sqrt{n}} \sum_{i=1}^{n} \frac{\partial^3}{\partial\theta^3} \log f(X_i, \theta_n^*) = \frac{a_n}{r} ,$$

where

$$|\theta_n^* - \hat{\theta}_n| < \frac{r}{n} .$$

Since $(1/n) \sum_{i=1}^{n} \left\{ (\partial^3/\partial\theta^3) \log f(X_i, \theta) \right\}$ converges in probability to $-3J(\theta)-K(\theta)$, it is seen that

$$(6.2.1)\quad Z_1(\hat{\theta}_n)+\frac{r}{2}\left\{-I(\hat{\theta}_n)+\frac{1}{n}Z_2(\hat{\theta}_n)\right\}-\frac{r^2}{6\sqrt{n}}\left\{3J(\hat{\theta}_n)+K(\hat{\theta}_n)\right\}=\frac{a_n}{r}\ .$$

On the other hand we have

$$Z_1(\hat{\theta}_n) = Z_1(\theta) + \frac{1}{\sqrt{n}}\left\{ Z_2(\theta) - \sqrt{n}\, I(\theta)\right\} T_n$$

$$- \frac{3J(\theta)+K(\theta)}{2\sqrt{n}} T_n^2 + o\left(\frac{1}{\sqrt{n}}\right) ,$$

where $T_n=\sqrt{n}(\hat{\theta}_n-\theta)$.

Since

$$Z_1(\hat{\theta}_n)=Z_1(\theta)+\frac{1}{\sqrt{n}}Z_2(\theta)T_n-I(\theta)T_n - \frac{3J(\theta)+K(\theta)}{2\sqrt{n}} T_n^2 + o\left(\frac{1}{\sqrt{n}}\right) ,$$

it follows that

$$(6.2.2)\quad T_n=\frac{1}{I(\theta)}\left\{ -Z_1(\hat{\theta}_n)+Z_1(\theta)+\frac{1}{\sqrt{n}}Z_2(\theta)T_n - \frac{3J(\theta)+K(\theta)}{2\sqrt{n}} T_n^2\right\}+o\left(\frac{1}{\sqrt{n}}\right)$$

Since

$$I(\hat{\theta}_n)=I(\theta)+\frac{1}{\sqrt{n}}\left\{2J(\theta)+K(\theta)\right\} T_n + o\left(\frac{1}{\sqrt{n}}\right) ;$$

$$J(\hat{\theta}_n)=J(\theta)+\frac{1}{\sqrt{n}}J'(\theta)T_n + o\left(\frac{1}{\sqrt{n}}\right) ;$$

$$K(\hat{\theta}_n)=K(\theta)+\frac{1}{\sqrt{n}}K'(\theta)T_n+o\left(\frac{1}{\sqrt{n}}\right) ;$$

$$Z_2(\hat{\theta}_n)=Z_2(\theta)-J(\theta)T_n+o(1) ,$$

it follows from (6.2.1) that

$$Z_1(\hat{\theta}_n) = \frac{a_n(\theta,r)}{r} + \frac{r}{2}I(\theta) + \frac{r}{6\sqrt{n}}\left\{ 3rJ(\theta) + rK(\theta) - 3Z_2(\theta)\right\}$$

$$+ \frac{r}{2\sqrt{n}} J(\theta)T_n + o_p\left(\frac{1}{n}\right)$$

up to order $n^{-1/2}$.

From (6.2.2) we have

$$T_n=-\frac{a_n^*}{rI} - \frac{r^2(3J+K)}{6I\sqrt{n}} + \frac{Z_1}{I} + \frac{1}{2I\sqrt{n}}\left(Z_2-\frac{J}{I}Z_1\right)+\frac{Z_1Z_2}{I^2\sqrt{n}} - \frac{(3J+K)Z_1^2}{2I^3\sqrt{n}}$$

$$+ o_p\left(\frac{1}{\sqrt{n}}\right)$$

up to order $n^{-1/2}$, where $a_n=-\frac{r^2 I}{2} + a_n^*$ with $a_n^*=o_p(1)$ so that $\hat{\theta}_n$ is AMU. In order to have second order asymptotic median unbiasedness of $\hat{\theta}_n$ we put

$$a_n^* = -\frac{r^3(3J+K)}{6\sqrt{n}} - \frac{rK}{6I\sqrt{n}} + o(\frac{1}{\sqrt{n}}) .$$

and we denote by T_n^* the T_n corresponding to this particular value of a_n. Then we may also denote $T_n^* = \sqrt{n}(\hat{\theta}_n^* - \theta)$.
Thus we see that $\hat{\theta}_n^*$ is second order AMU. Hence we have established the following theorem.

<u>Theorem 6.2.1.</u> Under conditions (A.3.1.1), (A.3.1.3), (A.4.1.1) $\sim$ (A.4.1.3) and (A.6.2.1), the DLE $\hat{\theta}_n^*$ with a_n^* defined above satisfies the following :

$$T_n^* = \sqrt{n}(\hat{\theta}_n^* - \theta)$$

$$= \frac{K(\theta)}{6I^2\sqrt{n}} + \frac{Z_1(\theta)}{I(\theta)} + \frac{Z_1(\theta)Z_2(\theta)}{\sqrt{n}\, I(\theta)^2} - \frac{3J(\theta)+K(\theta)}{2\sqrt{n}\, I(\theta)^3} Z_1(\theta)^2$$

$$+ \frac{r}{2I(\theta)\sqrt{n}} \left\{ Z_2(\theta) - \frac{J(\theta)}{I(\theta)} Z_1(\theta) \right\} + o_p(\frac{1}{\sqrt{n}})$$

up to order $n^{-1/2}$ as $n \to \infty$.

Remark : We have

$$E_\theta[\, Z_1(\theta) \left\{ Z_2(\theta) - \frac{J(\theta)}{I(\theta)} Z_1(\theta) \right\}] = 0 ;$$

$$V_\theta(T_n^*) = \frac{1}{I(\theta)} + o(\frac{1}{\sqrt{n}}) ,$$

and the third order cumulant is equal to that of the MLE up to the order $n^{-1/2}$. Hence the asymptotic expansion of the distribution of T_n^* is equal to that of $\sqrt{n}(\hat{\theta}^*_{ML} - \theta)$ up to the order $n^{-1/2}$.
Therefore the DLE $\hat{\theta}_n^*$ is asymptotically equivalent to the MLE $\hat{\theta}^*_{ML}$ up to that order. (See Theorems 4.1.2 and 4.1.3.)

6.3. Third order asymptotic efficiency of DLE.

We proceed to the problem of third order asymptotic efficiency. We further assume the following :

(A.6.3.1) For almost all $x[\mu]$, $f(x,\theta)$ is four times continuously differentiable in θ ;

(A.6.3.2) There exist

$$L(\theta) = E_\theta\left[\left\{\frac{\partial^3}{\partial\theta^3}\log f(X,\theta)\right\}\left\{\frac{\partial}{\partial\theta}\log f(X,\theta)\right\}\right] \quad ;$$

$$M(\theta) = E_\theta\left[\left\{\frac{\partial^2}{\partial\theta^2}\log f(X,\theta)\right\}^2\right] ;$$

$$N(\theta) = E_\theta\left[\left\{\frac{\partial^2}{\partial\theta^2}\log f(X,\theta)\right\}\left\{\frac{\partial}{\partial\theta}\log f(X,\theta)\right\}^2\right]$$

and

$$H(\theta) = E_\theta\left[\left\{\frac{\partial}{\partial\theta}\log f(X,\theta)\right\}^4\right]$$

and the following holds :

$$E_\theta\left[\frac{\partial^4}{\partial\theta^4}\log f(X,\theta)\right] = -4L(\theta) - 3M(\theta) - 6N(\theta) - H(\theta) .$$

Let $\hat{\theta}_n$ be an DLE.

Since

$$\sum_{i=1}^n \log f(X_i, \hat{\theta}_n + \frac{r}{\sqrt{n}}) - \sum_{i=1}^n \log f(X_i, \hat{\theta}_n) = a_n ,$$

it follows that

$$\frac{1}{\sqrt{n}}\sum_{i=1}^n \frac{\partial}{\partial\theta}\log f(X_i,\hat{\theta}_n) + \frac{r}{2n}\sum_{i=1}^n \frac{\partial^2}{\partial\theta^2}\log f(X_i,\hat{\theta}_n)$$
$$+\frac{r^2}{6n\sqrt{n}}\sum_{i=1}^n \frac{\partial^3}{\partial\theta^3}\log f(X_i,\hat{\theta}_n)+\frac{r^3}{24n^2}\sum_{i=1}^n \frac{\partial^4}{\partial\theta^4}\log f(X_i,\theta^*)=\frac{a_n}{r} ,$$

where $|\theta^* - \hat{\theta}_n| < \frac{r}{\sqrt{n}}$.

Since $(1/n)\sum_{i=1}^n (\partial^4/\partial\theta^4)\log f(X_i,\theta)$ converges in probability to $-4L(\theta)-3M(\theta)-6N(\theta)-H(\theta)$, it is seen that

$$Z_1(\hat{\theta}_n) + \frac{r}{2}\left\{-I(\hat{\theta}_n) + \frac{1}{\sqrt{n}}Z_2(\hat{\theta}_n)\right\} - \frac{r^2}{6\sqrt{n}}\left\{3J(\hat{\theta}_n) + K(\hat{\theta}_n)\right\}$$
$$+ \frac{r^2}{6n}Z_3(\hat{\theta}_n) - \frac{r^3}{24n}\left\{4L(\hat{\theta}_n)+3M(\hat{\theta}_n)+6N(\hat{\theta}_n)+H(\hat{\theta}_n)\right\} \sim \frac{a_n}{r} ,$$

where

$$Z_3(\theta) = \frac{1}{\sqrt{n}} \sum_{i=1}^{n} \left\{ \frac{\partial^3}{\partial\theta^3} \log f(X_i, \theta) + 3J(\theta) + K(\theta) \right\} .$$

Hence

$$(6.3.1) \qquad Z_1(\hat{\theta}_n) \sim \frac{a_n}{r} + \frac{1}{2} I(\hat{\theta}_n) - \frac{r}{2\sqrt{n}} Z_2(\hat{\theta}_n) + \frac{r^2}{6\sqrt{n}} \left\{ 3J(\hat{\theta}_n) + K(\hat{\theta}_n) \right\}$$

$$- \frac{r^2}{6n} Z_3(\hat{\theta}_n) + \frac{r^3}{24n} \left\{ 4L(\hat{\theta}_n) + 3M(\hat{\theta}_n) + 6N(\hat{\theta}_n) + H(\hat{\theta}_n) \right\} .$$

On the other hand we have

$$Z_1(\hat{\theta}_n) = Z_1(\theta) + \frac{1}{\sqrt{n}} \left\{ Z_2(\hat{\theta}_n) - \sqrt{n}\, I(\theta) \right\} T_n - \frac{3J(\theta)+K(\theta)}{2\sqrt{n}} T_n^2$$

$$- \frac{1}{6n} \left\{ 4L(\theta) + 3M(\theta) + 6N(\theta) + H(\theta) \right\} T_n^3 + o_p(\frac{1}{n}) ,$$

where $T_n = \sqrt{n}(\hat{\theta}_n - \theta)$.

We obtain

$$(6.3.2) \qquad T_n = \frac{1}{I(\theta)} [-Z_1(\hat{\theta}_n) + Z_1(\theta) + \frac{1}{\sqrt{n}} Z_2(\theta) T_n - \frac{3J(\theta)+K(\theta)}{2\sqrt{n}} T_n^2$$

$$- \frac{1}{6n} \left\{ 4L(\theta) + 3M(\theta) + 6N(\theta) + H(\theta) \right\} T_n^3] + o_p(\frac{1}{n}) .$$

Since

$$I(\hat{\theta}_n) = I(\theta) + \frac{1}{\sqrt{n}} \left\{ 2J(\theta) + K(\theta) \right\} T_n$$

$$+ \frac{1}{2n} \left\{ 2L(\theta) + 2M(\theta) + 5N(\theta) + H(\theta) \right\} T_n^2 + o_p(\frac{1}{n}) ;$$

$$J(\hat{\theta}_n) = J(\theta) + \frac{1}{\sqrt{n}} \left\{ L(\theta) + M(\theta) + N(\theta) \right\} T_n + o_p(\frac{1}{\sqrt{n}}) ;$$

$$K(\hat{\theta}_n) = K(\theta) + \frac{1}{\sqrt{n}} \left\{ 3N(\theta) + H(\theta) \right\} T_n + o_p(\frac{1}{\sqrt{n}}) ;$$

$$L(\hat{\theta}_n) = L(\theta) + \frac{1}{\sqrt{n}} L'(\theta) T_n + o_p(\frac{1}{\sqrt{n}}) ;$$

$$M(\hat{\theta}_n) = M(\theta) + \frac{1}{\sqrt{n}} M'(\theta) T_n + o_p(\frac{1}{\sqrt{n}}) ;$$

$$N(\hat{\theta}_n) = N(\theta) + \frac{1}{\sqrt{n}} N'(\theta) T_n + o_p(\frac{1}{\sqrt{n}}) ;$$

$$H(\hat{\theta}_n) = H(\theta) + \frac{1}{\sqrt{n}} H'(\theta) T_n + o_p(\frac{1}{\sqrt{n}}) ;$$

$$Z_2(\hat{\theta}_n) = Z_2(\theta) - J(\theta) T_n - \frac{1}{2\sqrt{n}} \left\{ 2L(\theta) + M(\theta) + N(\theta) \right\} T_n^2 + o_p(\frac{1}{\sqrt{n}}) ;$$

and

$$Z_3(\hat{\theta}_n) = \frac{1}{\sqrt{n}} Z_3'(\theta) T_n + o_p(\frac{1}{\sqrt{n}}) \ ,$$

it follows from (6.3.1) that

$$Z_1(\hat{\theta}_n) = \frac{a_n}{r} + \frac{r}{2} I - \frac{r}{2\sqrt{n}} + \frac{r^2(3J+K)}{6\sqrt{n}} - \frac{r^2}{6n} Z_3$$

$$+ \frac{r}{2\sqrt{n}} J \ T_n + \frac{r^3}{24n}(4L+3M+6N+H)$$

$$+ \frac{r}{4n} (4L+3M+6N+H) T_n^2 + \frac{r^2}{6n} (3L+3M+6N+H) T_n + o_p(\frac{1}{n})$$

up to order n^{-1} as $n \to \infty$.

Under conditions (A.3.1.1), (A.3.1.3), (A.4.1.2), (A.6.2.1), (A.6.3.1) and (A.6.3.2) we have from (6.3.2)

$$(6.3.3) \quad T_n = -\frac{a_n^{**}}{rI} + \frac{K}{6I^2\sqrt{n}} + \frac{3N+H}{6I^2 n} - \frac{r^3}{24In} + \frac{Z_1}{I}$$

$$+ \frac{r}{2I\sqrt{n}} Z_2 - \frac{r}{2I\sqrt{n}} J T_n + \frac{1}{I\sqrt{n}} Z_2 T_n - \frac{3J+K}{2I\sqrt{n}} T_n^2$$

$$- \frac{K(2J+K)}{6I^3 n} T_n - \frac{\mathcal{L}}{6In} T_n^3 + \frac{r^2 Z_3}{6In}$$

$$- \frac{1}{3In}(4L+4M+6N+H) T_n - \frac{r}{4In} L T_n^2 + o_p(\frac{1}{n})$$

$$= -\frac{a_n^{**}}{rI} + \frac{K}{6I^2\sqrt{n}} + \frac{3N+H}{6I^2 n} - \frac{r^3}{24In} + \frac{Z_1}{I}$$

$$+ \frac{1}{I^2\sqrt{n}}(Z_1 Z_2 - \frac{3J+K}{2I} Z_1^2) + \frac{r}{2I\sqrt{n}}(Z_2 - \frac{J}{I} Z_1)$$

$$+ \frac{1}{I^3 n}\left\{ \frac{KZ_2}{6} - \frac{3(3J+K)}{2I} Z_1^2 Z_2 + Z_1 Z_2^2 - \frac{K(2J+K)}{6I} Z_1 \right.$$

$$\left. - \frac{\mathcal{L}}{6I} Z_1^3 - \frac{1}{3}(4L+4M+6N+H) Z_1 \right\}$$

$$+ \frac{r}{2I^2 n} \Big[(Z_2 - \frac{3J+K}{I} Z_1)(Z_2 - \frac{J}{I} Z_1) - J \Big\{ \frac{K}{6I} + \frac{Z_1 Z_2}{I}$$

$$+ \frac{r}{2} Z_2 - \frac{rJ}{2I} Z_1 - \frac{3J+K}{2I^2} Z_1^2 + \frac{rI}{3} Z_3 - \frac{L}{2I} Z_1^2 \Big\} \Big]$$

$$+ o_p(\frac{1}{n})$$

up to order n^{-1} as $n \to \infty$, when $a_n = -\frac{r^2 I}{2} - \frac{r^3(3J+K)}{6\sqrt{n}} + \frac{rK}{6I\sqrt{n}} + a_n^{**}$

with $a_n^{**} = O(1/n)$ and $\mathcal{L} = 4L+3M+6N+H$.

We denote by T_n^* the T_n corresponding to this particular value of a_n. Then we may write $T_n^* = \sqrt{n}(\hat{\theta}_n^* - \theta)$.

On the other hand under conditions (A.3.1.1), (A.3.1.3), (A.4.1.2), (A.6.2.1), (A.6.3.1) and (A.6.3.2) we have obtained in section 4.2

$$(6.3.4) \quad \sqrt{n}(\hat{\theta}_{ML} - \theta) = \frac{Z_1}{I} + \frac{1}{I^2\sqrt{n}}\left\{Z_1 Z_2 - \frac{(3J+K)Z_1^2}{2I}\right\}$$
$$+ \frac{1}{I^3 n}\left\{Z_1 Z_2^2 + \frac{1}{2} Z_1^2 Z_3 - \frac{3J+K}{2I} Z_1^2 Z_2 \right.$$
$$\left. + \frac{(3J+K)^2}{2I^2} Z_1^3 - \frac{4L+3M+6N+H}{6I} Z_1^3\right\} + o_p(\frac{1}{n}) .$$

Let $\hat{\theta}_{ML}^*$ be a modified MLE so that it is third order AMU. Comparing (6.3.3) with (6.3.4), we see that $\sqrt{n}(\hat{\theta}_{ML}^* - \theta)$ and T_n^* are essentially different in the order n^{-1}. We put $T_{ML}^* = \sqrt{n}(\hat{\theta}_{ML}^* - \theta)$. Note that the difference of T_{ML}^* and T_n^* appears in the linear term of order $n^{-1/2}$ and in the other terms of order n^{-1}.

It was shown in sections 5.1 and 5.2 that the asymptotic cumulants are determined up to order n^{-1} by the terms of order up to $n^{-1/2}$ if the first term is equal to $Z_1(\theta)/I(\theta)$; and the first term of order n^{-1} of the fourth order cumulant is identical for all asymptotically efficient estimators. For the third order cumulants we have

$$K_3(T_{ML}^* - E_\theta(T_{ML}^*)) = \frac{\beta_3(\theta)}{\sqrt{n}} + o(\frac{1}{n}) ;$$

$$K_3(T_n^* - E_\theta(T_n^*)) = \frac{\beta_3(\theta)}{\sqrt{n}} + \frac{\gamma_3(\theta)}{n} + o(\frac{1}{n}) .$$

Therefore there is a difference between asymptotic distributions of $\sqrt{n}(\hat{\theta}_{ML}^* - \theta)$ and T_n^* in the term of order n^{-1} if $\gamma_3(\theta) \neq 0$.

Using Theorem 6.2.1 and (6.3.3) write T_n^* as :

$$T_n^* = \frac{Z_1}{I} - \frac{J+K}{6I^2\sqrt{n}} + \frac{r}{\sqrt{n}}L^* + \frac{1}{\sqrt{n}}(Q-c) + \frac{1}{n}R + o_p(\frac{1}{n}) ,$$

where

$$L^* = \frac{1}{2I}(Z_2 - \frac{J}{I}Z_1) ;$$

$$Q = \frac{1}{I^2}(Z_1Z_2 - \frac{3J+K}{2I} Z_1^2) ;$$

$$c = -\frac{J+K}{2I^2} .$$

We have

$$(6.3.5) \quad \frac{\beta_3(\theta)}{\sqrt{n}} + \frac{\gamma_3(\theta)}{n} + o(\frac{1}{n}) = -\frac{1}{2I^3}E_\theta(Z_1^3) + \frac{3}{2I}E_\theta[Z_1(T_n^* - E_\theta(T_n^*))^2]$$

$$+ \frac{3}{2In}E[Z_1(rL^*+Q-c)^2] + o(\frac{1}{n}) .$$

Note that $E(Q)=c$.

We have $E_\theta(Z_1{}^3) = K/\sqrt{n}$. For the second term of the right-hand side of (6.3.5) we use lemma 5.1.1.

By the Lemma we obtain

$$(6.3.6) \quad E_\theta[Z_1(T_n^* - E_\theta(T_n^*))^2]$$

$$= \frac{1}{\sqrt{n}}\frac{\partial}{\partial\theta}V_\theta(T_n^*) - \frac{2}{\sqrt{n}}E_\theta[(T_n^* - E_\theta(T_n^*))\left\{\frac{\partial}{\partial\theta}(T_n^* - E_\theta(T_n^*))\right\}]$$

$$= \frac{1}{\sqrt{n}}\frac{\partial}{\partial\theta}(\frac{1}{I}) + o(\frac{1}{n})$$

$$= -\frac{2J+K}{I^2\sqrt{n}} + o(\frac{1}{n}) .$$

We also obtain

$$(6.3.7) \quad E_\theta[Z_1(rL^*+Q-c)^2] = 2rE_\theta[Z_1L^*(Q-c)] + o(1)$$

$$= \frac{r}{I}\left\{\frac{M}{I} - \frac{(3J+K)K}{2I^2}\right\} + o(1) .$$

From (6.3.5), (6.3.6) and (6.3.7) we have

$$\gamma_3(\theta)= \frac{3r}{2I(\theta)^2}\left[\frac{M(\theta)}{I(\theta)} - \frac{\{3J(\theta)+K(\theta)\}K(\theta)}{2I(\theta)^2}\right].$$

Since the third order asymptotic distribution of T_n^* attains the bound of the third order asymptotic distributions of third order AMU estimators at r, there is no third order AMU estimator which uniformly attains the bound, if $\gamma_3(\theta)$ is not equal to zero. Hence we have established :

Theorem 6.3.1. Under conditions (A.3.1.1), (A.3.1.3), (A.4.1.2), (A.6.2.1), (A.6.3.1) and (A.6.3.2), $\hat{\theta}_{ML}^*$ is not third order asymptotically efficient if $\gamma_3(\theta)\neq 0$.

6.4. Maximum log-likelihood estimator

Instead of the equation (6.0.1) we may take a solution $\hat{\theta}_n$ of the discretized likelihood equation :

$$(6.4.1)\quad \sum_{i=1}^{n}\log f(X_i, \hat{\theta}_n+rc_n^{-1}) - \sum_{i=1}^{n}\log f(X_i, \hat{\theta}_n-rc_n^{-1}) = 0 .$$

The solution $\hat{\theta}_n$ is the θ which maximizes

$$\int_{-rc_n^{-1}}^{rc_n^{-1}} \sum_{i=1}^{n}\log f(X_i,\theta+t)dt = \int_{-rc_n^{-1}}^{rc_n^{-1}} \log L(\theta+t)dt .$$

where $L(\theta)$ denotes the likelihood function.

Then $\hat{\theta}_n$ is called the maximum log-likelihood estimator(MLLE).

If $\log L(\hat{\theta}_n+t)$ is locally linearized, $\hat{\theta}_n$ agrees with maximum probability estimator of Weiss and Wolfowitz [54].

Now we shall consider a location parameter case when the density function f satisfied the following assumption :

(A.6.4.1) $f(x,\theta)=f(x-\theta)$ and $f(x)$ is symmetric w.r.t. the origin.

It follows by the symmetry of f that the solution $\hat{\theta}_n$ of (6.4.1) is AMU.

We also have

$$J(\theta) = K(\theta) = 0 .$$

Let $\hat{\theta}_n$ be an MLLE.

Then

$$\sum_{i=1}^{n} \log f(X_i - \hat{\theta}_n - (r/\sqrt{n})) - \sum_{i=1}^{n} \log f(X_i - \hat{\theta}_n + (r/\sqrt{n})) = 0 .$$

Without loss of generality we assume that $\theta_0 = 0$.

Since

$$\sum_{i=1}^{n} \log f(X_i - \hat{\theta}_n - (r/\sqrt{n})) - \sum_{i=1}^{n} \log f(X_i)$$
$$- \left\{ \sum_{i=1}^{n} \log f(X_i - \hat{\theta}_n + (r/\sqrt{n})) - \sum_{i=1}^{n} \log f(X_i) \right\} = 0 ,$$

it follows by Taylor expansion around $\theta = 0$ that

$$\frac{2r}{\sqrt{n}} \left[\frac{\partial}{\partial\theta} \sum_{i=1}^{n} \log f(X_i - \theta) \right]_{\theta=0} + \frac{2r\hat{\theta}_n}{\sqrt{n}} \left[\frac{\partial^2}{\partial\theta^2} \sum_{i=1}^{n} \log f(X_i - \theta) \right]_{\theta=0}$$
$$+ \left(\frac{r\hat{\theta}_n^{\,2}}{\sqrt{n}} + \frac{r^3}{3n\sqrt{n}} \right) \left[\frac{\partial^3}{\partial\theta^3} \sum_{i=1}^{n} \log f(X_i - \theta) \right]_{\theta=0}$$
$$+ \frac{1}{3} \left(\frac{r\hat{\theta}_n}{\sqrt{n}} + \frac{r^3\hat{\theta}_n}{n\sqrt{n}} \right) \left[\frac{\partial^4}{\partial\theta^4} \sum_{i=1}^{n} \log f(X_i - \theta) \right]_{\theta=0} \sim 0 ,$$

where $|\theta^*| \leqq |\theta|$.

Putting $T_n = \sqrt{n}\,\hat{\theta}_n$, we obtain

$$2rZ_1 + 2r\left(\frac{Z_2}{\sqrt{n}} - I(0)\right)T_n + \left\{ Z_3 - \sqrt{n}(3J(0)+K(0)) \right\} \left(\frac{rT_n^{\,2}}{n} + \frac{r^3}{3n} \right)$$
$$- \frac{1}{3} \left\{ 4L(0)+3M(0)+6N(0)+H(0) \right\} \left(\frac{r}{n} T_n^{\,3} + \frac{r^3}{n} T_n \right) \sim 0$$

Hence it follows that

$$T_n = \frac{Z_1}{I} + \frac{Z_2}{I\sqrt{n}} \left(\frac{Z_1}{I} + \frac{Z_1 Z_2}{I^2 n} \right) + \frac{r^2}{6In} Z_3 + \frac{1}{2In} Z_3 \frac{Z_1^{\,2}}{I^2}$$
$$- \frac{1}{6In} (4L+3M+6N+H) \left(\frac{Z_1^{\,3}}{I^3} + r^2 \frac{Z_1}{I} \right) + o_p\left(\frac{1}{n}\right)$$

Under conditions (A.3.1.1), (A.3.1.3), (A.4.1.1), (A.4.1.2), (A.6.3.1), (A.6.3.2) and

$$(6.4.2) \quad T_n = \frac{Z_1}{I} + \frac{Z_1 Z_2}{I^2 n} + \frac{1}{I^3 n} \left\{ Z_1 Z_2^{\,2} + \frac{1}{2} Z_1^{\,2} Z_3 - \frac{1}{6I}(4L+3M+6N+H) Z_1^{\,3} \right\}$$
$$+ \frac{r^2 Z_3}{6In} - \frac{r^2(4L+3M+6N+H)}{6I^2 n} Z_1 + o_p\left(\frac{1}{n}\right)$$

up to order n^{-1} as $n \to \infty$.

As was stated previously the difference in the order n^{-1} term between (6.3.4) and (6.4.2) does not affect the asymptotic distribution up to the order n^{-1}. Hence we have established :

Theorem 6.4.1. Under conditions (A.3.1.1), (A.3.1.3), (A.4.1.1), (A.4.1.2), (A.6.3.1), (A.6.3.2) and (A.6.4.1), the MLLE $\hat{\theta}_n$ is asymptotically equivalent to the MLE $\hat{\theta}_{ML}$ up to order n^{-1}.

It was shown in sections 5.1 and 5.2 that $\hat{\theta}_{ML}$ maximizes the symmetric probability

$$P_{\theta,n}\left\{\sqrt{n}\,|\hat{\theta}_{ML} - \theta| < u\right\}$$

up to the order n^{-1} among all regular AMU estimators.

Therefore it is seen that

$$\varliminf_{n\to\infty} n\left[P_{\theta,n}\left\{\sqrt{n}\,|\hat{\theta}_{ML} - \theta| < u\right\} - P_{\theta,n}\left\{\sqrt{n}\,|\hat{\theta}_{DL} - \theta| < u\right\}\right] \geqq 0$$

for all u, where $\hat{\theta}_{DL}$ denotes DLE.

The asymptotic distribution of the MLLE $\hat{\theta}_{MLL}$ is equal to that of $\hat{\theta}_{ML}$ up to the order n^{-1}, but not up to order $n^{-3/2}$ and we can show that

$$\varliminf_{n\to\infty} n^{3/2}\left[P_{\theta,n}\left\{\sqrt{n}\,|\hat{\theta}_{MLL} - \theta| < r\right\} - P_{\theta,n}\left\{\sqrt{n}\,|\hat{\theta}_{ML} - \theta| < r\right\}\right] \geqq 0$$

in general situations. Hence for symmetric intervals $\hat{\theta}_{MLL}$ is not fourth order asymptotically efficient.

6.5. Remarks

If we also define an asymptotic efficient estimator as the $\hat{\theta}_n$ which maximizes $\lim_{n\to\infty} P_{\theta,n}\left\{ c_n |\hat{\theta}_n - \theta| < r\right\}$ among AMU estimators $\hat{\theta}_n$, then $\hat{\theta}_n^*$ satisfing the following equation (6.5.1) is asymptotically efficient :

$$(6.5.1)\quad \prod_{i=1}^{n} \frac{f(X_i, \hat{\theta}_n^* + rc_n^{-1})}{f(X_i, \hat{\theta}_n^*)} - \prod_{i=1}^{n} \frac{f(X_i, \hat{\theta}_n^* - rc_n^{-1})}{f(X_i, \hat{\theta}_n^*)} = a_n .$$

where a_n is chosen so that $\hat{\theta}_n^*$ is AMU.

Then (6.5.1) can also be expressed as

$$(6.5.2) \quad \exp\left\{\log L(\hat{\theta}_n^* + rc_n^{-1}) - \log L(\hat{\theta}_n^*)\right\}$$

$$- \exp\left\{\log L(\hat{\theta}_n^* - rc_n^{-1}) - \log L(\hat{\theta}_n^*)\right\} = a_n .$$

If $e^{\log L}$ is linearized then (6.5.2) is reduced to a form like (6.4.1), where in this case the right-hand side of (6.4.1) is not always zero.

Hence it is seen from the above that the asymptotic efficiency (including higher order cases) may be systematically discussed by discretized likelihood methods.

Further discretized likelihood methods may be also applied to the satistical estimation theory of fixed sample sizes.

Chapter 7

Higher Order Asymptotic Efficiency and Asymptotic Completeness

Summary :

In this chapter the concept of (higher order) asymptotic completeness of an estimator is introduced and it is shown that the maximum likelihood estimator is second order and third order asymptotically complete. Further the second order asymptotic efficiency of unbiased confidence intervals is discussed.

7.1. The concept of asymptotic completeness of an estimator

We consider first the case of a real-valued parameter. The concept of asymptotic efficiency and higher order asymptotic efficiency gives a very strong condition for the "goodness" of an estimator in that it requires it to uniformly maximize asymptotically the probability of its lying within a neighborhood of the true value, irrespective of the parameter value and the choice of neighborhood. But only one somewhat artificial restriction is that the estimators considered should be (higher order) asymptotically median unbiased. It is, however, easily seen from the context that the only requirement is that the class of estimators to be compared must have the same asymptotic "location" whatever the exact meaning of the word may be, without which nothing can be derived. Therefore one may consider different types of definitions of "location", and accordingly different types of "adjustment" of the estimator to satisfy the condition. Thus we proceed in the following way. Suppose that we want to establish asymptotic optimality of an estimator $\hat{\theta}_n^0$, e.g. MLE, in some class of estimators.

What has been virtually established is that for any estimator $\hat{\theta}_n$ in the class if both $\hat{\theta}_n$ and $\hat{\theta}_n^0$ are adjusted like

$$\hat{\theta}_n^* = \hat{\theta}_n + \frac{1}{n} g(\hat{\theta}_n) \ ;$$

$$\hat{\theta}_n^{0*} = \hat{\theta}_n^0 + \frac{1}{n} h(\hat{\theta}_n^0)$$

so that $\hat{\theta}_n^*$ and $\hat{\theta}_n^{0*}$ are both (higher order) asymptotically median unbiased, then $\hat{\theta}_n^{0*}$ is (higher order) asymptotically uniformly "better" than $\hat{\theta}_n^*$. If we change the definition of "location" the way of "adjustment" will be changed, but the same results hold for adjusted estimators $\hat{\theta}_n^{0*}$ and $\hat{\theta}_n^*$. This fact leads to the following question : Instead of comparing two adjusted estimators $\hat{\theta}_n^*$ and $\hat{\theta}_n^{0*}$, isn't it possible to compare $\hat{\theta}_n$ and the adjusted $\hat{\theta}_n^*$? Naturally, the way $\hat{\theta}_n^*$ is adjusted depends on the $\hat{\theta}_n$ to be compared, but is independent of the parameter. So we introduce the following concept.

An estimator $\hat{\theta}_n^*$ is called <u>asymptotically complete with respect to a class $\mathcal{E}$</u> of estimators if for any $\hat{\theta}_n \in \mathcal{E}$ there is an "adjusted" estimator

$$\hat{\theta}_n^{**} = \hat{\theta}_n^* + g_n(\hat{\theta}_n^*)$$

so that for any $a > 0$, $b > 0$ and θ we have

$$\underline{\lim}_{n\to\infty} [\ P_{\theta,n}\{-a < c_n(\hat{\theta}_n^{**}-\theta) < b\} - P_{\theta,n}\{-a < c_n(\hat{\theta}_n-\theta) < b\}\ ] \geqq 0\ . \tag{7.1.1}$$

Now let L be a loss function such that $L(0)=0$ and $L(u)$ is an increasing function for $u > 0$ and a decreasing function for $u < 0$ and it is bounded. If $\hat{\theta}_n^*$ is asymptotically complete in the above sense, then for any estimator $\hat{\theta}_n$ in $\mathcal{E}$ there exists $\hat{\theta}_n^{**}$ which is a function of $\hat{\theta}_n^*$, for which

$$\overline{\lim}_{n\to\infty} [E_{\theta,n}(L(c_n(\hat{\theta}_n^{**}-\theta))) - E_{\theta,n}(L(c_n(\hat{\theta}_n-\theta)))\ ] \leqq 0\ . \tag{7.1.2}$$

Higher order asymptotic completeness can be defined in a similar way. For each k=2,3,..., an estimator $\hat{\theta}_n^*$ is called <u>k-th order asymptotically complete with respect to a class $\mathcal{E}$</u> of estimators if for any $\hat{\theta}_n \in \mathcal{E}$ there is an "adjusted" estimator

$$\hat{\theta}_n^{**} = \hat{\theta}_n^* + g_n(\hat{\theta}_n^*)$$

so that for any $a > 0$, $b > 0$, and θ we have

$$\text{(7.1.3)} \quad \underline{\lim}_{n\to\infty} c_n^{k-1} [P_{\theta,n}\{-a<c_n(\hat{\theta}_n^{**}-\theta)<b\} - P_{\theta,n}\{-a<c_n(\hat{\theta}_n-\theta)<b\}] \geqq 0 .$$

Further $\hat{\theta}_n^*$ is called <u>k-th order asymptotically symmetrically complete with respect to a class $\mathcal{E}$</u> of estimators when (7.1.3) holds for a=b. If $\hat{\theta}_n^*$ is k-th order asymptotically complete with respect to $\mathcal{E}$, then for any estimator $\hat{\theta}_n$ in $\mathcal{E}$ there exists $\hat{\theta}_n^{**}$ which is a function of $\hat{\theta}_n^*$ for which

$$\text{(7.1.4)} \quad \overline{\lim}_{n\to\infty} c_n^{k-1}[E_{\theta,n}(L(c_n(\hat{\theta}_n^{**}-\theta))) - E_{\theta,n}(L(c_n(\hat{\theta}_n-\theta)))] \leqq 0.$$

If $\hat{\theta}_n^*$ is k-th order asymptotically symmetrically complete with respect to $\mathcal{E}$, then (7.1.4) holds for all loss function symmetric around the origin.

Unfortunately the above definition is meaningless as it stands, since no estimator can be asymptotically complete, which can be seen below. Let $\hat{\theta}_n^*$ be an asymptotically efficient estimator (e.g. MLE) and $\sqrt{n}(\hat{\theta}_n^* - \theta)$ be asymptotically normally distributed with mean 0 and variance 1/I, where I is the Fisher information. Let $\hat{\theta}_n$ be an AMU estimator with asymptotic variance larger than 1/I. Obviously $\hat{\theta}_n$ is not asymptotically efficient and inferior to $\hat{\theta}_n^*$. A naturally adjusted estimator of $\hat{\theta}_n^*$ corresponding to $\hat{\theta}_n$ will be $\hat{\theta}_n^{**} = \hat{\theta}_n^* + c/\sqrt{n}$ which has the same asymptotic mean as $\hat{\theta}_n$. Now take $(0<)b(<c)$ small enough and $a(>0)$ large enough.

Then we have

$$\overline{\lim_{n\to\infty}} \left[P_{\theta,n}\left\{ -a < \sqrt{n}(\hat{\theta}_n^{**} - \theta) < b \right\} - P_{\theta,n}\left\{ -a < \sqrt{n}(\hat{\theta}_n^{*} - \theta) < b \right\} \right] \leqq 0$$

which contradicts the condition of asymptotic completeness. However we may adjust $\hat{\theta}_n^*$ we can not get uniformity irrespective of a and b. It is, of course, possible to get the reverse inequality if we adjust $\hat{\theta}_n^*$ depending on a and b, but the nice uniformty would be lost.

There is nothing unnatural in this problem : If there is a bias in the estimators, the estimator with large variance may hit upon the true value by chance but one with small variance has little chance to get close to the true value. Therefore we must modify the concept of asymptotic completeness and call $\hat{\theta}_n^*$ asymptotically symmetrically complete if when (7.1.1) holds for $a=b$, then (7.1.2) holds for all symmetric loss functions.

The class $\mathbb{F}$ of estimators here is more specifically defined as the class of estimators $\hat{\theta}_n$ such that $c_n(\hat{\theta}_n - \theta)$ has an asymptotic distribution in the sense that

$$\lim_{n\to\infty} P_{\theta,n}\left\{ c_n(\hat{\theta}_n - \theta) < a \right\} = F_\theta(a) ,$$

where the convergence is locally uniform in θ and $F_\theta(a)$ is absolutely continuous in a and continuous in θ . We further assume that c_n is equal to the maximum order of consistency.
Now we have the following theorem, the proof of which is omitted.

<u>Theorem 7.1.1.</u> Let $\hat{\theta}_n^*$ be asymptotically efficient and the density of the asymptotic distribution of $\sqrt{c_n}(\hat{\theta}_n - \theta)$ be symmetric around the origin. Then $\hat{\theta}_n^*$ is asymptotically symmetrically complete with respect to the class $\mathbb{F}$.

In this case it is seen that $\hat{\theta}_n^*$ need not be modified.

Higher order asymptotic completeness has, however, a more meaningful implication. Now we consider the regular case with $c_n = \sqrt{n}$ and we define the classes $\mathbb{B}^*$, $\mathbb{C}^*$ and $\mathbb{D}^*$ in the same way as the classes $\mathbb{B}$, $\mathbb{C}$ and $\mathbb{D}$, respectively but omitting the condition of higher order asymptotic median unbiasedness.

Then the following theorems hold :

Theorem 7.1.2. The MLE is second order asymptotically complete with respect to the class $\mathbb{B}^*$.

Theorem 7.1.3. The MLE is third order asymptotically symmetrically complete with respect to the class $\mathbb{C}^*$.

Theorem 7.1.4. The MLE is third order asymptotically complete with respect to the class $\mathbb{D}^*$.

The outline of the proof of Theorem 7.1.4. is as follows :

Let $\hat{\theta}_n \in \mathbb{D}^*$ and the asymptotic bias of $\hat{\theta}_n$ be

$$E_{\theta,n}(\sqrt{n}(\hat{\theta}_n - \theta)) = \frac{1}{\sqrt{n}}\mu_1(\theta) + o(\frac{1}{n})$$

and that of $\hat{\theta}_n^*$ be

$$E_{\theta,n}(\sqrt{n}(\hat{\theta}_n^* - \theta)) = \frac{1}{\sqrt{n}}\mu_1^*(\theta) + o(\frac{1}{n}) .$$

Let the modified estimator $\hat{\theta}_n^{**}$ be defined as

$$\hat{\theta}_n^{**} = \hat{\theta}_n^* + \frac{1}{n}\{\mu_1(\hat{\theta}_n^*) - \mu_1^*(\hat{\theta}_n^*)\} .$$

Then we can prove exactly like in Theorem 5.1.6 that

$$\underline{\lim}_{n\to\infty} n[P_{\theta,n}\{-a<\sqrt{n}(\hat{\theta}_n^{**} - \theta)< b\} - P_{\theta,n}\{-a<\sqrt{n}(\hat{\theta}_n - \theta) < b\}] \geqq 0$$

for all $a,b > 0$ and θ , because the asymptotic moments of $\sqrt{n}(\hat{\theta}_n - \theta)$ and $\sqrt{n}(\hat{\theta}_n^{**} - \theta)$ are the same up to the order n^{-1} except for the second term in the variance which is smaller for $\hat{\theta}_n^*$.

The proofs of Theorems 7.1.2 and 7.1.3 are similar and even simpler.

The classes $\mathbb{C}^*$ and $\mathbb{D}^*$ are much more natural classes than classes $\mathbb{C}$ and $\mathbb{D}$. Theorems 7.1.2, 7.1.3 and 7.1.4 may be nearly the final conclusion of the asymptotic derivability of the MLE because we have the problem of the choice of the way of modification which should be based on the consideration of the asymptotic bias versus asymptotic variance, which can not be decided once and for all. What is established in the above implies that we only need to consider the estimators which are modifed MLE's and it is not necessary to take any other (regular) estimators into consideration.

The comparison of two estimators $\hat{\theta}_{1n}$ and $\hat{\theta}_{2n}$ of a real-valued parameter θ in the class $\mathbb{D}$ can be also made in terms of asymptotic deficiency as defined by Hodges and Lehmann [61].

$\hat{\theta}_{1n}$ and $\hat{\theta}_{2n}$ can be expressed as

$$\sqrt{n}(\hat{\theta}_{1n}-\theta) = U + \frac{1}{\sqrt{n}}Q_1 + o_p(\frac{1}{n}) \ ;$$

$$\sqrt{n}(\hat{\theta}_{2n}-\theta) = U + \frac{1}{\sqrt{n}}Q_2 + o_p(\frac{1}{n}) \ ,$$

where $U = \dfrac{1}{I\sqrt{n}} \sum_{i=1}^{n} \dfrac{\partial}{\partial\theta}\log f(X_i,\theta)$.

Now we may calculate "adjusted" estimators

$$\hat{\theta}^*_{1n} = \hat{\theta}_{1n} + \frac{1}{n}h_1(\hat{\theta}_{1n}) \ ;$$

$$\hat{\theta}^*_{2n} = \hat{\theta}_{2n} + \frac{1}{n}h_2(\hat{\theta}_{2n})$$

so that $\hat{\theta}^*_{1n}$ and $\hat{\theta}^*_{2n}$ have the same asymptotic bias up to the order $n^{-1/2}$. Then as have been repeatedly discussed, $\hat{\theta}^*_{1n}$ and $\hat{\theta}^*_{2n}$ have the same asymptotic cumulants up to the order n^{-1}, except for the order n^{-1} in the variance, which are given as

$$V(\sqrt{n}(\hat{\theta}^*_{1n}-\theta)) = \frac{1}{I} + \frac{\tau}{n} + \frac{1}{n}V(Q_1) + o(\frac{1}{n}) \ ;$$

$$V(\sqrt{n}(\hat{\theta}^*_{2n}-\theta)) = \frac{1}{I} + \frac{\tau}{n} + \frac{1}{n}V(Q_2) + o(\frac{1}{n}) \ .$$

Therefore in the Edgeworth expansion of the distribution of $\sqrt{n}(\hat{\theta}^*_{1n}-\theta)$ and $\sqrt{n}(\hat{\theta}^*_{2n}-\theta)$ up to the order n^{-1}, the only difference appearing in the term

$$\frac{1}{n}V(Q_i)t\,\phi(t) \qquad (i=1,2)\ ,$$

where $\phi(t)$ is the density function of the standard normal distribution. Now if we put $n'=n+V(Q_1)/I$ and $n''=n+V(Q_2)/I$, and consider the distributions of $\sqrt{n'}(\hat{\theta}_{1n'}-\theta)$ and $\sqrt{n''}(\hat{\theta}_{2n''}-\theta)$, that is, the estimators based on samples of sizes n' and n", respectively. Then it is seen that in the asymptotic expansion of the distributions of $\sqrt{n'}(\hat{\theta}_{1n'}-\theta)$ and $\sqrt{n''}(\hat{\theta}_{2n''}-\theta)$, the terms $V(Q_i)/n$ (i=1,2) vanish while all other terms up to the order n^{-1} remain the same as those of $\sqrt{n}(\hat{\theta}^*_{1i}-\theta)$ (i=1,2). Thus the following useful theorem is established.

<u>Theorem 7.1.5.</u> If $E_{\theta,n}(\hat{\theta}^*_{1n})-E_{\theta,n}(\hat{\theta}^*_{2n})=o(1/n)$, then the asymptotic distributions of $\sqrt{n'}(\hat{\theta}_{1n'}-\theta)$ and $\sqrt{n''}(\hat{\theta}_{2n''}-\theta)$ coincide with each other up to the order n^{-1}.

From the fact we may consider the amount

$$\frac{1}{I}\left\{V(Q_2) - V(Q_1)\right\}$$

as the (relative) deficiency of $\hat{\theta}_{1n}$ with respect to $\hat{\theta}_{2n}$. Especially when $\hat{\theta}_{1n}$ is the MLE, then

$$\frac{1}{I}\left\{V(Q_2)-V(Q_1)\right\} = \frac{1}{I}V(Q_2-Q_1) \geqq 0$$

which can be simply termed as the deficiency of $\hat{\theta}_{1n}$ (Akahira [6]).

7.2. Second order asymptotic efficiency of unbiased confidence intervals

Summary :

It was known in [30] that the confidence interval based on maximum likelihood (or asymptotically efficient) estimator $\hat{\theta}_n$ given as $[\hat{\theta}_n - \{u_{\alpha/2} / \sqrt{I(\hat{\theta}_n)}\}, \hat{\theta}_n + \{u_{\alpha/2} / \sqrt{I(\hat{\theta}_n)}\}]$ is asymptotically unbiased and asymptotically efficient.

The purpose of this section is to discuss the problem of the second order asymptotic efficiency of confidence intervals, that is, to consider the asymptotic power of the confidence interval up to the order $n^{-1/2}$ in the neighborhood of the true value of the parameter.

It will be shown that asymptotically unbiased, and second order asymptotically (most powerful) confidence interval at $\theta' = \theta + tn^{-1/2}$ depends on the specified value of t, thus proving that there exists no uniformly second order asymptotically most powerful unbiased confidence interval. The situation is in contrast to point estimation where maximum likelihood (or usually BAN) estimators are uniformly second order asymptotically efficient ([32], [44]). This is related to the curvature of the parameter space as was discussed by Efron [19].

Let c_n be the maximum order of convergence of consistent estimators.

A subinterval of Ⓗ, $[\underline{\theta}_n(\tilde{X}_n), \bar{\theta}_n(\tilde{X}_n)]$ whose limit depends on the observed random variables $\tilde{X}_n$ (properly measurable statistics) is called consistent of order c_n if for every $t(\neq 0)$

$$\lim_{|t|\to\infty} \underline{\lim}_{n\to\infty} P_{\theta,n}\left\{\theta + tc_n^{-1} \in [\underline{\theta}_n(\tilde{X}_n), \bar{\theta}_n(\tilde{X}_n)]\right\} = 1 .$$

A consistent interval $[\underline{\theta}_n(\tilde{X}_n), \bar{\theta}_n(\tilde{X}_n)]$ is called an <u>α-asymptotically unbiased confidence (α-AUC) interval</u>, $0<\alpha<1$, if for every $\vartheta \in \textcircled{H}$ and every $t(\neq 0)$

$$\overline{\lim_{n\to\infty}}\, P_{\theta,n}\{\theta + tc_n^{-1} \in [\underline{\theta}_n(\tilde{X}_n), \bar{\theta}_n(\tilde{X}_n)]\} \leqq 1-\alpha ;$$

and for every $\vartheta \in \textcircled{H}$ there exists a positive number δ such that

$$\lim_{n\to\infty} \inf_{\theta:|\theta-\vartheta|<\delta} P_{\theta,n}\{\theta \in [\underline{\theta}_n(\tilde{X}_n), \bar{\theta}_n(\tilde{X}_n)]\} \geqq 1-\alpha .$$

Further an α-AUC interval $[\underline{\theta}_n(\tilde{X}_n), \bar{\theta}_n(\tilde{X}_n)]$ is called <u>second order α-AUC</u> if for every $\theta \in \textcircled{H}$ and every $t(\neq 0)$

$$\overline{\lim_{n\to\infty}}\, c_n[P_{\theta,n}\{\theta + tc_n^{-1} \in [\underline{\theta}_n(\tilde{X}_n), \bar{\theta}_n(\tilde{X}_n)]\} - (1-\alpha)] \leqq 0 ;$$

and for every $\vartheta \in \textcircled{H}$ there exists a positive number δ such that

$$\lim_{n\to\infty} \inf_{\theta:|\theta-\vartheta|<\delta} c_n[P_{\theta,n}\{\theta \in [\underline{\theta}_n(\tilde{X}_n), \bar{\theta}_n(\tilde{X}_n)]\} - (1-\alpha)] \geqq 0 .$$

An α-AUC interval $[\underline{\theta}_n^*(\tilde{X}_n), \bar{\theta}_n^*(\tilde{X}_n)]$ is called <u>asymptotically efficient</u> if for every α-AUC interval $[\underline{\theta}_n(\tilde{X}_n), \bar{\theta}_n(\tilde{X}_n)]$, every $\theta \in \textcircled{H}$ and every $t(\neq 0)$

$$\overline{\lim_{n\to\infty}}\, [P_{\theta,n}\{\theta + tc_n^{-1} \in [\underline{\theta}_n^*(\tilde{X}_n), \bar{\theta}_n^*(\tilde{X}_n)]\} - P_{\theta,n}\{\theta + tc_n^{-1} \in [\underline{\theta}_n(\tilde{X}_n), \bar{\theta}_n(\tilde{X}_n)]\}] \leqq 0.$$

A second order α-AUC interval $[\underline{\theta}_n^*(\tilde{X}_n), \bar{\theta}_n^*(\tilde{X}_n)]$ is called <u>second order asymptotically efficient</u> if for every second order α-AUC interval $[\underline{\theta}_n(\tilde{X}_n), \bar{\theta}_n(\tilde{X}_n)]$, every $\theta \in \textcircled{H}$ and every $t(\neq 0)$

$$\overline{\lim_{n\to\infty}}\, c_n[P_{\theta,n}\{\theta + tc_n^{-1} \in [\underline{\theta}_n^*(\tilde{X}_n), \bar{\theta}_n^*(\tilde{X}_n)]\} - P_{\theta,n}\{\theta + tc_n^{-1} \in [\underline{\theta}_n(\tilde{X}_n), \bar{\theta}_n(\tilde{X}_n)]\}] \leqq 0$$

Note that our difinition of the aymptotically efficient confidence interval is different from that of Wald [53] (asymptotically shortest confidence interval in his terminology) in that a) he considered only exact confidence intervals, and b) he defined efficiency in terms of the supremum of the difference of powers of two intervals over all pairs of parameter values θ and θ' which implies uniformity of convergence of the power function, which in turn implies the bound from below of the Fisher information ; and both aspects limit the applicability of the theory.

Let Θ be an open set in a Euclidean R^1. We assume that for each $\theta \in \Theta$ P_θ is absolutely continuous with respect to a σ-finite measure μ. We denote a density $dP_\theta/d\mu$ by $f(x,\theta)$. Then the joint density is given by $\prod_{i=1}^{n} f(x_i,\theta)$. In the subsequent discussion we shall deal with the case when $c_n = \sqrt{n}$.

Suppose that $X_1, X_2, \ldots, X_n, \ldots$ is a sequence of i.i.d. random variables with the density $f(x,\theta)$ satisfying (A.3.1.1), (A.3.1.3), (A.4.1.1) $\sim$ (A.4.1.3).

We put

$$Z_1(\theta) = \frac{1}{\sqrt{n}} \sum_{i=1}^{n} \frac{\partial}{\partial\theta} \log f(X_i,\theta) \quad ;$$

$$Z_2(\theta) = \frac{1}{\sqrt{n}} \sum_{i=1}^{n} \left\{ \frac{\partial^2}{\partial\theta^2} \log f(X_i,\theta) + I(\theta) \right\} \quad ;$$

$$W(\theta) = \frac{1}{n} \sum_{i=1}^{n} \frac{\partial^3}{\partial\theta^3} \log f(X_i,\theta) \ .$$

Let θ_0 be arbitrary but fixed in Θ. Consider the problem of testing the hypothesis $H : \theta = \theta_0$ against the alternative $K : \theta = \theta_0 + tn^{-1/2}$ $(t \neq 0)$. By the fundamental lemma of Neyman and Pearson the most powerful locally unbiased test function has a rejection region of the following type ;

$$(7.2.1) \qquad \frac{\prod_{i=1}^{n} f(X_i, \theta_0 + tn^{-1/2})}{\prod_{i=1}^{n} f(X_i, \theta_0)} > \lambda_0 + \lambda_1 tn^{-1/2} \sum_{i=1}^{n} \frac{\partial}{\partial\theta} \log f(X_i,\theta) \ ,$$

where λ_0 and λ_1 are fixed real numbers, and if it is actually (globally) unbiased it is the most powerful unbiased test function. By Taylor expansion we obtain

$$\prod_{i=1}^{n} f(X_i, \theta_0+tn^{-1/2}) \Big/ \prod_{i=1}^{n} f(X_i, \theta_0)$$

$$= \exp \left[\sum_{i=1}^{n} \left\{ \log f(X_i, \theta_0+tn^{-1/2}) - \log f(X_i, \theta_0)\right\} \right]$$

$$= \exp \left\{ \frac{t}{\sqrt{n}} \sum_{i=1}^{n} \frac{\partial}{\partial\theta}\log f(X_i, \theta_0) + \frac{t^2}{2n} \sum_{i=1}^{n} \frac{\partial^2}{\partial\theta^2}\log f(X_i, \theta_0)\right.$$

$$\left. + \frac{t^3}{6n\sqrt{n}} \sum_{i=1}^{n} \frac{\partial^3}{\partial\theta^3}\log f(X_i, \theta^*) \right\},$$

$$= \exp \left[tZ_1(\theta_0) + \frac{t^2}{2\sqrt{n}}\left\{ Z_2(\theta_0) - \sqrt{n}\, I(\theta_0)\right\} + \frac{t^3}{6\sqrt{n}} W(\theta^*) \right]$$

$$= \exp\left\{- \frac{t^2}{2} I(\theta_0) + tZ_1(\theta_0) + \frac{t^2}{2\sqrt{n}} Z_2(\theta_0) + \frac{t^3}{6\sqrt{n}} W(\theta^*)\right\},$$

where $|\theta^* - \theta| < tn^{-1/2}$.

Since $W(\theta)$ converges in probability to $-3J(\theta)-K(\theta)$, it follows that

$$(7.2.2)\quad \prod_{i=1}^{n} f(X_i, \theta_0+tn^{-1/2}) \Big/ \prod_{i=1}^{n} f(X_i, \theta_0)$$

$$=\exp\left[-\frac{t^2}{2}I(\theta_0)+tZ_1(\theta_0)+\frac{t^2}{2\sqrt{n}} Z_2(\theta_0)-\frac{t^3}{6\sqrt{n}}\left\{3J(\theta_0)+K(\theta_0)\right\}+o_p\left(\frac{1}{\sqrt{n}}\right) \right]$$

$$=e^{-\frac{t^2}{I}(\theta_0)}.\ e^{tZ_1(\theta_0)} \left[1+\frac{t^2}{2\sqrt{n}}Z_2(\theta_0)-\frac{t^3}{6\sqrt{n}}\left\{3J(\theta_0)+K(\theta_0)\right\} +o_p\left(\frac{1}{\sqrt{n}}\right)\right] .$$

First we consider the problem up to constant order. It follows by (7.2.1) and (7.2.2) that there exist solutions $\overline{Z}_1$ and $\underline{Z_1}$ ($\underline{Z_1} < \overline{Z}_1$) of the following equation in $Z_1(\theta_0)$:

$$(7.2.3)\quad e^{-\frac{t^2}{2} I(\theta_0)}\, e^{tZ_1(\theta_0)} = \lambda_0 + \lambda_1 t Z_1(\theta_0) .$$

Then it follows that

$$e^{-\frac{t^2}{2} I(\theta_0)} \cdot e^{tZ_1(\theta_0)} > \lambda_0 + \lambda_1 t Z_1(\theta_0)$$

if and only if

$$Z_1(\theta_0) < \underline{Z_1}\ ,\quad Z_1(\theta_0) > \overline{Z}_1\ ,$$

It can be easily seen that in order to make the α-AUC interval we have to take $\underline{Z}_1 = -u_{\alpha/2}\sqrt{I(\theta_0)}$ and $\overline{Z}_1 = u_{\alpha/2}\sqrt{I(\theta_0)}$ (See [30]), where $u_{\alpha/2}$ denotes upper $100\alpha/2$-percent point of the standard normal distribution. In order to determine λ_0 and λ_1 we substitute $\underline{Z}_1 = -u_{\alpha/2}\sqrt{I(\theta_0)}$ and $\overline{Z}_1 = u_{\alpha/2}\sqrt{I(\theta_0)}$ in the equations

$$e^{-\frac{t^2}{2}I(\theta_0)}\, e^{t\underline{Z}_1} = \lambda_0 + \lambda_1 t\underline{Z}_1 \ :$$

$$e^{-\frac{t^2}{2}I(\theta_0)}\, e^{t\overline{Z}_1} = \lambda_0 + \lambda_1 t\overline{Z}_1 \ .$$

Then we have

$$e^{-\frac{t^2}{2}I(\theta_0)} e^{-tu_{\alpha/2}\sqrt{I(\theta_0)}} = \lambda_0 + \lambda_1 t u_{\alpha/2}\sqrt{I(\theta_0)} \ :$$

$$e^{-\frac{t^2}{2}I(\theta_0)} e^{tu_{\alpha/2}\sqrt{I(\theta_0)}} = \lambda_0 + \lambda_1 t u_{\alpha/2}\sqrt{I(\theta_0)} \ .$$

Hence we obtain

$$\lambda_0 = e^{-\frac{t^2}{2}I(\theta_0)} \cosh\left(t u_{\alpha/2}\sqrt{I(\theta_0)}\right) \ ;$$

$$\lambda_1 = \frac{1}{u_{\alpha/2}\sqrt{I(\theta_0)}}\, e^{-\frac{t^2}{2}I(\theta_0)} \sinh\left(t u_{\alpha/2}\sqrt{I(\theta_0)}\right) \ .$$

If there exist $\underline{\theta}_n^*(\tilde{X}_n)$ and $\overline{\theta}_n^*(\tilde{X}_n)$ such that

$$Z_1(\underline{\theta}_n^*(\tilde{X}_n)) = u_{\alpha/2}\sqrt{I(\underline{\theta}_n^*(\tilde{X}_n))} \ ;$$

$$Z_1(\overline{\theta}_n^*(\tilde{X}_n)) = -u_{\alpha/2}\sqrt{I(\overline{\theta}_n^*(\tilde{X}_n))} \ ,$$

then the α-AUC interval $[\underline{\theta}_n^*(\tilde{X}_n), \overline{\theta}_n^*(\tilde{X}_n)]$ is asymptotically efficient. Hence we have established the following :

Theorem 7.2.1. Under the conditions (A.3.1.1) $\sim$ (A.3.1.3), the α-AUC interval $[\underline{\theta}_n^*(\tilde{X}_n), \overline{\theta}_n^*(\tilde{X}_n)]$ is asymptotically efficient.

We can actually construct the asymptotically efficient α-AUC interval using the maximum likelihood estimator (MLE) $\hat{\theta}_{ML}$.

We put

$$\underline{\theta}_n^*(\tilde{X}_n) = \hat{\theta}_{ML} + \frac{\gamma'}{\sqrt{n}} \quad ;$$

$$\overline{\theta}_n^*(\tilde{X}_n) = \hat{\theta}_{ML} + \frac{\gamma}{\sqrt{n}} \quad .$$

We have

$$Z_1(\underline{\theta}_n^*) = Z_1(\hat{\theta}_{ML} + \frac{\gamma'}{\sqrt{n}})$$

$$= \frac{\gamma'}{\sqrt{n}} Z_1'(\hat{\theta}_{ML}) + o_p(\frac{1}{\sqrt{n}})$$

$$= \frac{\gamma'}{\sqrt{n}} \left\{ Z_2(\hat{\theta}_{ML}) - I(\hat{\theta}_{ML})\sqrt{n} \right\} + o_p(\frac{1}{\sqrt{n}})$$

$$= -\gamma' I(\hat{\theta}_{ML}) + o_p(\frac{1}{\sqrt{n}})$$

Since

$$Z_1(\underline{\theta}_n^*) = u_{\alpha/2}\sqrt{I(\underline{\theta}_n^*)} \quad ,$$

it follows that

$$-\gamma' I(\hat{\theta}_{ML}) + o_p(\frac{1}{\sqrt{n}})$$

$$= u_{\alpha/2}\sqrt{I(\hat{\theta}_{ML})} + o_p(\frac{1}{\sqrt{n}}) \quad .$$

Hence we have

$$\gamma' = -\frac{u_{\alpha/2}}{\sqrt{I(\hat{\theta}_{ML})}} \quad .$$

In a way similar the above we obtain

$$\gamma = \frac{u_{\alpha/2}}{\sqrt{I(\hat{\theta}_{ML})}} \quad .$$

Hence we have established the following :

<u>Theorem 7.2.2.</u> Under the conditions (A.3.1.1) $\sim$ (A.3.1.3), the α-AUC interval

$$\left[\hat{\theta}_{ML} - \frac{u_{\alpha/2}}{\sqrt{I(\hat{\theta}_{ML})n}} \quad , \quad \hat{\theta}_{ML} + \frac{u_{\alpha/2}}{\sqrt{I(\hat{\theta}_{ML})n}} \right]$$

is asymptotically efficient.

Next we consider the case up to the order of $n^{-1/2}$, i.e., second order case ([43]).

We put $Z_1(\theta_0)=\bar{Z}_1+\frac{\Delta}{\sqrt{n}}$ in the following equation :

(7.2.4) $e^{-\frac{t^2}{2}I(\theta_0)}.e^{tZ_1(\theta_0)}[1+\frac{t^2}{2\sqrt{n}}Z_2(\theta_0)-\frac{t^3}{6\sqrt{n}}\{3J(\theta_0)+K(\theta_0)\}]=\lambda_0+\lambda_1 tZ_1(\theta_0)$

We have

$$\lambda_0 + \lambda_1 t(\bar{Z}_1 + \frac{\Delta}{\sqrt{n}})$$
$$= e^{-\frac{t^2}{2}I}\, e^{t(\bar{Z}_1+\frac{\Delta}{\sqrt{n}})} \left\{ 1 + \frac{t^2}{2\sqrt{n}}Z_2 - \frac{t^3}{6\sqrt{n}}(3J+K) \right\}$$
$$= e^{-\frac{t^2}{2}I}\, e^{t\bar{Z}_1}\, (1+\frac{t}{\sqrt{n}}\Delta) \left\{ 1 + \frac{t^2}{2\sqrt{n}}Z_2 - \frac{t^3}{6\sqrt{n}}(3J+K) \right\} .$$

It follows by (7.2.3) and (7.2.4) that

$$1+\lambda_1 t\, \frac{\Delta}{\sqrt{n}}\, e^{-t\bar{Z}_1+\frac{t^2}{2}I} = 1 + \frac{t\Delta}{\sqrt{n}} + \frac{t^2}{2\sqrt{n}}Z_2 - \frac{t^3}{6\sqrt{n}}(3J+K) .$$

Hence we obtain

(7.2.5) $$\Delta = - \frac{1}{t(1-\lambda_1 e^{-t\bar{Z}_1+\frac{t^2}{2}I})} \left\{ \frac{t^2}{2}Z_2 - \frac{t^3}{6}(3J+K) \right\} .$$

Putting $Z_1(\theta_0)=\underline{Z_1} + \frac{\Delta'}{\sqrt{n}}$ in the equation (7.2.4) we similarly have

(7.2.6) $$\Delta' = - \frac{1}{t(1-\lambda_1 e^{-t\underline{Z_1}+\frac{t^2}{2}I})} \left\{ \frac{t^2}{2}Z_2 - \frac{t^3}{6}(3J+K) \right\} .$$

The rejection region like the type of (7.2.1) is given by

(7.2.7) $$Z_1(\theta_0) - \frac{\Delta'}{\sqrt{n}} < \underline{Z_1} , \quad Z_1(\theta_0) - \frac{\Delta}{\sqrt{n}} > \bar{Z}_1 ,$$

where Δ and Δ' are given by (7.2.5) and (7.2.6), respectively.

We have

$$Z_1(\theta_0) - \frac{\Delta}{\sqrt{n}} = Z_1(\theta_0) + \frac{c_1}{\sqrt{n}} Z_2(\theta_0) - \frac{c_2}{\sqrt{n}} ;$$
$$Z_1(\theta_0) - \frac{\Delta'}{\sqrt{n}} = Z_1(\theta_0) + \frac{c_1'}{\sqrt{n}} Z_2(\theta_0) - \frac{c_2'}{\sqrt{n}} ,$$

where

$$c_1 = \frac{t}{2(1-\lambda_1 e^{-t\overline{Z}_1} + \frac{t^2}{2}I)}, \qquad c_2 = \frac{t^2(3J+K)}{6(1-\lambda_1 e^{-t\overline{Z}_1} + \frac{t^2}{2}I)},$$

$$c_1' = \frac{t}{2(1-\lambda_1 e^{-t\underline{Z}_1} + \frac{t^2}{2}I)}, \qquad c_2' = \frac{t^2(3J+K)}{6(1-\lambda_1 e^{-t\overline{Z}_1} + \frac{t^2}{2}I)}.$$

It follows from (7.2.7) that the rejection region like the type (7.2.1) is given by

$$Z_1(\theta_0) + \frac{c_1'}{\sqrt{n}} Z_2(\theta_0) < \underline{Z}_1 + \frac{c_2'}{\sqrt{n}}, \qquad Z_1(\theta_0) + \frac{c_1}{\sqrt{n}} Z_2(\theta_0) > \overline{Z}_1 + \frac{c_2}{\sqrt{n}},$$

wheere c_1, c_1', c_2 and c_2' are given above.

We put

$$A_1 = \left\{ \tilde{X}_n : Z_1(\theta_0) + \frac{c_1}{\sqrt{n}} Z_2(\theta_0) > \overline{Z}_1 + \frac{c_2}{\sqrt{n}} \right\}.$$

$$A_2 = \left\{ \tilde{X}_n : Z_1(\theta_0) + \frac{c_1'}{\sqrt{n}} Z_2(\theta_0) < \underline{Z}_1 + \frac{c_2'}{\sqrt{n}} \right\}.$$

In order to construct second order α-AUC interval we have to determine $\underline{Z}_1$ and $\overline{Z}_1$ such that

$$(7.2.8) \qquad P_{\theta_0,n}(A_1) + P_{\theta_0,n}(A_2) = \alpha + o\left(\frac{1}{n}\right);$$

$$(7.2.9) \qquad \frac{\partial}{\partial\theta} P_{\theta_0,n}(A_1) + \frac{\partial}{\partial\theta} P_{\theta_0,n}(A_2) = o\left(\frac{1}{n}\right).$$

If χ_A denotes the indicator of the set A, then (7.2.8) is given by

$$(7.2.10) \qquad E_{\theta_0,n}(\chi_{A_1}) + E_{\theta_0,n}(\chi_{A_2}) = \alpha + o\left(\frac{1}{n}\right).$$

Since

$$\begin{aligned}
\frac{\partial}{\partial\theta} P_{\theta,n}(A) &= \int \chi_A(\tilde{x}_n) \left\{ \frac{\partial}{\partial\theta} \prod_{i=1}^{n} f(x_i,\theta) \right\} d\mu^{(n)} \\
&= \int \chi_A(\tilde{x}_n) \left\{ \frac{\partial}{\partial\theta} \log \prod_{i=1}^{n} f(x_i,\theta) \right\} \prod_{i=1}^{n} f(x_i,\theta)\, d\mu^{(n)} \\
&= \int \chi_A(\hat{x}_n) \left\{ \sum_{i=1}^{n} \frac{\partial}{\partial\theta} \log f(x_i,\theta) \right\} \prod_{i=1}^{n} f(x_i,\theta)\, d\mu^{(n)} \\
&= E_{\theta,n}\left[\chi_A(\tilde{x}_n) \sum_{i=1}^{n} \frac{\partial}{\partial\theta} \log f(x_i,\theta) \right],
\end{aligned}$$

where $\mu^{(n)}$ denotes the n-fold direct product measure of μ, it follows that (7.2.9) is given by

(7.2.11) $$E_{\theta_0,n}[\chi_{A_1}(\tilde{x}_n)Z_1(\theta_0)]+E_{\theta_0,n}[\chi_{A_2}(\tilde{x}_n)Z_1(\theta_0)]=o(\frac{1}{n}) .$$

Putting

$$Z_1^*(\theta_0)=Z_1(\theta_0)+\frac{c_1}{\sqrt{n}}Z_2(\theta_0) ; \quad Z_1^{**}(\theta_0)=Z_1(\theta_0)+\frac{c_1'}{\sqrt{n}}Z_2(\theta_0) ,$$

we have from (7.2.10)

(7.2.12) $$\begin{aligned} o(\frac{1}{n}) &= E_{\theta_1,n}[\chi_{A_1}(\tilde{x}_n)\left\{Z_1^*(\theta_0)-\frac{c_1}{\sqrt{n}}Z_2(\theta_0)\right\}] \\ &\quad + E_{\theta_0,n}[\chi_{A_2}(\tilde{x}_n)\left\{Z_1^{**}(\theta_0)-\frac{c_1}{\sqrt{n}}Z_2(\theta_0)\right\}] \\ &= E_{\theta_0,n}[\chi_{A_1}(\tilde{x}_n)Z_1^*(\theta_0)] - \frac{c_1}{\sqrt{n}}E_{\theta_0,n}[\chi_{A_1}(\tilde{x}_n)Z_2(\theta_0)] \\ &\quad + E_{\theta_0,n}[\chi_{A_2}(\tilde{x}_n)Z_1^{**}(\theta_0)] - \frac{c_1'}{\sqrt{n}}E_{\theta_0,n}[\chi_{A_2}(\tilde{x}_n)Z_2(\theta_0)] . \end{aligned}$$

Since

$$E_{\theta_0,n}[Z_1^2(\theta_0)]=I(\theta_0) ; \quad E_{\theta_0,n}[Z_1(\theta_0)Z_2(\theta_0)]=J(\theta_0) ;$$

$$E_{\theta_0,n}[Z_2^2(\theta_0)]=M(\theta_0)-I^2(\theta_0) ,$$

where $M(\theta)=E_\theta[\left\{\frac{\partial^2}{\partial\theta^2}\log f(x,\theta)\right\}^2]$,

it follows that

(7.2.13) $$\begin{aligned} V_{\theta_0,n}(Z_1^*(\theta_0)) &= E_{\theta_0,n}[\left\{Z_1(\theta_0)+\frac{c_1}{\sqrt{n}}Z_2(\theta_0)\right\}^2] \\ &= E_{\theta_0,n}[Z_1^2(\theta_0)]+\frac{2c_1}{\sqrt{n}}E_{\theta_0,n}[Z_1(\theta_0)Z_2(\theta_0)]+o(\frac{1}{\sqrt{n}}) \\ &= I(\theta_0)+\frac{2c_1}{\sqrt{n}}J(\theta_0)+o(\frac{1}{\sqrt{n}}) ; \end{aligned}$$

(7.2.14) $$\begin{aligned} \mathrm{Cov}_{\theta_0,n}(Z_1^*(\theta_0), Z_2(\theta_0)) &= E_{\theta_0,n}[Z_1^*(\theta_0)Z_2(\theta_0)]-E_{\theta_0,n}[Z_1^*(\theta_0)] \\ &\quad E_{\theta_0,n}[Z_2(\theta_0)] \\ &= E_{\theta_0,n}[\left\{Z_1(\theta_0)+\frac{c_1}{\sqrt{n}}Z_2(\theta_0)\right\}Z_2(\theta_0)] \\ &= E_{\theta_0,n}[Z_1(\theta_0)Z_2(\theta_0)] + \frac{c_1}{\sqrt{n}}E_{\theta_0,n}[Z_2^2(\theta_0)] \end{aligned}$$

$$= J(\theta_0) + \frac{c_1}{\sqrt{n}}\left\{M(\theta_0) - I^2(\theta_0)\right\} + o_p(\frac{1}{\sqrt{n}}) \quad .$$

From (7.2.13) and (7.2.14) we have

$$E_{\theta_0,n}[Z_2(\theta_0) \mid Z_1^*(\theta_0)] = \frac{Cov_{\theta_0,n}(Z_1^*(\theta_0), Z_2(\theta_0))}{V_{\theta_0,n}(Z_1^*(\theta_0))} Z_1^*(\theta_0)$$

$$= \frac{J(\theta_0)}{I(\theta_0)} Z_1^*(\theta_0) + o(\frac{1}{\sqrt{n}}) \quad .$$

Hence we obtain

$$(7.2.15) \quad E_{\theta_0,n}[\chi_{A_1}(\tilde{x}_n)Z_2(\theta_0)] = E_{\theta_0,n}[\chi_{A_1}(\tilde{x}_n)E_{\theta_0,n}(Z_2(\theta_0) \mid Z_1^*(\theta_0))]$$

$$= \frac{J(\theta_0)}{I(\theta_0)} E_{\theta_0,n}[\chi_{A_1}(\tilde{x}_n)Z_1^*(\theta_0)] + o(\frac{1}{\sqrt{n}}) \quad .$$

In a way similar way to the above we obtain

$$(7.2.16) \quad E_{\theta_0,n}[\chi_{A_2}(\tilde{x}_n)Z_2(\theta_0)] = \frac{J(\theta_0)}{I(\theta_0)} E_{\theta_0,n}[\chi_{A_2}(\tilde{x}_n)Z_1^{**}(\theta_0)] + o(\frac{1}{\sqrt{n}}) \quad .$$

From (7.2.12), (7.2.15) and (7.2.16) it follows that

$$(7.2.17) \quad \left\{1 - \frac{c_1 J(\theta_0)}{I(\theta_0)\sqrt{n}}\right\} E_{\theta_0,n}[\chi_{A_1}(\tilde{x}_n)Z_1^*(\theta_0)] + \left\{1 - \frac{c_1' J(\theta_0)}{I(\theta_0)\sqrt{n}}\right\} E_{\theta_0,n}$$

$$[\chi_{A_2}(\tilde{x}_n)Z_1^{**}(\theta_0)] = o(\frac{1}{n}) \quad .$$

It follows by Edgeworth expansion that the probability density function $f(x)$ of $Z_1^*(\theta_0)/\sigma$, where $\sigma^2 = V_{\theta_0,n}(Z_1^*(\theta_0))$((7.2.13)), is given by

$$f(x) = \phi(x) + \frac{\beta_3}{6\sqrt{n}} H_3(x)\phi(x) + o(\frac{1}{\sqrt{n}}) \quad ,$$

where $\phi(x) = \frac{1}{\sqrt{2\pi}} e^{-x^2/2}$ and β_3 is the third order cumulant of Z_1^*/σ and $H_3(x)$ is a Hermite polynomial, i.e. $H_3(x) = x^3 - 3x$.

Since

$$\int xH_3(x)\phi(x)dx = \int x(x^3-3x)\phi(x)dx$$

$$= \int (x^4-3x^2)\phi(x)dx$$

$$= \int \left\{(x^4-6x^2+3)+3(x^2-1)\right\}\phi(x)dx$$

$$=\int\{H_4(x)+3H_2(x)\}\ \phi(x)dx$$
$$=\int H_4(x)\phi(x)dx+3\int H_2(x)\phi(x)dx$$
$$=\{-H_3(x)-3H_1(x)\}\phi(x)$$
$$=(-x^3+3x-3x)\ \phi(x)$$
$$=-x^3\phi(x)\quad ,$$

where $H_i(x)$ $(i=1,2,3,4)$ are Hermite polynomials, it follows that

$$(7.2.18)\quad \int xf(x)dx=\int x\{\phi(x)+\frac{\beta_3}{6\sqrt{n}}H_3(x)\phi(x)\}\ dx+o(\frac{1}{\sqrt{n}})$$
$$=\int x\,\phi(x)dx-\frac{\beta_3}{6\sqrt{n}}\int xH_3(x)\phi(x)dx+o(\frac{1}{\sqrt{n}})$$
$$=-\phi(x)-\frac{\beta_3}{6\sqrt{n}}x^3\phi(x)+o(\frac{1}{\sqrt{n}})\quad .$$

Since by a way similar to the above, the probability density function $g(x)$ of $Z_1^{**}(\theta_0)/\sigma'$, where $\sigma'^2=V_{\theta_0,n}(Z_1^{**}(\theta_0))=I(\theta_0)+\frac{2c_1'}{\sqrt{n}}J(\theta_0)+o(\frac{1}{\sqrt{n}})$, is given by

$$g(x)=\phi(x)+\frac{\beta_3'}{6\sqrt{n}}H_3(x)\phi(x)+o(\frac{1}{\sqrt{n}})\quad ,$$

where β_3' is the third order cumulant of $Z_1^{**}(\theta_0)/\sigma'$, it follows that

$$(7.2.19)\qquad \int xg(x)dx=-\phi(x)-\frac{\beta_3'}{6\sqrt{n}}x^3\phi(x)+o(\frac{1}{\sqrt{n}})\quad .$$

Putting

$$k_1=\overline{Z}_1+\frac{c_2}{\sqrt{n}}\quad ;\quad k_1'=\underline{Z}_1+\frac{c_2'}{\sqrt{n}}$$

we have

$$A_1=\{Z_1^*>k_1\}\quad ;\quad A_2=\{Z_1^{**}<k_1'\}\ .$$

By (7.2.18) we obtain

$$(7.2.20)\qquad E_{\theta_0,n}[\chi_{A_1}(\tilde{x}_n)Z_1^*(\theta_0)]=[-\phi(\frac{x}{\sigma})-\frac{\beta_3}{6\sqrt{n}}(\frac{x}{\sigma})^3\phi(\frac{x}{\sigma})\ ]_{k_1}^{\infty}$$
$$=\phi(\frac{k_1}{\sigma})+\frac{\beta_3}{6\sqrt{n}}(\frac{k_1}{\sigma})^3\phi(\frac{k_1}{\sigma})\quad .$$

By (7.2.19) we also have

$$(7.2.21)\quad E_{\theta_0,n}[\chi_{A_2}(\tilde{x}_n)Z_1^{**}(\theta_0)]=\left[-\Phi(\frac{x}{\sigma'})-\frac{\beta_3'}{6\sqrt{n}}(\frac{x}{\sigma'})^3\phi(\frac{x}{\sigma'})\right]_{-\infty}^{k_1'}$$

$$=-\Phi(\frac{k_1'}{\sigma'})-\frac{\beta_3'}{6\sqrt{n}}(\frac{k_1'}{\sigma'})^3\phi(\frac{k_1'}{\sigma'}) .$$

Since $\beta_3=\beta_3'=K(\theta_0)$, it follows by (7.2.17), (7.2.20) and (7.2.21) that

$$(7.2.22)\quad o(\frac{1}{n})=(1-\frac{c_1J}{I\sqrt{n}})E_{\theta_0,n}(\chi_{A_1}Z_1^*)+(1-\frac{c_1'J}{I\sqrt{n}})E_{\theta_0,n}(\chi_A Z_1^{**})$$

$$=(1-\frac{c_1J}{I\sqrt{n}})\left\{\Phi(\frac{k_1}{\sigma})+\frac{K}{6\sqrt{n}}(\frac{k_1}{\sigma})^3\phi(\frac{k_1}{\sigma})\right\}$$

$$-(1-\frac{c_1'J}{I\sqrt{n}})\left\{\Phi(\frac{k_1'}{\sigma'})+\frac{K}{6\sqrt{n}}(\frac{k_1'}{\sigma'})^3\phi(\frac{k_1'}{\sigma'})\right\}$$

$$=(1-\frac{c_1J}{I\sqrt{n}})\left\{1+\frac{K}{6\sqrt{n}}(\frac{k_1}{\sigma})^3\right\}\phi\left(\frac{\bar{Z}_1+\frac{c_2}{\sqrt{n}}}{\sigma}\right)$$

$$-(1-\frac{c_1'J}{I\sqrt{n}})\left\{1+\frac{K}{6\sqrt{n}}(\frac{k_1'}{\sigma'})^3\right\}\phi\left(\frac{\underline{Z}_1+\frac{c_2'}{\sqrt{n}}}{\sigma'}\right) .$$

From (7.2.22) we have

$$(7.2.23)\quad \phi\left(\frac{\bar{Z}_1+\frac{c_2}{\sqrt{n}}}{\sigma}\right)\Big/\phi\left(\frac{\underline{Z}_1+\frac{c_2'}{\sqrt{n}}}{\sigma'}\right)=\frac{(1-\frac{c_1'J}{I\sqrt{n}})\left\{1+\frac{K}{6\sqrt{n}}(\frac{k_1'}{\sigma'})^3\right\}}{(1-\frac{c_1J}{I\sqrt{n}})\left\{1+\frac{K}{6\sqrt{n}}(\frac{k_1}{\sigma})^3\right\}}$$

$$=\left\{1-\frac{c_1'J}{I\sqrt{n}}+\frac{K}{6\sqrt{n}}(\frac{k_1'}{\sigma'})^3\right\}\left\{1+\frac{c_1J}{I\sqrt{n}}-\frac{K}{6\sqrt{n}}(\frac{k_1}{\sigma})^3\right\}$$

$$=1+\frac{c_1-c_1'}{I\sqrt{n}}J+\frac{K}{6\sqrt{n}}\left\{(\frac{k_1'}{\sigma'})^3-(\frac{k_1}{\sigma})^3\right\}+o(\frac{1}{\sqrt{n}}) .$$

Since $\sigma^2=\sigma'^2=I(\theta_0)+o(\frac{1}{\sqrt{n}})$, it follows from (7.2.23) that

$$(7.2.24)\quad -\frac{1}{2I}\left\{(\bar{Z}_1+\frac{c_2}{\sqrt{n}})^2-(\underline{Z}_1+\frac{c_2'}{\sqrt{n}})^2\right\}$$

$$=\log\left\{1+\frac{(c_1-c_1')J}{I\sqrt{n}}+\frac{K}{6I^{3/2}\sqrt{n}}(\underline{Z}_1{}^3-\bar{Z}_1{}^3)\right\} ,$$

$$= \frac{(c_1-c_1')J}{I\sqrt{n}} + \frac{K}{6I^{3/2}\sqrt{n}}(\underline{Z}_1{}^3 - \overline{Z}_1{}^3) + o(\frac{1}{\sqrt{n}}) .$$

Putting $\overline{Z}_1 = u_{\alpha/2}\sqrt{I(\theta_0)} + \frac{\eta}{\sqrt{n}}$ and $\underline{Z}_1 = -u_{\alpha/2}\sqrt{I(\theta_0)} + \frac{\eta'}{\sqrt{n}}$,

we have from (7.2.24)

$$- \frac{u_{\alpha/2}}{\sqrt{I}} \frac{\eta+\eta'+c_2+c_2'}{\sqrt{n}}$$

$$= \frac{(c_1-c_1')J}{I\sqrt{n}} + \frac{K}{6I^{3/2}\sqrt{n}} (I^{3/2}u_{\alpha/2}{}^3 + I^{3/2}u_{\alpha/2}{}^3) .$$

Hence we have

$$(7.2.25) \quad \eta+\eta' = - \frac{(c_1-c_1')J}{u_{\alpha/2}\sqrt{I}} - \frac{k\sqrt{I}}{3} u^2_{\alpha/2} - c_2 - c_2' .$$

Since

$$\int f(x)dx = \int \left\{ \phi(x) + \frac{K}{6\sqrt{n}} H_3(x)\,\phi(x) \right\} dx$$

$$= \Phi(x) - \frac{K}{6\sqrt{n}} \int H_3(x)\,\phi(x)dx$$

$$= \Phi(x) - \frac{K}{6\sqrt{n}} H_2(x)\,\phi(x)$$

$$= \Phi(x) - \frac{K}{6\sqrt{n}}(x^2-1)\,\phi(x) \quad ;$$

$$\int g(x)dx = \Phi(x) - \frac{K}{6\sqrt{n}}(x^2-1)\,\phi(x) ,$$

where $\Phi(x)$ is a standard normal distribution function, it follows from (7.2.10) that

$$\alpha + o(\frac{1}{n}) = E_{\theta_0,n}(\chi_{A_1}) + E_{\theta_0,n}(\chi_{A_2})$$

$$= 1 - \Phi(\frac{k_1}{\sigma}) + \frac{K}{6\sqrt{n}}(\frac{k_2^2}{\sigma^2} - 1)\,\phi(\frac{k_1}{\sigma})$$

$$+ \Phi(\frac{k_1'}{\sigma'}) - \frac{K}{6\sqrt{n}}(\frac{k_1'^2}{\sigma'^2} - 1)\,\phi(\frac{k_1'}{\sigma'}) .$$

Hence we have

$$(7.2.26)\qquad 1-\alpha+o(\tfrac{1}{n}) = \Phi\left(\frac{\bar{Z}_1+\frac{c_2}{\sqrt{n}}}{\sigma}\right) - \Phi\left(\frac{\underline{Z}_1+\frac{c_2'}{\sqrt{n}}}{\sigma'}\right)$$

$$+\frac{K}{6\sqrt{n}}\left\{\left(\frac{\underline{Z}_1^2}{\sigma'^2}-1\right)\phi\left(\frac{\underline{Z}_1}{\sigma'}\right)-\left(\frac{\bar{Z}_1^2}{\sigma^2}-1\right)\phi\left(\frac{\bar{Z}_1}{\sigma}\right)\right\}$$

Since we put

$$\bar{Z}_1 = u_{\alpha/2}\sqrt{I} + \frac{\eta}{\sqrt{n}}\ ;\ \underline{Z}_1 = -u_{\alpha/2}\sqrt{I} + \frac{\eta'}{\sqrt{n}}\ ,$$

we have from (7.2.26)

$$(7.2.27)\quad \Phi\left(\frac{u_{\alpha/2}\sqrt{I} + \frac{c_2+\eta}{\sqrt{n}}}{\sigma}\right) - \Phi\left(\frac{-u_{\alpha/2}\sqrt{I} + \frac{c_2'+\eta'}{\sqrt{n}}}{\sigma'}\right) = 1-\alpha+o(\tfrac{1}{n})\ .$$

Since

$$\sigma = \sqrt{I(\theta_0)} + \frac{c_1 J(\theta_0)}{\sqrt{I(\theta_0)n}} + o(\tfrac{1}{\sqrt{n}})\ ;$$

$$\sigma' = \sqrt{I(\theta_0)} + \frac{c_1 J(\theta_0)}{\sqrt{I(\theta_0)n}} + o(\tfrac{1}{\sqrt{n}})\ ,$$

it follows that

$$\frac{u_{\alpha/2}\sqrt{I} + \frac{c_2+\eta}{\sqrt{n}}}{\sigma} = \left(u_{\alpha/2} + \frac{c_2+\eta}{\sqrt{I\,n}}\right)\left\{1 - \frac{c_1 J}{I\sqrt{n}} + o(\tfrac{1}{\sqrt{n}})\right\}$$

$$= u_{\alpha/2} - u_{\alpha/2}\frac{c_1 J}{I\sqrt{n}} + \frac{c_2+\eta}{\sqrt{I\,n}} + o(\tfrac{1}{\sqrt{n}})\ ;$$

$$\frac{-u_{\alpha/2}\ I + \frac{c_2'+\eta'}{\sqrt{n}}}{\sigma'} = -u_{\alpha/2} + u_{\alpha/2}\frac{c_1' J}{I\sqrt{n}} + \frac{c_2'+\eta'}{\sqrt{I\,n}} + o(\tfrac{1}{\sqrt{n}})\ .$$

Hence we have

$$\Phi\left(\frac{u_{\alpha/2}\ I + \frac{c_2+\eta}{\sqrt{n}}}{\sigma}\right) = 1 - \frac{\alpha}{2} + \frac{1}{\sqrt{n}}\phi(u_{\alpha/2})\left(-\frac{u_{\alpha/2}c_1 J}{I} + \frac{c_2+\eta}{\sqrt{I}}\right) + o(\tfrac{1}{\sqrt{n}})$$

$$\Phi\left(\frac{-u_{\alpha/2}I + \frac{c_2'+\eta'}{\sqrt{n}}}{\sigma'}\right) = \frac{\alpha}{2} + \frac{1}{\sqrt{n}}\phi(u_{\alpha/2})\left(\frac{u_{\alpha/2}c_1' J}{I} + \frac{c_2'+\eta'}{\sqrt{I}}\right) + o(\tfrac{1}{\sqrt{n}}).$$

From (7.2.27) we obtain

$$(7.2.28)\quad \frac{1}{\sqrt{n}}\,\phi(u_{\alpha/2})\left(-\frac{u_{\alpha/2}(c_1+c_1')J}{I}+\frac{c_2-c_2'+\eta-\eta'}{\sqrt{I}}\right)=o\left(\frac{1}{n}\right).$$

Since by (7.2.28)

$$\frac{c_2-c_2'+\eta-\eta'}{\sqrt{I}}=\frac{u_{\alpha/2}(c_1+c_1')J}{I},$$

It follows that

$$(7.2.29)\quad \eta-\eta'=\frac{u_{\alpha/2}(c_1+c_1')J}{\sqrt{I}}-c_2+c_2'.$$

From (7.2.25) and (7.2.29) we obtain

$$(7.2.30)\quad \eta=\frac{u_{\alpha/2}(c_1+c_1')J}{2\sqrt{I}}-\frac{(c_1-c_1')J\sqrt{I}}{2u_{\alpha/2}}-\frac{K\sqrt{I}}{6}u^2_{\alpha/2}-c_2;$$

$$(7.2.31)\quad \eta'=-\frac{u_{\alpha/2}(c_1+c_1')J}{2\sqrt{I}}-\frac{(c_1-c_1')}{2u_{\alpha/2}}-\frac{K\sqrt{I}}{6}u^2_{\alpha/2}-c_2'$$

where

$$c_1=\frac{t}{2(1-\lambda_1 e^{-tu_{\alpha/2}\sqrt{I}+\frac{t^2}{2}I})}\ ;\quad c_2=\frac{t^2(3J+K)}{6(1-\lambda_1 e^{-tu_{\alpha/2}\sqrt{I}+\frac{t^2}{2}I})}\ ;$$

$$c_1'=\frac{t}{2(1-\lambda_1 e^{tu_{\alpha/2}\sqrt{I}+\frac{t^2}{2}I})}\ ;\quad c_2=\frac{t^2(3J+K)}{6(1-\lambda_1 e^{tu_{\alpha/2}\sqrt{I}+\frac{t^2}{2}I})}\ :$$

with $\lambda_1=\frac{1}{u_{\alpha/2}\,I(\theta_0)}e^{-\frac{t^2}{2}I}\sinh\ (tu_{\alpha/2}\sqrt{I})+O(1/\sqrt{n})$.

Hence we have

$$\bar{Z}_1=u_{\alpha/2}\sqrt{I}+\frac{\eta}{\sqrt{n}}\ ;\quad \underline{Z}_1=-u_{\alpha/2}\sqrt{I}+\frac{\eta'}{\sqrt{n}},$$

where η and η' are given by (7.2.30) and (7.2.31), respectively.

We also obtain

$$k_1=u_{\alpha/2}\sqrt{I}+\frac{c_2+\eta}{\sqrt{n}}\ ;\quad k_1'=-u_{\alpha/2}\sqrt{I}+\frac{c_2'+\eta'}{\sqrt{n}}.$$

We put

$$u(\theta\ ;\ t)=c_2+\eta\ ;\ l(\theta\ ;\ t)=c_2'+\eta'.$$

If there exist $\underline{\theta}_n^{**}(\tilde{x}_n)$ and $\overline{\theta}_n^{**}(\tilde{x}_n)$ in the neighborhood of the maximum likelihood estimator (MLE) $\hat{\theta}_{ML}$ such that

$$(7.2.32)\quad Z_1(\underline{\theta}_n^{**}(\tilde{x}_n))+\frac{c_1}{\sqrt{n}}\,Z_2(\underline{\theta}_n^{**}(\tilde{x}_n))=u_{\alpha/2}\sqrt{I(\underline{\theta}_n^{**}(\tilde{x}_n))}+\frac{u(\underline{\theta}_n^{**}(\tilde{x}_n);t)}{\sqrt{n}}$$

$$(7.2.33)\quad Z_1(\overline{\theta}_n^{**}(\tilde{x}_n))+\frac{c_1'}{\sqrt{n}}Z_2(\overline{\theta}_n^{**}(\tilde{x}_n))=-u_{\alpha/2}\sqrt{I(\overline{\theta}_n^{**}(\tilde{x}_n))}+\frac{l(\overline{\theta}_n^{**}(\tilde{x}_n);t)}{\sqrt{n}}$$

then the second order α-AUC interval $[\underline{\theta}_n^{**}(\tilde{x}_n),\ \overline{\theta}_n^{**}(\tilde{x}_n)]$ is second order asymptotically efficient.

Hence we have established the following :

Theorem 7.2.3. Under conditions (A.3.1.1), (A.3.1.3), (A.4.1.1) $\sim$ (A.4.1.3), the second order α-AUC interval $[\underline{\theta}_n^{**}(\tilde{x}_n),\overline{\theta}_n^{**}(\tilde{x}_n)]$ is second order asymptotically efficient.

We shall actually construct the second order asymptotically efficient α-AUC interval using the MLE $\hat{\theta}_{ML}$.

We put

$$\underline{\theta}_n^{**}(\tilde{x}_n)=\hat{\theta}_{ML}-\frac{u_{\alpha/2}}{\sqrt{I(\hat{\theta}_{ML})n}}+\frac{\delta'}{n}\quad;$$

$$\overline{\theta}_n^{**}(\tilde{x}_n)=\hat{\theta}_{ML}-\frac{u_{\alpha/2}}{\sqrt{I(\hat{\theta}_{ML})n}}+\frac{\delta}{n}\quad.$$

we have

$$Z_1(\underline{\theta}_n^{**})+\frac{c_1}{\sqrt{n}}\,Z_2(\underline{\theta}_n^{**})$$

$$=Z_1\Big(\hat{\theta}_{ML}-\frac{u_{\alpha/2}}{\sqrt{I(\hat{\theta}_{ML})n}}+\frac{\delta'}{n}\Big)+\frac{c_1}{\sqrt{n}}Z_2\Big(\hat{\theta}_{ML}-\frac{u_{\alpha/2}}{\sqrt{I(\hat{\theta}_{ML})n}}+\frac{\delta'}{n}\Big)+o_p\Big(\frac{1}{n}\Big)$$

$$=\Big(-\frac{u_{\alpha/2}}{\sqrt{I(\hat{\theta}_{ML})n}}+\frac{\delta'}{n}\Big)Z_1'(\hat{\theta}_{ML})+\frac{1}{2}\Big(-\frac{u_{\alpha/2}}{\sqrt{I(\hat{\theta}_{ML})n}}+\frac{\delta'}{n}\Big)^2 Z_1''(\hat{\theta}_{ML})$$

$$+\frac{c_1}{\sqrt{n}}\,Z_2(\hat{\theta}_{ML})+\frac{c_1}{\sqrt{n}}\Big(-\frac{u_{\alpha/2}}{\sqrt{I(\hat{\theta}_{ML})n}}+\frac{\delta'}{n}\Big)Z_2'(\hat{\theta}_{ML})+o_p\Big(\frac{1}{n}\Big)\ .$$

Since

$$Z_1'(\theta_0)=Z_2(\theta_0)-\sqrt{n}\,I(\theta_0)\quad;$$

$$I'(\theta_0) = 2J(\theta_0) + K(\theta_0) \quad ,$$

it follows that

$$Z_1''(\theta_0) = Z_2'(\theta_0) - \sqrt{n}\, I'(\theta_0)$$

$$\sim -\sqrt{n}\, J(\theta_0) - \sqrt{n}\left\{2J(\theta_0) + K(\theta_0)\right\}$$

$$= -\sqrt{n}\left\{3J(\theta_0)+K(\theta_0)\right\} \quad .$$

We also obtain

$$Z_2'(\theta_0) = \frac{1}{\sqrt{n}} \sum_{i=1}^{n} \frac{\partial^3}{\partial\theta^3}\log f(x_i,\theta_0) + \sqrt{n}\, I'(\theta_0)$$

$$= \frac{1}{\sqrt{n}} \sum_{i=1}^{n} \frac{\partial^3}{\partial\theta^3}\log f(x_i,\theta_0) + \sqrt{n}\left\{2J(\theta_0) + K(\theta_0)\right\}$$

$$\sim \sqrt{n}\left\{-3J(\theta_0) - K(\theta_0)\right\} + \sqrt{n}\left\{2J(\theta_0) + K(\theta_0)\right\}$$

$$= -J(\theta_0)\sqrt{n} \quad .$$

Hence we have

$$Z_1(\underline{\theta}_n^{**}) + \frac{c_1}{\sqrt{n}} Z_2(\underline{\theta}_n^{**})$$

$$= \left(-\frac{u_{\alpha/2}}{\sqrt{I(\hat\theta_{ML})n}} + \frac{\delta'}{n}\right)\left\{Z_2(\hat\theta_{ML}) - \sqrt{n}\, I(\hat\theta_{ML})\right\}$$

$$+ \frac{u_{\alpha/2}^2}{2I(\hat\theta_{ML})n}\left\{-\sqrt{n}(3J(\hat\theta_{ML}) + K(\hat\theta_{ML}))\right\} + \frac{c_1}{\sqrt{n}} Z_2(\hat\theta_{ML})$$

$$- \frac{c_1 u_{\alpha/2}}{\sqrt{I(\hat\theta_{ML})n}}(-\sqrt{n}\, J(\hat\theta_{ML})) + o_p\left(\frac{1}{\sqrt{n}}\right)$$

$$= u_{\alpha/2}\sqrt{I(\hat\theta_{ML})} - \frac{u_{\alpha/2}}{\sqrt{I(\hat\theta_{ML})n}} Z_2(\hat\theta_{ML}) + \frac{c_1}{\sqrt{n}} Z_2(\hat\theta_{ML})$$

$$- \frac{u_{\alpha/2}^2}{2I(\hat\theta_{ML})\sqrt{n}}\left\{3J(\hat\theta_{ML})+K(\hat\theta_{ML})\right\} + \frac{c_1 u_{\alpha/2} J(\hat\theta_{ML})}{I(\hat\theta_{ML})} + o_p\left(\frac{1}{\sqrt{n}}\right) .$$

Since

$$\sqrt{I(\underline{\theta}_n^{**})} = \sqrt{I(\hat{\theta}_{ML})} + o_p(1)$$

$$u(\underline{\theta}_n^{**}, t) = u(\hat{\theta}_{ML}\ ;\ t) + o_p(\frac{1}{\sqrt{n}}) \quad ,$$

it follows by (7.2.32) that

$$-\frac{u_{\alpha/2}}{\sqrt{I(\hat{\theta}_{ML})n}} Z_2(\hat{\theta}_{ML}) - \frac{\delta'}{\sqrt{n}} I(\hat{\theta}_{ML}) + \frac{c_1}{\sqrt{n}} Z_2(\hat{\theta}_{ML}) - \frac{u_{\alpha/2}^2}{2I(\hat{\theta}_{ML})\sqrt{n}} \left\{3J(\hat{\theta}_{ML}) + K(\hat{\theta}_{ML})\right\}$$

$$+ \frac{c_1 u_{\alpha/2} J(\hat{\theta}_{ML})}{\sqrt{I(\hat{\theta}_{ML})\ n}}$$

$$= -\frac{u_{\alpha/2}^2 \left\{2J(\hat{\theta}_{ML}) + K(\hat{\theta}_{ML})\right\}}{2\sqrt{I(\hat{\theta}_{ML})n}} \quad .$$

Then we have

$$(7.2.34) \quad \delta' = (c_1 - \frac{u_{\alpha/2}}{\sqrt{I(\hat{\theta}_{ML})}}) \frac{Z_2(\hat{\theta}_{ML})}{I(\hat{\theta}_{ML})} + c_1 u_{\alpha/2} \frac{J(\hat{\theta}_{ML})}{I(\hat{\theta}_{ML})^{3/2}}$$
$$- \frac{u_{\alpha/2}^2}{2I(\hat{\theta}_{ML})} \left\{ \frac{3J(\hat{\theta}_{ML}) + K(\hat{\theta}_{ML})}{I(\hat{\theta}_{ML})} + \frac{2J(\hat{\theta}_{ML}) + K(\hat{\theta}_{ML})}{\sqrt{I(\hat{\theta}_{ML})}} \right\} .$$

In a way similar to the above we obtain

$$(7.2.35) \quad \delta = (c_1 + \frac{u_{\alpha/2}}{\sqrt{I(\hat{\theta}_{ML})}}) \frac{Z_2(\hat{\theta}_{ML})}{I(\hat{\theta}_{ML})} - c_1 u_{\alpha/2} \frac{J(\hat{\theta}_{ML})}{I(\hat{\theta}_{ML})^{3/2}}$$
$$- \frac{u_{\alpha/2}^2}{2I(\hat{\theta}_{ML})} \left\{ \frac{3J(\hat{\theta}_{ML}) + K(\hat{\theta}_{ML})}{I(\hat{\theta}_{ML})} + \frac{2J(\hat{\theta}_{ML}) + K(\hat{\theta}_{ML})}{\sqrt{I(\hat{\theta}_{ML})}} \right\} .$$

Hence we have established the following :

Theorem 7.2.4. Under conditions (A.3.1.1) $\sim$ (A.3.1.3), (A.4.1.1) $\sim$ (A.4.1.3), the second order α-AUC interval

$$(7.2.36) \quad [\hat{\theta}_{ML} - \frac{u_{\alpha/2}}{\sqrt{I(\hat{\theta}_{ML})n}} + \frac{\delta'}{n}\ ,\ \hat{\theta}_{ML} + \frac{u_{\alpha/2}}{\sqrt{I(\hat{\theta}_{ML})n}} + \frac{\delta}{n}\ ]$$

is second order asymptotically efficient, where δ' and δ are given by (7.2.34) and (7.2.35), respectively.

Remark. It should be noted that both δ' and δ depend on t and $Z_2(\cdot)$. The second order α-AUC interval (7.2.36) is most powerful but not uniformly most powerful because of the dependence on t. This testifies the fact that the asymptotically sufficient statistic up to the order $n^{-1/2}$ is given by the pair $(\hat{\theta}_{ML}, Z_2(\hat{\theta}_{ML}))$ ([41], also [31]), while in the case of point estimation $\hat{\theta}_{ML}$ is alone asymptotically efficient up to order $n^{-1/2}$, i.e. second order.

7.3. Concluding Remarks

The modern theory of asymptotic efficiency of estimators was begun 60 years ago by the work of R.A.Fisher. Its development has not always been steady, but the works of R.A.Fisher, J.Neyman, H.Cramér, A.Wald, C.R.Rao, R.R.Bahadur and L.LeCam, among others, contributed to the formulation of a systematic theory. In the past five years it has reached a stage where some of the basic problems have been properly formulated and given general solutions.
The concepts and properties of higher order asymptotic efficiency especially, have been given definitive forms. We have tried to present the basic logical structure of the recently built theory, focussing on the more important aspects of the results, but not necessarily with the widest generality or with strict rigour.

There still remains, however, some important theoretical problems to be explored : One is the relationship between the concept of higher order asymptotic efficiency of estimators and that of higher order asymptotic sufficiency of statistics, as has recently been developed by R.Michel, T.Suzuki and others.
It appears to the authors that there must be some close relation between the concepts, but a clear picture has yet to emerge.
In connection with this, it is also necessary to investigate the relationship between the asymptotic theory of estimation and other

forms of inference, particularly hypothesis testing.

Another problem is that of asymptotic efficiency versus robustness of estimators. Here, there is the problem of asymptotic efficiency where the underlying probability distribution may differ from, but be close to that postulated. We discuss this in a paper now in preparation.

Finally, it should be pointed out that higher order asymptotic efficiency can be reviewed in the framework of differential geometry, as was done by B.Efron. Recently S.Amari gave a differential geometric treatment of the third order asymptotic efficiency of the MLE. Higher order asymptotic efficiency is also closely related to the concept of asymptotic loss of information which was discussed by R.A.Fisher, C.R.Rao, J.K.Ghosh and others. Still other problems could be mentioned, but now, at least one act of the drama is closed.

References

[1] Akahira,M. : Asymptotic theory for estimation of location in non-regular cases, I : Order of convergence of consistent estimators. Rep. Stat. Appl. Res., JUSE, 22 (1975), 8-26.

[2] Akahira,M. : Asymptotic theory for estimation of location in non-regular cases, II : Bounds of asymptotic distributions of consistent estimators. Rep. Stat. Appl. Res., JUSE 22 (1975), 99-115.

[3] Akahira,M. : A note on the second order asymptotic efficiency of estimators in an autoregressive process. Rep. Univ. Electro-Comm., 26 (1975), 143-149.

[4] Akahira,M. : On the asymptotic efficiency of estimators in an autoregressive process. Ann. Inst. Statist. Math., 28 (1976), 35-48.

[5] Akahira,M. : A remark on asymptotic sufficiency of statistics in non-regular cases. Rep. Univ. Electro-Comm., 27 (1976), 125-128.

[6] Akahira,M. : On asymptotic deficiency of estimators. To be published in Austral. J. Statist. 23 (1981).

[7] Akahira,M and Takeuchi,K : On the second order asymptotic efficiency of estimators in multiparameter cases. Rep. Univ. Electro-Comm., 26 (1976), 261-269.

[8] Akahira,M and Takeuchi,K : Discretized likelihood methods —— Asymptotic properties of discretized likelihood estimators (DLE's). Ann. Inst. Statist. Math. 31 (1979), Part A, 39-56.

[9] Akahira,M and Takeuchi,K. : On the second order asymptotic efficiency of unbiased confidence intervals. Rep. Stat. Appl. Res., JUSE, 26 (1979).

[10] Akahira,M and Takeuchi,K. : Remarks on asymptotic efficiency and inefficiency of maximum probability estimators. Rep. Stat. Appl. Res., JUSE, 26 (1979).

[11] Anderson,T.W. : On asymptotic distributions of estimates of stochastic difference equations. Ann. Math. Statist., 30 (1959), 676-687.

[12] Bahadur,R.R. : On Fisher's bound for asymptotic variances. Ann. Math. Statist., 35 (1964), 1545-1552.

[13] Bhattacharya,R.N. and Ghosh, J.K. : On the validity of the formal Edgeworth expansion. Ann. Statist., 6 (1978), 434-451.

[14] Blom,G. : Statistical Estimates and Transformed Beta-Variables. John Wiley & Sons (1958).

[15] Chibisov,D.M. : On the normal approximation for a certain class of statistics. Proc. Sixth Berkeley Symp. on Math. Statist. and Prob., 1 (1972), 153-174.

[16] Chibisov, D.M. : Asymptotic expansions for Neyman's C(α) tests. Proc. Second Japan-USSR Symp. on Probability Theory. Lecture Notes in Math., 330 (1973) Springer Verlag, Berlin , 16-45.

[17] Cramér,H. : Mathematical Methods of Statistics. Princeton University Press. (1946).

[18] Diananda,P.H. : Some Probability limit theorems with satistical applications. Proc. Camb. Phil. Soc., 49 (1953), 239-246.

[19] Efron,B. : Defining the curvature of a statistical problem (with application to second order efficiency). Ann. Statist., 3 (1975), 1189-1242.

[20] Fisher,R.A. : Theory of statistical estimation. Proc. Camb. Phil. Soc., 22 (1925), 700-725.

[21] Ghosh,J.K. and Subramanyam,K. : Second order efficiency of maximum likelihood estimators. Sankhyā Ser.A, 36 (1974), 325-358.

[22] Gnedenko,B.V. and Kolmogorov,A.N. : Limit Distributions for Sums of Independent Random Variables. Addison-Wesley, Cambridge, Massachussets (Translated from Russian), (1954).

[23] Gusev,S.I. : Asymptotic expansions associated with some statistical estimators in the smooth case I. Expansions of random variables. Theory Prob. Applications. 20 (1975), 470-498.

[24] Hoeffding,W. and Robbins,H. : The central limit theorem for depending random variables. Duke Math.J., 15 (1948), 773-780.

[25] Hoeffding,W. and Wolfowitz,J. : Distinguishability of sets of distributions. Ann. Math. Statist. 29 (1958), 700-718.

[26] Kendall,M.G. and Stuart, A. : The Advanced Theory of Statistics. Vol.1, Charles Griffin (1969).

[27] LeCam,L. : On the asymptotic theory of estimation and testing hypothesis. Proc. third Berkeley Symp. on Math. Statist. and Prob., 1 (1956), 129-156.

[28] LeCam,L. and Schwarts,L. : A necessary and sufficient condition for the existence of consistent estimates. Ann. Math. Statist., 31 (1960), 140-150.

[29] LeCam,L. : Notes on Asymptotic Methods in Statistical Decision Theory. Lecture Note, (1974).

[30] Lehmann,E.L. : Testing Statistical Hypotheses. John Wiley & Sons (1959).

[31] Michel,R. : Higher order asymptotic sufficiency. Sankhyā 40 (1978), 76-84.

[32] Pfanzagl,J. : Asymptotic expansions related to minimum contrast estimators. Ann. Statist., 1 (1973), 993-1026.

[33] Pfanzagl,J, : On asymptotically complete classes, Proc. Summer Research Institute of Statistical Inference for Stochastic Processes, 2 (1975), 1-43.

[34] Pfanzagl,J. and Wefelmeyer.W. : A third order optimum property of maximum likelihood estimator. J.Multivariate Anal. 8(1978), 1-29.

[35] Pfanzagl,J and Wefelmeyer,W. : Addendum to "A third-order optimum property of the maximum likelihood estimator," J.Multivariate Anal., 9 (1979), 179-182.

[36] Polfeldt,T. : The order of the minimum variance in a non-regular case, Ann. Math. Statist., 41 (1970), 667-672.

[37] Rao, C.R. : Asymptotic efficiency and limiting information. Proc. Fourth Berkeley Symp. on Math. Statist. and Prob., 1 (1961), 531-545.

[38] Rao, C.R. : Efficient estimates and optimum inference procedures in large sample. J.Roy. Statist. Soc. 24(B), (1962), 46-72.

[39] Sargan, J.D. : Econometric estimators and the Edgeworth approximation. Econometrica 14 (1976), 421-448.

[40] Strasser,H. : Asymptotic expansions for Bayes procedures. In Recent Development of Statistics (J.R.Barra et al., Ed.) North-Holland (1977), 9-35.

[41] Suzuki,T. : Asymptotic sufficiency up to higher orders and its applications to statistical tests and estimates. Osaka J. Math., 15 (1978), 575-588.

[42] Takeuchi,K. : Tôkei-teki suitei no Zenkinriron (Asymptotic Theory of Statistical Estimation). (in Japanese) Kyôiku-Shuppan, Tokyo, (1974).

[43] Takeuchi,K. : Asymptotic expansions for the critical limits of unbiased tests. Discussion paper No.131, Kyoto Institute of Economic Research, Kyoto University (1978).

[44] Takeuchi,K. and Akahira,M. : On the second order asymptotic efficiencies of estimators. Proceedings of the Third Japan-USSR Symposium on Probability Theory (G.Maruyama and J.V.

Prokhorov, Eds.). Lecture Notes in Math., 550 (1976), Springer Verlag, Berlin, 604-638.

[45] Takeuchi,K. and Akahira,M. : On the asymptotic properties of statistical estimators. (in Japanese) Sûgaku, 29 (1977), 110-123.

[46] Takeuchi,K. and Akahira,M. : Third order asymptotic efficiency of maximum likelihood estimator for multiparameater exponential case. Rep. Univ. Electro-Comm. 28 (1978), 271-293.

[47] Takeuchi,K. and Akahira,M. : On the asymptotic efficiency of estimators. (in Japanese). A report of the Symposium on Several Problems of Asymptotic Theory, Annual Meeting of the Mathmatical Society of Japan (1978), 1-24.

[48] Takeuchi,K. and Akahira,M. : Asymptotic optimality of the generalized Bayes estimator. Rep. Univ. Electro-Comm. 29 (1978), 37-45.

[49] Takeuchi,K. and Akahira,M. : Asymptotic optimality of the generalized Bayes estimator in multiparameter cases. Ann. Inst. Statist. Math., 31 (1979), Part A, 403-415.

[50] Takeuchi,K. and Akahira,M. : Third order asymptotic efficiency of maximum likelihood estimator in general case. (to appear).

[51] Takeuchi,K. and Akahira,M. : Asymptotic efficeincy of estimators, with order $\{c_n\}$ consistency, in regular and non-regular cases. (to appear).

[52] Wald,A. : Note on the consistency of the maximum likelihood estimator. Ann. Math. Statist., 20 (1949), 595-601.

[53] Wald, A. : Asymptotically shortest confidence intervals. Ann. Math. Statist., 13 (1942), 127-137.

[54] Weiss,L. and Wolfowitz,J. : Maximum probability estimators. Ann. Inst. Statist. Math., 19 (1967), 193-206.

[55] Weiss,L. and Wolfowitz,J. : Generalized maximum likelihood estimators in a particular case. Theor. Probability Appl. 13 (1968), 622-627.

[56] Weiss,L. and Wolfowitz,J. : Maximum Probability Estimators and Related Topics. Lecture Notes in Math., 424 (1974).

[57] Woodroofe,M. : Maximum likelihood estimation of a translation parameter of a truncated distribution. Ann. Math. Statist., 43 (1972), 113-122.

References in addendum

[58] Amari,S. : Theory of information space : A differential-geometrical foundation of statistics. RAAG Reports, 106 (1980).

[59] Berkson,J : Minimum chi-square, not maximum likelihood! Ann. Statist., 8(1980), 457-487.

[60] Ghosh,J.K., Sinha,B.K. and Wieand,H.S. : Second order efficiency of the mle with respect to any bounded bowl-shaped loss function. Ann. Statist., 8 (1980), 506-521.

[61] Hodges,J.L. and Lehmann, E.L. : Deficiency. Ann. Math. Statist., 41 (1970), 783-801.

[62] Takeuchi,K. and Akahira,M. : Third order asymptotic efficiency and asymptotic completeness of estimators. Rep. Univ. Electro-Comm., 31 (1980), 89-96.

Subject Index

Lecture Notes in Statistics

Vol. 1: R. A. Fisher: An Appreciation. Edited by S. E. Fienberg and D. V. Hinkley. xi, 208 pages, 1980.

Vol. 2: Mathematical Statistics and Probability Theory. Proceedings 1978. Edited by W. Klonecki, A. Kozek, and J. Rosiński. xxiv, 373 pages, 1980.

Vol. 3: B. D. Spencer, Benefit-Cost Analysis of Data Used to Allocate Funds. viii, 296 pages, 1980.

Vol. 4: E. A. van Doorn: Stochastic Monotonicity and Queueing Applications of Birth-Death Processes. vi, 118 pages, 1981.

Vol. 5: T. Rolski, Stationary Random Processes Associated with Point Processes. vi, 139 pages, 1981.

Vol. 6: S. S. Gupta and D.-Y. Huang, Multiple Statistical Decision Theory: Recent Developments. viii, 104 pages, 1981.

Vol. 7: M. Akahira and K. Takeuchi, Asymptotic Efficiency of Statistical Estimators. viii, 242 pages, 1981.

Vol. 8: The First Pannonian Symposium on Mathematical Statistics. Edited by P. Révész, L. Schmetterer, and V. M. Zolotarev. vi, 308 pages, 1981.

Springer Series in Statistics

L. A. Goodman and W. H. Kruskal, Measures of Association for Cross Classifications. x, 146 pages, 1979.

J. O. Berger, Statistical Decision Theory: Foundations, Concepts, and Methods. xiv, 420 pages, 1980.

R. G. Miller, Jr., Simultaneous Statistical Inference, 2nd edition. 300 pages, 1981.

P. Brémaud, Point Processes and Queues: Martingale Dynamics. 352 pages, 1981.

Lecture Notes in Mathematics

Selected volumes of interest to statisticians and probabilists:

Vol. 532: Théorie Ergodique. Actes des Journées Ergodiques, Rennes 1973/1974. Edited by J.-P. Conze, M. S. Keane. 227 pages, 1976.

Vol. 539: Ecole d'Ete de Probabilités de Saint-Flour V–1975. A. Badrikian, J. F. C. Kingman, J. Kuelbs. Edited by P.-L. Hennequin. 314 pages, 1976.

Vol. 550: Proceedings of the Third Japan-USSR Symposium on Probability Theory. Edited by G. Maruyama, J. V. Prokhorov. 722 pages, 1976.

Vol. 566: Empirical Distributions and Processes. Selected Papers from a Meeting at Oberwolfach, March 28–April 3, 1976. Edited by P. Gänssler, P. Revesz. 146 pages, 1976.

Vol. 581: Séminaire de Probabilités XI. Université de Strasbourg. Edited by C. Dellacherie, P. A. Meyer, and M. Weil. 573 pages, 1977.

Vol. 595: W. Hazod, Stetige Faltungshalbgruppen von Warscheinlichkeitsmassen und erzeugende Distributionen. 157 pages, 1977.

Vol. 598: Ecole d'Ete de Probabilités de Saint-Flour VI–1976. J. Hoffmann-Jorgensen, T. M. Ligett, J. Neveu. Edited by P.-L. Hennequin. 447 pages, 1977.

Vol. 636: Journées de Statistique des Processus Stochastiques, Grenoble 1977. Edited by Didier Dacunha-Castelle, Bernard Van Cutsem. 202 pages, 1978.

Vol. 649: Séminaire de Probabilités XII. Strasboug 1976–1977. Edited by C. Dellacherie, P. A. Meyer, and M. Weil. 805 pages, 1978.

Vol. 656: Probability Theory on Vector Spaces, Proceedings 1977. Edited by A. Weron. 274 pages. 1978.

Vol. 672: R. L. Taylor, Stochastic Convergence of Weighted Sums of Random Elements in Linear Spaces. 216 pages, 1978.

Vol. 675: J. Galambos, S. Kotz, Characterizations of Probability Distributions. 169 pages, 1978.

Vol. 678: Ecole d'Ete de Probabilités de Saint-Flour VII–1977. D. Dacunha-Castelle, H. Heyer, and B. Roynette. Edited by P.-L. Hennequin. 379 pages, 1978.

Vol. 690: W. J. J. Rey, Robust Statistical Methods. 128 pages, 1978.

Vol. 706: Probability Measures on Groups, Proceedings 1978. Edited by H. Heyer. 348 pages, 1979.

Vol. 714: J. Jacod, Calcul Stochastique et Problemes de Martingales. 539 pages, 1979.

Vol. 721: Seminairé de Probabilités XIII. Proceedings, Strasbourg, 1977/78. Edited by C. Dellacherie, P. A. Meyer, and M. Weil. 647 pages, 1979.

Vol. 794: Measure Theory, Oberwolfach 1979, Proceedings, 1979. Edited by D. Kolzöw. 573 pages, 1980.

Vol. 796: C. Constantinescu, Duality in Measure Theory. 197 pages, 1980.

Vol. 821: Statistique non Paramétrique Asymptotique, Proceedings 1979. Edited by J.-P. Raoult. 175 pages, 1980.

Vol. 828: Probability Theory on Vector Spaces II, Proceedings 1979. Edited by A. Weron. 324 pages, 1980.